Bones, Stones and Molecules

"Out of Africa" and Human Origins

**Reconstruction of *Paranthropus*
(Adapted from Matternes [Isaac & McCown, 1976])**

Bones, Stones and Molecules
"Out of Africa" and Human Origins

David W. Cameron

Department of Anatomy and Histology
The University of Sydney

and

Colin P. Groves

School of Archaeology and Anthropology (Faculties)
Australian National University

Amsterdam • Boston • Heidelberg • London • New York • Oxford
Paris • San Diego • San Francisco • Singapore • Sydney • Tokyo

ELSEVIER
ACADEMIC
PRESS

Acquisition Editor: *David Cella*
Project Manager: *Troy Lilly*
Editorial Coordinator: *Kelly Sonnack*
Marketing Manager: *Linda Beattie*
Marketing Manager: *Clare Fleming*
Cover Design: *Cate Rickard Barr*
Interior Design: *Julio Esperas*
Composition: *Newgen Imaging Systems*
Interior Printer: *The Maple-Vail Book Manufacturing Group*
Cover Printer: *Phoenix Color Corporation*

Elsevier Academic Press
200 Wheeler Road, Burlington, MA 01803, USA
525 B Street, Suite 1900, San Diego, California 92101-4495, USA
84 Theobald's Road, London WC1X 8RR, UK

This book is printed on acid-free paper. (∞)

Library of Congress Cataloging-in-Publication Data
Cameron, David W.
  Bones, stones and molecules: "Out of Africa" and human origins / David W. Cameron and
Colin P. Groves.
    p. cm.
Includes bibliographical references and index.
  ISBN 0-12-156933-0 (pbk. : alk. paper)
  1. Paleoanthropology. 2. Human evolution. I. Groves, Colin P. II. Title.
  GN282.C36 2004
  569.9—dc22
                                                                              2003022774

British Library Cataloguing in Publication Data
A catalogue record for this book is available from the British Library

ISBN: 0-12-156933-0

For all information on all Academic Press publications
visit our Web site at www.academicpressbooks.com

Printed in the United States of America
04   05   06   07   08   09   9   8   7   6   5   4   3   2   1

David dedicates this book to Debbie and our little Cameron clan, Emma, Anita, and Lloyd Jr.

Colin dedicates it to his long-suffering wife, the inspiration of the last thirty years, Phyll.

CONTENTS

ACKNOWLEDGMENTS

We want to thank all our colleagues, mentors, students, and friends who have contributed to our thought-processes on human evolution. Some are no longer with us: John Napier, Vratja Mazák, Peter Whybrow, Allan Wilson. Most are still around, and bickering in a very lively fashion: Peter Andrews, Debbie Argue, Ray Bernor, Peter Brown, David Bulbeck, Judith Caton, Ron Clarke, Denise Donlon, Peter Grubb, Vu The Long, Maciej Henneberg, Jacob Hogarth, Bill Howells, Jean-Jaques Hublin, Dan Lieberman, Theya Molson, Charles Oxnard, Rajeev Patnaik, David Pilbeam, Brett Still, Chris Stringer, Phillip Tobias, Alan Thorne, Peter Ungar, Alan Walker, Michael Westaway, Milford Wolpoff, Benard Wood, Wu Xinzhi. We also thank Troy Lilly and Kelly Sonnack at Elsevier, whose nurturing actions have been above and beyond the call of duty.

We thank, but do not blame them. We finally thank each other, for putting up with each other's foibles, pig-headedness and exasperating habits.

PREFACE

Within the last decade there have been a number of truly significant discoveries relating to the evolution of humans and their ancestors. Most recent have been the discovery and publication of the late Miocene fossil specimen from Chad allocated to *Sahelanthropus* and the mid-Pliocene fossils from Kenya allocated to *Kenyanthropus*. Ongoing discoveries of more recent human remains, especially from the Pleistocene of Africa, Europe, and Australia are also forcing us to reassess our views of modern human origins. Discoveries by archaeologists over the last decade have not only pushed back the earliest dates for stone tool manufacture, but are also challenging our current view of past human behavior. New methods of collecting, analyzing, and interpreting molecular evidence have also had considerable impact on the way we interpret the evolution of our species. Molecular biology has enabled us to identify the likely period when proto-chimpanzees and proto-humans last shared a common ancestor (around 6 million years ago), and the most recent contribution from this field to the study of human evolution has been the extraction and analysis of Neanderthal mtDNA. All of this evidence supports the idea that human evolution over the last few million years is a complex story, defined by considerable species diversity.

It is becoming increasingly clear to both authors that the "Out of Africa" model for recent human origins is supported by the available fossil, archaeological and molecular evidence, though, as we will also argue, there was more than one "Out of Africa," and in some cases there were dispersals into Africa during the early Pleistocene by some human species. That is not to say that we both agree on the details of human evolution over the last 5 million years or so. As the reader will see, we agree to disagree, which is shown most markedly in our differing taxonomies of the hominids, both of which suggest distinct relationships within the more recent members of our own family, the Hominidae.

CHAPTER 1

Introduction

Is the evolution of modern humans an African genesis followed by prehistoric worldwide genocide of earlier pre-*sapiens*, or is it a slow progression
from pre-*sapiens* to modern humans? Theories concerned with modern
human evolution have been polarized by these extreme views. These two
basic positions have been referred to, respectively, as the "Out of Africa" and
the "Multiregional" hypotheses. Does the paleontological, archaeological,
and molecular evidence support the mass extinction of earlier humans, the
last of all being the Neanderthals, or did these diverse pre-*sapiens* interbreed
with the more "successful," modern *H. sapiens,* thus being swamped genetically and physically? Indeed, are Neanderthals just an extreme version of
the one species *H. sapiens* — is there still a little Neanderthal left in us all?

Any understanding of human evolution, undoubtedly, must be based on
an interpretation of human physical (bone) and cultural (stone) remains.
This is particularly true of the remains that predate the origins of our own
species, *H. sapiens,* whose earliest representatives appear around 250,000–
150,000 years ago (Bräuer, 1984, 1989; Rightmire, 1984, 1993; Stringer &
Andrews, 1988; Groves, 1989a; F.H. Smith *et al.,* 1989; Stringer, 1989,
2003; Stringer & McKie, 1996; F.H. Smith, 2002; T.D. White *et al.,* 2003).
With the late emergence of our own species, we are also able to invoke
molecular evidence from preserved human mitochondrial DNA (mtDNA).
The molecular evidence, if assessed cautiously, provides a date for the origins of our own species, which is independent of other "hard" evidence,
such as bones and stones, and also suggests likely evolutionary relationships

between different human groups. Unlike bones and stones, however, the molecular evidence does not provide a picture of what our ancestors looked like or how they adapted physically and behaviorally to their seasonally fluctuating environments.

The overall tempo and mode of evolution best fits in with long periods of morphological stasis followed by rapid speciation. While this was suggested by Haldane as long ago as 1932 (and even earlier by Huxley in correspondence to Charles Darwin), it was Eldredge and Gould (1972) who first popularized this theory of evolution, most commonly referred to as *punctuated equilibrium* (Figure 1.1) (see also Gould & Eldredge, 1977;

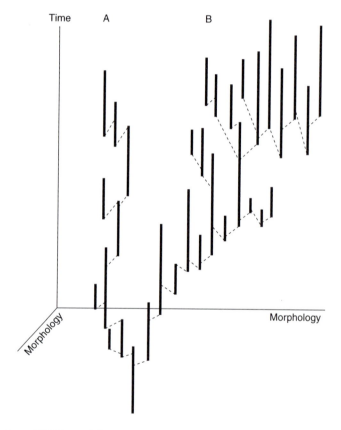

Figure 1.1 ▶   Eldridge and Gould's original diagrammatic model of punctuated equilibrium tied to rapid speciation via cladogenesis, with numerous extinctions along the way. This tempo and mode of evolution best fits the "Out of Africa" hypothesis for modern human origins.

From Eldredge and Gould (1972), p. 113.

Stanley, 1978, 1979; Tattersall, 1986; Eldredge, 1989). Under this model, the many gaps in the fossil record are not merely annoying hiatuses, they are actually data: they are informing us about the tempo of evolution, that in many cases these gaps are the result of rapid speciation (rapid in geological time, that is, about 100,000 years!). Given this rapid turnover, then, the transitional forms were unlikely to be fossilized; or if they were fossilized, they are unlikely ever to be discovered, given their small population size and occupation of a restricted geographical region. While there certainly are many, many gaps in the hominid fossil record, it is perhaps the Miocene hominid record from 23 to 6 million years ago that has been most clearly shown to be characterized by a tempo and mode of evolution that best fits a punctuationist model (Cameron, in press a). This will surely prove to be the case for the hominids and hominins of the Old World, for they are marked by a sudden explosion of contemporary species, many of which appear to have left no direct descendants. This is further emphasized because the fossil record will always underestimate the number of species, and we will never have fossils representing all of the species that have ever existed.

The theory of punctuated equilibrium argues that the mode of speciation is the result of reproductive isolation at the periphery of a species' range, the emphasis being on cladogenesis as opposed to anagenesis (see Eldredge & Gould, 1972; Gould & Eldredge, 1977; Stanley, 1978, 1979, 1996; Eldredge, 1989; Gould, 2002). *Cladogenesis* is the splitting of a single species into two reproductively isolated or genetically distinct lineages so that species remain relatively unchanged for long periods of time, occasionally interrupted by rapid or short bursts of evolutionary change resulting in speciation. The isolation of a marginalized population results in a rapid rate of speciation, which may be accompanied by the new daughter species taking over the parent species' territory. If this does occur, it is at this stage that we find the new species within the paleontological record. The daughter species is of course much more likely to be competitively inferior to the parent species and so to become extinct; but very occasionally it may outcompete or coexist with the parent species and become successful and abundant enough to become visible to us in the fossil record, having found its own niche, distinct from that of its parent species. This tempo and mode of evolution best fits the model of evolution espoused by those who support an "Out of Africa" origin for the hominins.

Anagenesis, the alternative to cladogenesis, is slow evolutionary transformation over a long period of time within a single lineage so that an

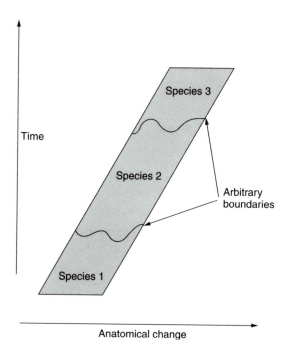

Figure 1.2 ▶ Evolution via anagensis, in which there is limited or no cladogenesis. One species is considered to have evolved into another through gradual evolution, resulting in "chronospecies."

ancestral species blends insensibly into its immediate descendants (Figure 1.2). Whether anagenesis actually exists, at least as a form of gradual change, is controversial. The existence of it depends both on whether selection pressures can remain the same over long periods of time and on there being a constant stream of mutations for selection to work on. This model of evolution and "speciation" tends to be supported by those advocating the Multiregional hypothesis for human origins. Van Valen's "Red Queen effect" assumed a pattern of evolution defined by anagenesis because it argued that a species or population has to keep changing to keep up with the changes that it wreaks in its environment (Van Valen, 1973).

It is undeniable that a species does change its environment, and keeps doing so, and there are of course other species in the same environment that are busy doing the same thing. It is arguable to what extent such changes may be progressive or may be cyclic. Perhaps the animals that cause the most havoc to their environment — after humans! — are elephants. The African savannah elephant (*Loxodonta africana*), the best studied of the three living species, bulldozes whole stands of trees and turns bush and forest into savannah or even desert, affecting the livelihood and abundance of the other mammals that live in the same habitat; so every year hundreds of elephants are shot in southern African game reserves and national parks, based on the premise that uncontrolled populations of elephants will destroy the whole ecosystem. Yet what sounds like a clear-cut Red Queen scenario has been challenged. On a large geographic scale, the effect may be cyclical (a stable limit cycle): The elephants eat themselves out of house and home, their populations plummet and the survivors emigrate, the vegetation recovers, the elephants increase again, the circle is closed.

If there is no sustained Red Queen effect, there is no anagenesis, at least in its traditional (gradualistic) form; or else it must depend solely on gradual, continuous nonbiological changes such as long-term, unidirectional climate or sea-level change. But these seem to have been episodic, not sustained uninterruptedly. At most there is the possibility, even likelihood, that a local environment is somewhat altered after each cycle so that the cumulative effect of a long chain of cycles is really noticeable. But this begins to stretch the concept of anagenesis as gradualism. The Red Queen, when she operates, is a downwardly directed oscillation, not an inexorable slope.

A much more likely scenario is the "Effect" hypothesis of Vrba (1980). A parent species splits into two; one of the daughters (A_1) is somewhat better adapted to the changed environment than the other (B_1) and flourishes while B_1 declines to extinction. Stasis is restored. Meanwhile the environment continues to fluctuate and undergo its stable limit cycles, but the extremes of the cycle change directionally over time — the open-country phase of the cycle gets more open over time; when the forest returns, it is less dense or less widespread. After some time, A_1 itself speciates. Of the two daughter species, A_2 is the one better adapted to the now-changed environment, and it flourishes in its turn while B_2 declines to extinction. And so it goes on. Over a long period of time, the differential survival of the daughter species that each time is better adapted to the now-changed environment is the one that survives, and the effect mimics anagenesis. In the main, the fossil record is too coarse-grained to differentiate the two processes, and prior to 1980, evolutionists would assume that it was anagenesis that was taking place. Maybe it was not.

The importance of speciation has been promoted many times in the fossil record. Groves (1989a) argued that, if it is true that evolutionary change is concentrated at the point of speciation, we can predict that, of two sister species, the one that is more changed (highly autapomorphic) from the common ancestor will have undergone more cladogenesis (its lineage has gone through more speciation events) than the one that is less changed. Unfortunately, the record of human evolution offers only a partial test of this. The human species is much more different than is the chimpanzee from our common ancestor, and the human fossil record is certainly enormously speciose, but the chimpanzee fossil record is empty. All we can do is predict that, when paleontologists start prospecting in the right place to find proto-chimpanzees, they will not be very speciose. Chimpanzee evolution will prove to be, let us say, as nearly unlinear in reality as human evolution was held to be up until the 1970s, when the single-species model finally became untenable. But, as we will see presently, the single-species hypothesis has reared its head again, though not through an analysis of fossil material but, rather, by an abstract discussion of the molecular evidence.

If any statement regarding our own origins is correct, it is that humans originally evolved in Africa. We can all trace our prehistoric roots back to the African continent around 6 million years ago. It was at this time that populations of proto-chimpanzees and proto-humans split from a common ancestor and each started its own evolutionary journey. The recently described fossils allocated to *Sahelanthropus* from Chad, dating to between 6–7 million years ago, and *Orrorin* from Kenya, dating to around 6.1–5.8 million years ago, are close to the point of separation (Brunet *et al.*, 2002; Senut *et al.*, 2001; Pickford *et al.*, 2002), as is the earlier hominid discovery from Lothagam, dated to between 5.0–5.2 million years ago (see M.G. Leakey & Walker, 2003).

Following on from these late Miocene genera comes *Ardipithecus*, which occurs at the Miocene/Pliocene transition of Ethiopia between 5.8 and 4.4 millions of years ago (Ma) (T.D. White *et al.*, 1995; Haile-Selassie, 2001; White, T.D. 2002). *Ardipithecus* displays a mixture of features, some of which are chimpanzee-like while others are human-like. What traditionally marks *Ardipithecus* as being on the human line is that they, unlike chimpanzees, seem to have walked upright. It is from *Ardipithecus* or an *Ardipithecus*-like hominid that the later proto-australopithecines are thought by most to have emerged (Figure 1.3).

The proto-australopithecines are represented by a number of species commonly allocated to the genus *Australopithecus* even though they do not form a monophyletic group, meaning that they do not share an exclusive

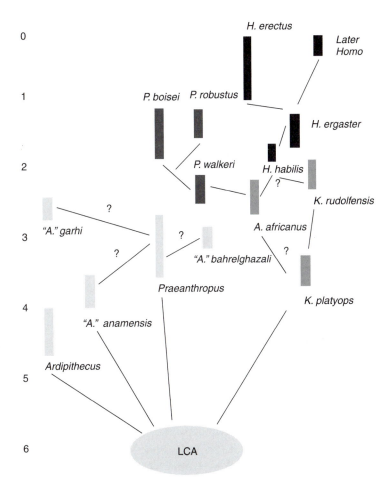

Figure 1.3 ▶   Proposed evolutionary scheme for the Plio-/Pleistocene hominids and hominins. LCA = last common ancestor (Miocene, e.g., *Sahelanthropus* and/or *Orrorion*?).

common ancestor (discussed in detail in Chapter 5). Given their distinct evolutionary histories, they cannot be allocated to the same genus, at least not to a genus that does not include modern humans too; rather, they represent a pattern of hominin diversity, each eventually leading to extinction. Following the scheme proposed by Strait *et al.* (1997), Strait and Grine (2001), and Cameron (in press b), we agree that "*A.*" *anamensis*, "*A.*" *afarensis*, and "*A.*" *garhi* either represent distinct genera (Cameron's preference) or, like all Plio-/Pleistocene hominids, should be subsumed into *Homo* (Groves's preference). Recently, Strait *et al.* (1997) and Strait and Grine (1998, 2001) have reallocated "*A*." *afarensis* (which contains the famous "Lucy" skeleton) to the genus *Praeanthropus*. This genus was first described in the 1950s (see also Harrison, 1993). Thus they and Cameron would argue that only one species, *A. africanus* (the type species), exists within the genus *Australopithecus*.

The evolution of the later, more derived hominins, *Paranthropus*, the "*rudolfensis* group" (represented by the famous 1470 skull), and early *Homo*, appear to be distinct from that of the proto-australopithecines, suggesting that these lineages have a relatively longer history than currently recognized. There are two candidates for the last common ancestor of these later hominins: *Australopithecus africanus* and *Kenyanthropus platyops* (see Dart, 1925; M.G. Leakey *et al.*, 2001; D.E. Lieberman, 2001; Cameron, in press a & b). Indeed, it is likely that both of these "basal hominins" branched off the line before the emergence of the proto-australopithecines. It is possible that their success occurred at the expense of the proto-australopithecines in competition for available resources. Species of *Paranthropus*, early *Homo*, and the "*rudolfensis* group" occupied the same habitats in time and space, so some form of competition must also have occurred between these various groups in the African forests and savannas. If the earliest representatives of *Homo* had succumbed to the competitive pressures of these other groups, then the world as we know it would be very different indeed!

Homo represents the first hominin to disperse out of Africa (though we will see in the next chapter that the original hominid "out of Africa" occurred during the early/middle Miocene transition). Species of *Homo* were in both far southeastern Europe (Georgia) and Asia (Java) by 1.6 million years ago, while *Kenyanthropus, Paranthropus*, and members of the "*rudolfensis* group" remained restricted to Africa (Strait & Wood, 1999; Gabunia *et al.*, 2000a; Dunsworth & Walker, 2002). About 1 million years before *Homo* was extending its range outside of Africa, *K. platyops* disappeared from the fossil record. By the time early *Homo* were occupying a number of diverse habitats in Africa, Europe, and Asia, the last relict *Paranthropus* populations disappeared from the fossil record.

Later *Homo* were a diverse lot: Those populations from different parts of the Old World can all be distinguished easily from one another based on a number of distinct facial features. Some authorities (the "Out of Africa" school) regard them as belonging to a number of different species (*Homo erectus* in Java, *Homo pekinensis* in China, *Homo heidelbergensis* in Africa and Europe). It is true that a general likeness of skull shape is maintained over vast eons of time — hundreds of thousands of years — within each of these regions, though this is to be expected given the similar rate of encephalization. Only in Europe, however, was there a measurable change within one of these species: After about 400,000 years ago, *Homo heidelbergensis*, which had entered Europe from Africa a few hundred thousand years before, had by 120,000 years ago become *Homo neanderthalensis*, the famous Neanderthal people (Stringer, 1989, 1994; Stringer & McKie, 1996), whereas the deme that remained in Africa had by 160,000 years ago emerged into near modern *H. sapiens*, as defined by the recent significant discoveries of the Herto specimens from Ethiopia (T.D. White *et al.*, 2003; Clark *et al.*, 2003; see also Stringer, 2003).

It has also been suggested by some, however, that the lineage leading to *H. neanderthalensis* had already been established as early as 780,000 years ago, as represented by the hominins from Atapuerca (Gran Dolina), Spain, sometimes referred to as *H. antecessor* (Bermúdez *et al.*, 1997). They suggest that *H. heidelbergensis* was already a part of the Neanderthal lineage, and as such the African hominins usually allocated to the same species must be a different species because they are not part of the Neanderthal lineage. Thus a separate and parallel line in Africa (*H. rhodesiensis*?) may have led to the evolution of *H. sapiens* via African populations, as represented by the Herto, Elandsfontein, and Kabwe specimens (see Stringer, 1998, 2003; Clark *et al.*, 2003; T.D. White *et al.*, 2003), so having nothing to do with the emergence of the Neanderthals.

Other authorities (multiregionalists) disagree with these interpretations. These are not different species, they say, but races of early *Homo sapiens*; just as modern *Homo sapiens* has somewhat different geographic varieties, which we sometimes refer to as "races," so did ancient *Homo sapiens* (Wolpoff, 1989, 1999; Wolpoff & Caspari, 1997; Wolpoff *et al.*, 1984, 2001). This minor semantic difference makes all the difference. If they were different species, then they were genetically discontinuous, and if there was any interbreeding between them it was marginal, and their distinct genetic makeup remained unaffected. If they were demes ("races") of the same species, then they were fuzzy at the edges, and new genes from one of them would flow easily into the others.

Despite what some molecular biologists might say, fossils are still the most informative pieces of information available to us when trying to interpret evolutionary relationships among extant and fossil species. They enable us to recognize distinct and common anatomical features, which provide clues to the evolutionary relationship between the species being examined and other fossils and living organisms. Fossils also enable us to identify adaptive strategies employed by these extinct organisms. For example, the identification of large robust mandibles and molars (marked by hyperthick molar enamel) in *Paranthropus* species suggests that they consumed very tough food types (Tobias, 1967; Rak, 1983; Hylander, 1988; White, 2002). Using the bones and the archaeological record, we can identify, through time, how species evolved as a result of their environmental conditions and how they adapted to take advantage of new opportunities.

The study of fossils is largely an anatomical pursuit. Paleontologists spend much of their time examining fossils and comparing them to other fossils and to living organisms thought to share a close evolutionary relationship. One of the most important keys in the reconstruction of evolutionary relationships between species is the identification of *polarity* — those anatomical features that are *primitive* and those that are *derived.*

Primitive features are characters that are often commonly observed and widespread and are considered to have evolved at a very early stage in the group's evolution. Derived features are characters that are less widespread, often unique to a particular group, and so are likely to have evolved only recently in that group. For example, quadrupedal locomotion is a primitive character of the primates (we know this because almost all other mammals are quadrupedal), which tells us little about the evolutionary relationships within this large group. Habitual bipedal locomotion, however, is a derived feature linking humans and the proto-australopithecines and their immediate ancestors, to the exclusion of most other primates (see next chapter). In summary, fossils enable us to identify evolutionary relationships among species and likely physical adaptive trends through time and space.

Stone tools, and an interpretation of their immediate context, are an important source of information when trying to reconstruct past human behavior and cultural evolution. While early humans undoubtedly used other materials (such as wood and animal skins), these are not usually preserved in the archaeological record. The development of ever more sophisticated stone "tool kits" by early humans enabled them to adapt more readily to and extract new food resources from their ever-changing environments and habitats. It also allowed them to defend themselves from much larger and more ferocious animals, and it enabled them to hunt and thus to develop an increased sense of community. In developing this technology, early humans started their long journey on the road to reshaping their environment, rather than simply being shaped by it. Through time, a number of different tool traditions were developed. Archaeologists have been able to associate some of these tool traditions with particular human groups (Bordes, 1950, 1961, 1969; Bordes & Sonneville-Bordes, 1970; Foley & Lar, 1997), while other tool kits are clearly designed for specific functions and not related to differing "cultural" traditions (Binford & Binford, 1966; Binford, 1983). Interpreting how these tools were used has enabled archaeologists to help reconstruct aspects of past human behavior.

   The recent application of molecular biology to human evolutionary studies has greatly influenced current interpretations of human origins. Our genes contain all of the relevant information pertaining to our genetic makeup; they are the core of our being (Figure 1.4). These genes are made

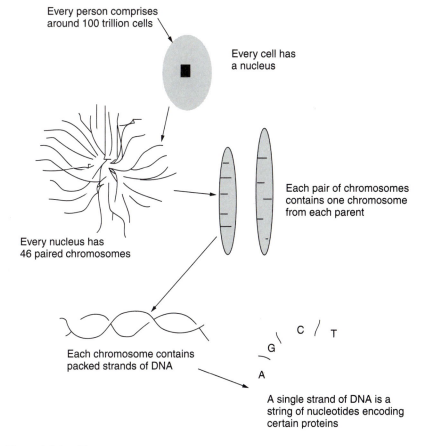

Figure 1.4 ▶   From person to gene.
Adapted from Kingdon (2003).

of deoxyribose nucleic acid (DNA for short). DNA itself consists of two long spiral strands, which form the chromosomes. Each of these strands is made up of four types of small molecules (coded A, G, C, and T). The sequence in these strands forms a code, which carries all of the genetic information transmitted from parents to offspring. The chromosomes are present in the nucleus of every cell; the DNA they contain is called nuclear DNA (nDNA). It is important to realize that genes actually make up only a very small part of nDNA; the rest does not code for anything and is (rightly or wrongly) often referred to as "junk DNA." There are *pseudogenes* (segments of DNA that used to be genes in the distant, evolutionary past but that have been "switched off" over time); *introns* (meaningless segments inserted in the middle of genes); and *repetitive DNA* (varying from long sequences repeated thousands of times to short sequences repeated hundreds of thousands of times, called *microsatellites*). Between them, these "junk" bits make up 90% or more of the complement of nDNA (Pilbeam, 1996; Dover, 1999; Relethford, 2001).

Outside the cell nucleus, in the body of the cell itself (the cytoplasm), are thousands of tiny bodies called *mitochondria*, which provide the energy on which the body's metabolism runs. The mitochondria have their own DNA, mitochondrial DNA (mtDNA). Because mtDNA mutates without any of the "correction" mechanisms operating in nDNA, it changes much faster, and so its variation is an important source of information with regard to the timing of a speciation event among species, as well as identifying likely evolutionary relationships within and between groups. Importantly, mtDNA is inherited, to all intents and purposes, solely from our mothers, for the contribution from the sperm is minute compared to that from the ovum; so mtDNA traces the path of genetic development for our female ancestors in the evolutionary past. If we want to trace where male ancestors went, we have to look at the nDNA of the chromosome that is unique to males: the Y chromosome (Sykes, 2001; Relethford, 2001).

For mtDNA, as for much of DNA, a constant rate of mutation has been assumed. Whether this assumption is always justified is another matter. Certainly mtDNA includes some genes that provide energy for the cell. But because of the way in which the genetic code operates, most mutations do not seem to affect the functioning of the organism, so the assumption of a constant rate of change is, overall, quite reasonable. Accepting that the mutation rates are constant, we can examine the number of shared and unique bases along any strand of mtDNA within a given population and then calculate the molecular distance between populations. The molecular distance between species, therefore, should also be proportional to their separation in time, that is, the time when they last shared a common ancestor. (It may not be exactly the same: The DNA has to become differentiated before the populations do.)

Among the apes, the greatest distance in mtDNA is between the gibbon and the others (orangutan, gorilla, chimpanzee, and human), with a difference of around 5%, and this suggests that the earliest divergence date is between gibbons and the other apes. Next is the orangutan, which differs in mtDNA by 3.6% from the gorilla, chimpanzee, and human; and then the gorilla, at 2.3% difference from chimpanzee and human. The two chimpanzee species (the common and pygmy chimpanzees) differ in only 0.7% of their mtDNA. Chimpanzees and humans are relatively close and differ in only 1.6% of their mtDNA (Ruvolo, 1994, 1997; Pilbeam, 1996, 1997; Stringer & McKie, 1996). Because our own mtDNA differs from that of the chimpanzee by 1.6% (which is about half the distance of the orangutan from the chimpanzee), and because we know, or think we know, that the orangutan split from the other apes 12–16 million years ago (based on fossil evidence), we can use simple mathematics to calculate that the proto-chimpanzees and proto-humans diverged 4.2–6.2 million years ago, the gorilla lineage split around 6.2–8.4 million years ago, while the gibbons were the first to diverge, around 18 million years ago (Chen & Li, 2001).

It was the German paleoanthropologist Franz Weidenreich who originally argued, in the 1930s and 1940s, for a theory of regional continuity. He suggested that the Chinese *Homo erectus* (or what we would call *Homo pekinensis*) fossils, commonly referred to as "Peking Man," gave rise to the modern Chinese, while *Homo erectus* from Java was the ancestor to the original Australians, and Neanderthals gave rise to modern Europeans (Weidenreich, 1946, 1949). The problem with this original scheme was this: How did individual and isolated human groups manage to evolve in the same direction at around the same time through similar successive stages? Weidenreich skirted this question and never successfully addressed the contradiction. Weidenreich (1943:88–89) merely stated that

the fact remains that the Paleolithic population of western France already showed a considerable variety of types. Of no less importance is the fact that these types lived close together in a relatively small area and that there are no signs of a strict separation by geographical barriers. All the facts available indicate that racial characters made their appearance as individual variations ...

and, furthermore, that they started with a great range of variations in a relatively small population. The kind of isolation mechanism which prevented the breakdown of the gene system remains to be studied. It cannot differ much from that which causes the persistence and stability of the nongeographical differentiations of modern mankind. However, this is a problem, not for physical anthropologists alone, but also for geneticists and sociologists.

The Multiregional hypothesis (Figure 1.5) was later revised to emphasize gene flow between groups to help explain a similar rate and "direction" within the evolution of all modern humans (Wolpoff *et al.*, 1984, 2001; Wolpoff, 1989; Wolpoff & Caspari, 1997). It should be noted, however, that while Weidenreich's theory also invoked gene flow, the revised version of Weidenreich's scheme used gene flow between groups at their overlapping peripheries as its central platform to help explain how human groups evolved through similar successive stages. The multiregionalists have proposed that there was sexual contact between different human groups, at least along the fringes of certain regional communities, that enabled traits to be spread by a sequential process of passing and receiving genetic information. Some anatomical features are said to have developed in a particular region as a result of the need to cope with new and unique environmental conditions encountered within that region and to have been maintained through time (to the present day) within those regions.

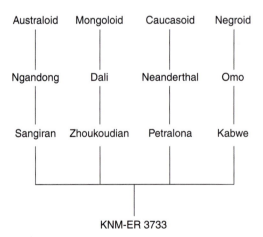

Figure 1.5 ►   A model of the Multiregional hypothesis, with long-existing human regional lineages shown to be established within parts of the Old World by 1.8 million years ago. From Groves (1994), p. 30.

Wolpoff *et al.* (1984) argued that in both Europe and Australia, peripheral groups absorbed genetic material from the main population centers of Asia and Africa. In Australia, they maintain, gene flow was mainly from the southern and eastern parts of East Asia, while Europe is thought to have been influenced more by the major centers of Africa and western Asia. Therefore, some regional continuity in fossil anatomy can be shown, especially in the peripheral regions, and anatomy is still linked to the ongoing evolution of our species by gene flow between the centers and the peripheral regions. For example, a continuation of anatomical form, they suggest, can be seen between the *H. erectus* populations of Java and the Pleistocene Australians. Both groups are said to have a large supraorbital torus, a flat frontal bone, a developed occipital torus, and facial prognathism. The Pleistocene *Homo pekinensis* populations of China are linked, they argue, to the modern populations of northeast Asia and the Americas by possession of large and shovel-shaped incisors (incisor cutting edge is curved, not straight, at the lateral margins) and other features (Wolpoff *et al.*, 1984). Conversely, the peripheral European populations, allocated here to *H. heidelbergensis*, are said to maintain certain Neanderthal features, including strong midfacial prognathism and a backward projection (bunning) of the occipital. Only Africa is said to lack any evidence for regional continuity features. Of course, multiregionalists do not recognize *Homo erectus*, *Homo pekinensis*, and *Homo heidelbergensis* as different species. For multiregionalists, they are all archaic versions of *Homo sapiens*.

Many of these "unique" features, however, appear to be no more than primitive retentions, passed on from a common ancestor, for they can be identified in numerous human populations, not just in the regions where they are claimed (rightly or wrongly) to predominate. Some of these regional "transitional" fossils are characterized by a mixture of primitive and derived features, which say little about their evolutionary past (Groves, 1989a). Other features, which may be considered regionally distinct (Neanderthal populations with large nasal cavities and sinuses), are as likely to be related to environmental conditions (part of an exaptation that enables greater warming of the freezing "Ice Age" air before it reaches the brain [see Chapter 8; also partly Coon, 1962]) as they are to regional continuity based on close evolutionary relationships. Indeed, modern Africans, who are said to lack a list of regionally unique features, have a number of derived features commonly observed in all modern human populations throughout the world, including a high forehead positioned directly above a vertical face, a chin, a rounded occipital, and a short, flexed braincase

(D.E. Lieberman, 1995). This would tend to support the idea that modern humans really did originate in Africa.

Recent studies and interpretations of fossil *H. sapiens* and Neanderthal mtDNA suggest to some multiregionalists that interpretations based on living human mtDNA may be oversimplifying the picture of modern human origins. It is suggested that mtDNA from a Willandra Lakes Australian fossil skeleton (Mungo 3), dating from between 40,000 and 60,000 years ago, is moderately different from mtDNA observed in living modern humans (Adcock *et al.*, 2001). No one denies that Mungo 3 represents a modern human, so the difference in mtDNA must be the result of the "extinction" of a modern mtDNA lineage from a prehistoric modern human population. This, the describers suggest, creates a problem for previous molecular interpretations of modern human origins. For example, when examining living human mtDNA, the deepest branch is African, but when examining fossil human mtDNA (Mungo 3), the deepest branch is Australian. This does not mean that modern humans originated in Australia any more than extant mtDNA means they originated in Africa. Indeed, the difference observed between Neanderthal and modern human mtDNA (which is even more distinct) does not necessarily mean that the Neanderthals did not play a direct role in our own evolution. Rather, the absence of the modern mtDNA type, as in the case of Mungo 3, is the result of their long prehistory, and as such it has become extinct through the vagaries of time. We would argue, however, that this study when interpreted correctly actually supports the "Out of Africa" hypothesis. For example, the Mungo specimen is shown by this same analysis to be closer in its mtDNA to the modern human range than to the Neanderthal samples, which are later in time, thus confirming the distinctiveness of the Neanderthals not only from living humans, but also from earlier fossil modern human populations. This interpretation supports the "Out of Africa" model and reflects the distinctiveness of all modern humans (in time and space) compared to our near contemporaries, the Neanderthals.

Recently, Curnoe and Thorne (2003) have provided a revision of the Multiregional hypothesis based on their interpretation of extant ranges of genetic distance. They propose that the human lineage consists of one genus, *Homo*, spanning a period of around 6 million years. In addition to this, they suggest that only four or five species of *Homo* have ever existed over this long temporal span, with the last species, *H. sapiens*, having emerged around 2 million years ago. This is a revised version of the "Single Species Hypothesis," originally rejected in the 1970s when the fossil evidence made it clear to all that a number of different species had to be recognized, given the great degree of variability observed with the available hominid fossil samples. The continued recovery of more fossil specimens over the last 25 years or so has provided even greater evidence that a number of hominid species were contemporary in time and space. The new version of the single species hypothesis, however, ignores the fossil evidence and is based on an abstract interpretation of the available molecular data. Those of us who work with the fossil record tend to recognize that the current situation, in which there is only one species of *Homo*, is unique in the history of our own lineage. Cladogenesis is recognized by almost all as the mode defining evolution, and multiple species of hominids are to be expected. This is best summed up perhaps by Arsuaga (2002:36), who states that

> In reality, a species' complete disappearance from the world does not necessarily have to coincide with the appearance of its descendant species in any given place. This would be a theoretical prerequisite only if one species evolved into another species throughout its entire geographical range, in a process that affected each and every one of its separate populations. In most cases though, a descendant species evolves in a specific geographical location and from a specific population of its ancestral species. Thus the two may coexist over long periods of time within different geographical ranges. . . . In fact, if a descendant species extends its range to other areas still inhabited by its ancestral species, the mother and daughter species could even coexist within one geographical range. Eventually, if the two species occupy the same ecological niche, they compete with each other and the ancestral species could finally disappear.

This is demonstrated not only in the Pliocene and early Pleistocene hominid fossil record, but also in the later Pleistocene hominins. It was only around 40,000 years ago that at least three species of *Homo* existed, *H. neanderthalensis* (the famous Neanderthal people), who occupied parts of Eurasia, *H. erectus* in Indonesia (if we can believe the recent dates for this species), and modern *H. sapiens*, who had by then occupied most parts of Africa, Europe, and Asia (including Australia). Following the emergence of modern humans in Africa between 250,000 and 150,000 years ago, and their later dispersal into Europe and Asia, the more archaic human populations became extinct, not through a form of genocide, but as a result of losing in a competition for finite resources to the *sapiens*.

Curnoe and Thorne (2003), however, recognize only four species within *Homo* that tend to be time successive. The four species that they acknowledge are, starting from the earliest, *Homo ramidus, H. africanus, H. habilis,* and finally, modern humans, *H. sapiens.* They suggest that the chimpanzee should be considered a species of *Homo.*

They dismiss the idea that the pygmy chimpanzee is a distinct species, *Pan paniscus,* and recognize only one species, *Homo troglodytes.* Their evidence for lumping these two species together is that hybrids have been born in captivity. This is a misunderstanding of what "species" are: Ernst Mayr, the biologist who first fully articulated the so-called Biological Species Concept, was very clear that two putative species should be reproductively isolated in nature, and it does not matter what happens in captivity. In fact, as Common and Pygmy Chimpanzees do not overlap in the wild, there is no way of deciding whether or not they rank as distinct species under the Biological Species Concept, so primate specialists have turned to the Phylogenetic Species Concept, under which two putative species differ absolutely (no individual can ever be mistaken for the "wrong" species).

Curnoe and Thorne argue that, given that the DNA differs by 1% between *Homo troglodytes* and *Homo sapiens,* and there is a minimum genetic difference that can support a species distinction (0.25%, in their estimation), only 4 or 5 species can be supported in the human fossil record (see also Eckhardt, 2000). This is paleoanthropology by short division. Their tacit assumption seems to be that hominin evolution is based on anagenesis: Most or all fossil hominin species are directly ancestral to *H. sapiens,* with no contemporary speciation events and no extinctions. This would be truly remarkable for any mammal group over a 5 to 6 million year period. Second, we must accept a concept of "generic ranges of variability," a concept that no other biologist would seriously entertain.

In any case, this preoccupation with genetic variation betrays the fundamental flaw in the revised Multiregional hypothesis — they confuse species and genera. A species is a real biological unit, while a genus is merely a system of biological classification — it is not a blown-up species. There is no automatic relationship between a species and genus, except that a genus will normally consist of a number of species, but this number is not fixed or constrained by genetic variation. Mayr defined living species by their propensity to interbreed in the wild and produce offspring that can themselves reproduce; in paleontology, we cannot possibly determine whether fossil A could interbreed with fossil B, let alone produce offspring that can themselves reproduce. It is because we can rarely make this determination, even in the case of living animals, that many, perhaps most, taxonomists nowadays reject the "interbreeding" criterion altogether, and instead use the Phylogenetic Species Concept. Usually, paleontologists measure degrees of anatomical variability in living species (especially the ones that are thought to be closely related to their chosen fossil group) in order to determine whether a fossil sample can reasonably be considered to fall within or outside an acceptable range of anatomical variability. This is nothing new and has been endorsed by practicing paleontologists for the last 100 years or so.

A genus, however, is *not* directly related to any species concept — that is, it does not presume to define populations within given genetic bounds, or whether members can successfully reproduce together; rather, genus is part of a human-made system of classification. While the concept of the genus has biological implications, it is a category, not a real biological entity like that of a species. This important and crucial distinction appears to lie at the heart of the Curnoe/Thorne confusion: They believe that a genus has a finite number of species that it can contain; i.e., over a 6 million year period, a maximum of only four or five species can exist. This is incorrect: A genus can potentially contain 1, 5, 20, 30, or 50 species. In living Old World monkeys, for example, there are at least 18 well recognized species within the genus *Macaca*, and at least 19 in the genus *Cercopithecus*; these numbers do not include the fossil species of these genera (see Fleagle, 1999; Groves, 2001, has 19 species in *Macaca* and 24 in *Cercopithecus*, both probably underestimates). There are certainly rules in the formulation and recognition of genera, though they have nothing to do with concepts of anatomical or genetic variability. Genera are groups of organisms recognized as sharing an immediate common ancestor, partly defined by all species sharing a number of unique anatomical features, usually associated with specific derived adaptations which help define the group. Thus, a genus is defined by evolutionary relationships between species and is totally unrelated to concepts of fixed degrees of anatomical or molecular variability (see Figure 1.6). Attempts to objectify the concept relate to giving it a standard time depth, and have nothing to do with the number of species allowed per genus; thus, Groves (2001) urged that a genus should have separated from its sister genera by the Miocene-Pliocene boundary, and this, if adopted, would indeed make it thinkable to unite humans and chimpanzees in the same genus. Ironically, therefore, Curnoe and Thorne (2003) might be doing the right things, but for totally the wrong reason!

Even ignoring this basic flaw in their model, if Curnoe and Thorne (2003) wish to propose such a fundamental revision of the human family

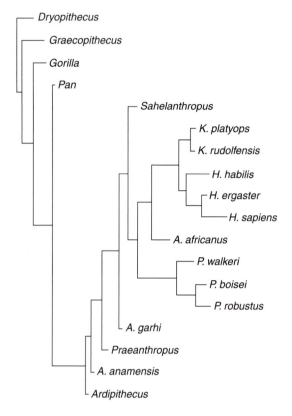

Figure 1.6 ▶   Unlike a species, which represents real biological entities, a genus is a unit of classification and is not defined by variability but by evolutionary relationships. It can be seen that the species within *Paranthropus, Homo,* and *Kenyanthropus* are each defined by a common ancestor, to the exclusion of all other taxa (monophyletic group). They are also defined by a number of unique adaptations. For example, *Paranthropus* species are defined by a large robust face, neuro-orbital disjunction (brain set back from the face with a low frontal), with large, grinding, stonelike molars, while *Homo* species have a large brain, marked neuro-orbital convergence (brain set above the face — high frontal), small face and dental complex, and a more efficient mode of bipedal locomotion. It can also be seen that the species usually allocated to *Australopithecus* (e.g., *A. africanus, A. garhi, A. anamensis*) do not share a common ancestor. Thus they are paraphyletic, and each can be considered as representing a distinct genus, with only the type species *africanus* representing a species of *Australopithecus*. A genus is defined by phylogenetic relationships and *not* by concepts of anatomical and/or genetic variability.

tree, then the onus is on them to provide anatomical definitions of their species, and this is conspicuously absent. So far, they have been working in a fossil-free zone. Species descriptions are crucial because any paleoanthropologist who finds a new fossil needs to be able to allocate his or her

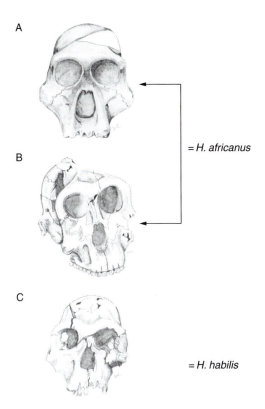

A

B   = *H. africanus*

C   = *H. habilis*

Figure 1.7 ▶   According to the revised Multiregional hypothesis, specimens A and B would belong to "*H. africanus*" while specimen C would be allocated to *H. habilis*. Almost all experts, however, consider specimens A and C to be more closely related to each other in phylogenetic terms (indeed some allocate both species to *Australopithecus*), while specimen B is considered by almost all paleoanthropologists to be very distinctive from all other hominins and as such has been allocated to its own genus *Paranthropus*.

new specimen to a species via the only material preserved — its anatomy. Curnoe and Thorne do not look at the fossil record at all; they say what the species ought to be, not what they actually are according to a detailed analysis of the evidence.

Their version of *H. africanus* would include the five species currently allocated to *Australopithecus* (one of which of course is *A. africanus*, specimen A in Figure 1.7) and the three species allocated to *Paranthropus* (one of which is *P. boisei*, specimen B in Figure 1.7). As such, this single species would include one "sub-population," defined by a high frontal (forehead), well-developed snout (similar to that observed in living chimpanzees), with a relatively narrow and gracile facial structure, undeveloped bony ridges along its braincase associated with reduced musculature, large front teeth, small back teeth, and relatively thin molar enamel; and another subpopulation with the exact opposite condition, totally lacking a forehead, a flat face lacking a snout, a broad and heavily built facial structure, massive bony ridges along its braincase for strong musculature attachments, extremely small front teeth, massive back teeth, and hyperthick molar enamel. Clearly these "subpopulations" are defined by a number of differing evolutionary adaptive trends (see Chapter 5). We would not expect to see such distinct trajectories within one species. In terms of adaptive trends, a number of patterns are present, again refuting the idea that these taxa represent one species. Let us also emphasize this most strongly: No living ape species, or monkey species for that matter, even comes close to such extremes in anatomical variability. It makes the large degree of anatomical variability observed in gorillas (male and females combined!) look tiny by comparison.

Indeed, if Curnoe and Thorne were to try and produce a description of their species *H. africanus*, we believe that the degree of anatomical variability expressed would swallow up the anatomical condition present within at least two of their four other species, namely *H. ramidus* and *H. habilis*, and possibly *H. troglodytes* as well. *Homo sapiens* would probably be the only species to survive this pruning and remain a distinct species. There would now be just two species, *H. africanus* and *H. sapiens*. It is also not unimportant that what most people today would call *Australopithecus africanus* (specimen A in Figure 1.7) and *Homo habilis* (specimen C in Figure 1.7) are increasingly thought to be closely related. Indeed, some consider *H. habilis* to represent a species of *Australopithecus* — *A. habilis* (Wood & Richmond, 2000). Infact, it is surprising that Curnoe and Thorne (2003) maintain the species distinction between *africanus* and *habilis*, which are overall very "similar," while at the same time lumping the very different species within *Paranthropus* and *Australopithecus* into just one species, *H. africanus*.

Let us remind ourselves that paleontologists who have conducted "blind" studies on samples of skeletal remains of living primates tell us that we are liable in almost all cases to vastly underestimate the number of species present in any given sample; the exact opposite of what the Multiregionalists would have us believe (Cope, 1988, 1993; Cope & Lacy, 1992; Plavcan, 1993; Shea *et al.*, 1993; see also other papers in Kimbel & Martin, 1993). This is to say that skeletal remains appear to underrepresent actual speciation: Variation is more prolific than skeletal or fossil anatomical variability. For example, Figure 1.8 presents a principal component

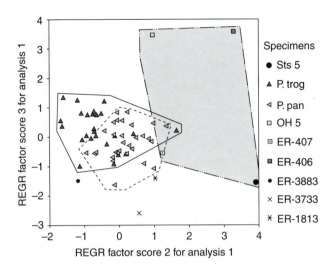

Figure 1.8 ▶ Principal components analysis of extant *Pan paniscus* and *Pan troglodytes* specimens as well as fossil hominins.

Extant *Pan* data is from Cameron (unpublished raw data), while fossil hominin data is from B.A. Wood (1991) (see text for details).

analysis (metric characters) of *Pan paniscus* and *Pan troglodytes* specimens as well as fossil hominins. It can be seen that under the "revised multiregionalist" definition, *Homo africanus* (shaded area) is almost twice as great in range as the combined range of the two species of *Pan*. The molecular clocks tell us that this combined range of variability represents around 2.5 million years of evolution (i.e., the two species of *Pan* split around 2.5 million years ago). The time depth of the revised *H. africanus* represents just half of this (the range from oldest to youngest specimens is around 1 to 1.5 million years), though the specimens have twice the range of variability. Clearly at least two species of hominin are represented within the "revised Multiregionalist" *Homo africanus*.

Again, non-metric variability is vastly greater between extant and fossil hominid groups than between the two chimpanzee species. Table 1.1 presents a breakdown on phenotypic characters (non-metric anatomical features) used by Cameron (in press [b], submitted) in his analysis of fossil hominin systematics. The first is based on an analysis of 72 characters, the second uses 92 characters. While we stress that these values are crude "yardsticks" and alone cannot be used to determine taxonomic allocations, they clearly support the idea that a number of species (even, in

TABLE 1.1 ► Phenotypic Differences Between Taxa

Specific Percentage: Based on 72 phenotypic characters	
(Cameron, in press b)	
Pan troglodytes and *Pan paniscus*	3%
Australopithecus afarensis and *Australopithecus africanus*	53%
Paranthropus walkeri and *Paranthropus boisei*	25%
Paranthropus boisei and *Paranthropus robustus*	13%
Kenyanthropus platyops and *Kenyanthropus rudolfensis*	13%
Homo habilis and *Homo ergaster*	26%
Homo ergaster and *Homo erectus*	13%
Homo erectus and *Homo sapiens*	26%
Specific Percentage: Based on 92 phenotypic characters	
(Cameron, submitted)	
Homo habilis and *Homo sapiens*	41%
Australopithecus afarensis and *Australopithecus africanus*	33%
Paranthropus walkeri and *Paranthropus boisei*	32%
Paranthropus boisei and *Paranthropus robustus*	21%
Kenyanthropus platyops and *Kenyanthropus rudolfensis*	3%
Homo ergaster and *Homo sapiens*	21%
Homo ergaster and *Homo sapiens*	21%
Homo habilis and *Homo ergaster*	26%
Homo habilis and *Australopithecus africanus*	47%

Missing variables counted as the same character state; thus there will be a tendency to underestimate phenotypic variability in fossil groups (see text for details).

conventional assessments, within differing genera) are present, as their degree of metric and non-metric anatomical variability are both *beyond* the species range observed in the species of *Pan*.

We suggest that the complexity we see in the evolution of the human lineage, and the rest of the vertebrates and invertebrates for that matter, is real and cannot be wished away by those seeking to short-circuit the well-documented burgeoning diversity of the natural world. They have produced a model which is not only unworkable for practicing paleoanthropologists, but is also tautological in its construction.

Those proposing an African origin for modern humans argue that, after the original human dispersion from Africa around 1.8 million years ago, a number of populations settled within specific regions and followed their own evolutionary course. This eventually resulted in the evolution of modern human populations in Africa around 250,000–150,000 years ago

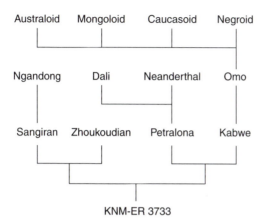

Figure 1.9 ▶ The "Out of Africa" hypothesis, with some key fossil specimens representing distinct hominin species. Note that all end in extinction and that only the later Omo African hominin populations (or populations very much like them) survive to give rise to modern *H. sapiens* around 200,000 years ago.

(Figure 1.9). In Europe, however, the earliest representatives of a Neanderthal lineage were starting to adapt to the freezing conditions of an "Ice Age" northern hemisphere, while in mainland Asia, relict populations such as *Homo erectus* and *Homo pekinensis* lived on in isolation (see Stringer *et al.*, 1984; Stringer & Andrews, 1988; Stringer, 1989; Groves, 1989; Stringer & McKie, 1997).

By 120,000 years ago, the modern humans of Africa began a second dispersal out of Africa into Europe and Asia. They eventually replaced the Neanderthal and Asian populations without much or any interbreeding. The "archaic" indigenous populations quickly succumbed to competition for the available resources by the more modern arrivals from Africa. According to a less extreme form, however, some paleoanthropologists who agree with much of the "Out of Africa" hypothesis suggest that there may have been some sexual contact between the moderns and the more primitive indigenous populations. However, given their suggested specific status (if correct) this would result in no offspring or in offspring that were unable to reproduce, though this may be a misunderstanding of the "reproductive isolation" model of species. Given the spatial and temporal overlap of Neanderthals and modern *H. sapiens*, as well as recent molecular studies (see Chapter 9), is it reasonable to suggest that there was little or no interaction between these "distinct" roaming groups, other than violence?

Initially, the hypothesis that modern humans originated in Africa was based on the available fossil and archaeological evidence (Bräuer, 1984, 1989). For example, while "classic" Neanderthal populations were beginning to dominate Europe around 120,000–80,000 years ago, more modern-looking people occupied parts of Africa, as represented in South Africa by fossil remains from Klasies River Mouth, Border Cave, and Die Kelders Cave; in northeast Africa by fossils from Omo-Kibish in Ethiopia; and in northwest Africa by the fossils from Jebel Irhoud in Morocco (F.H. Smith, 2002). The recent significant discoveries of modern *H. sapiens* from 160,000-year-old deposits in Ethiopia (T.D. White *et al.*, 2003; Clark *et al.*, 2003) have finally bridged the temporal gap between the more archaic and modern *sapiens*. As T.D. White *et al.* (2003:742) state:

> The Herto hominids are morphologically and chronologically intermediate between archaic African fossils and later anatomically modern Late Pleistocene humans. They therefore represent the probable immediate ancestors of anatomically modern humans. Their anatomy and antiquity constitute strong evidence of modern-human emergence in Africa.

Fossil specimens currently considered as representing early *H. sapiens* are defined by a relatively short, high braincase and reduced supraorbital torus. Associated with this physical change within the African populations are signs of a change in tool technology. The long history of a stone hand axe technology gave way to lighter and more refined toolkits, which included sharp stone flakes for more precision cutting, wooden spear shafts with attached spear points, bone fishhooks, and other specialized tools to assist in woodworking and in butchering carcasses (see Schick & Toth, 1993; Deacon & Deacon, 1999). There is also, in the case of the Herto hominin, evidence of postmortuary cultural modification (Clark *et al.*, 2003), similar to that observed in the Willandra Lakes people, who occupied Australia around 100,000 years later. It is suggested that two of the three crania so far discovered have evidence of cut marks on the zygomatic and parietal associated with selective defleshing (Clark *et al.*, 2003), though unlike the Willandra Lakes people there is no evidence of cremation.

The development of molecular biology and its application to the question of modern human origins has to some degree supported these paleontological and archaeological interpretations. Cann *et al.* (1987) in their now-classic study, took the placentas from 147 women from numerous ethnic backgrounds and analyzed their mtDNA. They concluded that the African populations were more variable than those of other groups, suggesting that their mtDNA had evolved for a slightly longer time than that from other groups. This in turn suggested that the first modern humans originated in Africa and that all present-day humans are descendants of the original African ancestral group. Calculating that two samples would differ by 20–40 base mutations every million years suggested a divergence mutation rate of 2–4% per million years. This rate is partially based on the fact that the greatest degree of divergence in mtDNA types within modern human populations is about one-twentieth as great as human mtDNA is from the chimpanzee. Because the last common ancestor between humans and chimpanzees occurred around 5–6 million years ago, the last common ancestor of all modern humans was estimated at around 200,000 years ago (see Ruvolo, 1994; Pilbeam, 1996; Dover, 1999; Sykes, 2001; Relethford, 2001). Finally, as discussed previously, the extraction of ancient mtDNA from a number of Australian Pleistocene modern human remains (Adcock *et al.*, 2001) has enabled us for the first time to examine and compare early modern human mtDNA, dating from between 60,000 and 8,000 years ago, with recent modern humans. It has been demonstrated that the preserved mtDNA extracted from Mungo 3 (dating to between 60,000 and 40,000 years ago) and from later specimens from Kow Swamp and Mungo (about 15,000–10,000 years ago) are all relatively similar to one another as well as to modern humans, especially when compared to the mtDNA of Neanderthals. The extracted Neanderthal mtDNA comes from specimens dated to around 35,000 years ago and is distinct not only from modern humans but also from Mungo 3; that is, the older Australian mtDNA is closer to that of modern humans than is the later Neanderthal mtDNA. This clearly supports the "Out of Africa" hypothesis. This as well as other issues discussed in this chapter will be considered in greater depth within the forthcoming chapters.

The next chapter will review the emergence of the earliest Miocene apes around 23 million years ago and finish just before the emergence of the earliest proto-humans in Africa from between 7 and 5 million years ago, which may (or may not) represent the earliest members of the human lineage. It is around 17 million years ago, we theorize, that the first hominid arose "Out of Africa." It is also from one of these early primitive hominid groups that the earliest members of our own lineage, *Homo*, evolved. Without them there would be no story to tell.

INTERLUDE I
Creationism and Other Brainstorms

Biologists are not always looking anxiously over their shoulders in case the general public should discover that evolution is all a confidence trick, though some creationist writings seem to imply that. But biologists have become a bit distressed over the past 20 years or so that the message is not getting through. In Australia, Mike Archer's probes seem to suggest that about 15% of the population would rather believe in the literal truth of the Bible than in science; in the United Kingdom and Canada, the figure is lower, perhaps 5%; but in the United States it is much, much higher — over 40%.

Very curious. Here is the world's most powerful country, with its most eminent scientists, yet nearly half of its population simply does not believe what these eminent scientists are finding out. In fact, science seems to be subject to democracy. In the United States, school boards — those who decide on the curriculum for schools in local districts — are elected by popular vote, and some members of some school boards have won election by promising that they will deny children the right to be taught about evolution. Actually, very few of them go quite that far; mostly, they push a wedge in the door by saying that fair's fair: Where "evolution science" is taught, then something called "creation science" must be taught as well. What could be more balanced than that?

Science is a process of finding out. It is not a matter for democracy. You can't vote on the truth.

Creationists write lots of books. Usually they are full of bright pictures and cartoons, and their arguments against evolution consist in the main of quoting scientists verbatim but out of context so that it looks as if the scientists are admitting some dreadful secret, that the evidence for evolution is actually pretty sparse, maybe even doctored. The books are not actually science books at all, though their adherents treat them as if they are. You can read some funny things there. For instance, there is a strange, intemperate book by one Father Patrick O'Connell called *Science of Today and the Problems of Genesis*, first published in 1959 and reprinted in 1993. Father O'Connell was in China during the 1930s and read the reports of the discoveries at Zhoukoudian as they came out in the newspapers. This made him, in his opinion and that of the creationists who quote him devotedly, an expert on "Peking Man." His woeful understanding of anatomy and geology and his ignorance of the process of casting (which he thought was just making models or copies) led him to accuse Franz Weidenreich (after the latter's death, of course) of falsifying the records. The tragedy that the original 1930s specimens were lost during the war, enabling him to propose a truly libelous hypothesis: that the eminent Chinese paleoanthropologist Bei Wenzhong, who had been involved in the Zhoukoudian discoveries, "may have destroyed the fossils [during the war] before the Chinese government returned to Peking in order to conceal the fact that the models did not correspond to the fossils." In 1981 a truly disgraceful book by J.W.G. Johnson, called *The Crumbling Theory of Evolution*, was published. Johnson had read no science at all, only other creationists. Treating O'Connell's hypothesis as if it were fact, he wrote, "That, to me, was the masterstroke. Get rid of the incriminating evidence; and let Peking Man live on as our immediate ancestor."

Creationists have let their ignorance give them free rein to accuse Eugene Dubois, the discoverer of the first *Homo erectus* at Trinil, of dishonesty as well. Both Malcolm Bowden's (1977) *Ape-men: Fact or Fallacy?* and Duane T. Gish's (1978) *Evolution: the Fossils Say No!* imply that, while he had been promoting his "Pithecanthropus" discovery, Dubois had all the time been hiding the fact that he had uncovered evidence that the ape-man had been contemporary with real human beings ("Dubois concealed the fact . . .""). They mean the Wajak *Homo sapiens* skulls — which of course had never been thought to be contemporary with Trinil at all. Needless to say, in the hands of J.W.G. Johnson, never one to avoid a barefaced lie for a good cause, this veiled implication becomes fact: "However, Dr Dubois had not told the whole truth. He had not told the most important part of the story. He did not tell that he had also found two human skulls in the same stratum as the skull-cap. To have told this would have spoiled his case because those human skulls, the Wadjak skulls, as they are called, showed that real human beings did live in Java at the same time as the supposed ape-men."

Of course, it is quite true that paleoanthropology has not been lacking in its embarrassing mistakes. One that is brought up without fail in creationist writings is Nebraska Man. A storm-in-a-teacup brewed by Osborn, who mistook a fossil peccary tooth for a primate one and described *Hesperopithecus haroldcooki* on it, was stirred by an overzealous artist in the *Illustrated London News*, who "reconstructed" a whole proto-human on that slender basis, before being finally laid to rest by Osborn himself when he realized his error — a nice example of the self-correcting nature of science, which creationists would do well to emulate.

Another creationist mainstay is the notorious Piltdown forgery, in which (in the early 20th century) parts of a human skull and an orangutan jaw were fraudulently modified to make it appear that they were one creature. Many scientists were fooled, and in 1913 the taxon *Eoanthropus dawsoni* was erected for the composite. Not until 1952 was the fraud uncovered, by first using fluorine content analysis and then carbon-14. The point creationists always overlook when retelling this tale is that, as more and more genuine fossils were discovered in the 1920s to 1940s, Piltdown came to look more and more anomalous. Specialists increasingly questioned whether the skull and jaw really did belong together. They also overlook that it was scientists — in creationist jargon, "evolutionists" — who exposed the fraud in the end. This very precisely illustrates the way science works; if some paleoanthropologists had been gullible, that is just human nature and has nothing to do with the study of human evolution as such.

Creationists also tend to trot out a batch of modern human remains that, they claim, are ignored or covered up by "evolutionists" because they are from very early strata and so don't fit. Actually the reason these are nowadays ignored (and appear in few modern textbooks) is not because of a cover-up but because they are rubbish. Calaveras (California) is a supposedly Miocene skull claimed by gold miners in the 1880s to have been found in gold-bearing deposits; it was in fact a notorious hoax — the creationists'

Piltdown, one might call it. The Castenedolo (Italy) skeletons and other remains were found in Pliocene marine deposits between the 1860s and 1880s; chemical and radio-carbon tests have shown them to be recent burials. The Foxhall (England) jaw, found in a sand quarry by workmen in 1855, was purchased by a pharmacist and sold to an American, Dr. Collyer, who claimed for some obscure reason that it came from the base of the Suffolk Red Crags, of Pliocene age. The Abbeville or Moulin Quignon (France) jaw was another fraud, placed in the ground by workmen, in 1863, to be found by them when the archaeologist Boucher de Perthes was watching. And so the sorry list goes on. The creationists who love to accuse paleoanthropologists of fraud (almost invariably dead paleoanthropologists, who can't sue) fall over themselves in the rush to resurrect discredited specimens as showing modern humans back in the Dark Ages, when only australopithecines cased the joint.

Oh, there are other brainstorms. There is the guy who is sure that our ancestors were bipedal and had huge globular heads from way back — we just haven't found their remains, that's all. There are the followers of Erich von Däniken, who wrote that human beings were not as bright as all that and that visitors from outer space had to come along and teach them to build pyramids. Why, they may even have tinkered a bit with our DNA to make us really truly human. Then there were Fred Hoyle and Chandra Wickramsinghe, two eminent astronomers, who maintained that there have been peri-odic influxes of viruses, dropping as the gentle rain from heaven, and that these invaded the DNA of earthbound organisms and caused evolutionary boosts. And sundry folk are convinced that Neanderthals, or *Homo erectus*, or *Gigantopithecus* at the very least, still roam the mountain fastnesses of the Himalayas, or the Altay, or the Pacific coast of North America, peering out at earnest souls who mount expeditions to look for them and giving them the slip. The Bigfoot seekers don't make as many millions from their efforts as did von Däniken. (Indeed, the world would be a poorer place without them.) And none of the minor brainstormers have anything like the insidious influence on peo-ple's minds as do the creationists.

Most creationists like to give the impression that the entire edifice of human evolu-tion is based on nothing more than a handful of fragments. So Malcolm Bowden in his 1977 book (see previous) wrote that "the fossil links between man and the animals con-sist only of fragments of jaws, some broken skull pieces, part of a foot, etc., no com-plete skeleton or even a reasonable proportion of one ever having been discovered." This was crap even at the time he wrote it. But of course most creationists read only other creationists, so it is no surprise to read Unfred and Mackay in 1986 telling the world how little fossil evidence there is for australopithecines: "*Australopithecus africanus*, . . . a nearly complete skull, several jaws, numerous teeth, portions of pelvis and fragments of long bones; and *Australopithecus robustus*, . . . a small portion of the left side of a skull, ends of a few limb bones and a young lower jaw." Their source for this devastatingly incomplete catalogue? Bowden.

And so it goes on. The dismal catalogue of creationist ignorance, half-truths, and utter dishonesty goes on. There are a few — a very, very few — creationists who do seem to be more honest. The Hindu creationists Michael Cremo and Richard Thompson in 1993 wrote a vast tome called *Forbidden Archaeology*, claiming that humans are billions of years old, just as the Hindu scriptures say they are, and that there are lots of traces of them in Palaeozoic deposits that have been suppressed by orthodox archaeology and paleoanthropology. Marvin Lubenow, a mainstream Christian creationist, wrote *Bones of Contention* in 1992. Both books at least don't try to hide how abundant human fossils are or attempt to accuse honest scientists of fraud; but, like the rest, Cremo and Thompson too drag up long-discredited "ancient" *Homo sapiens*, and both of them bust a gut to fit the facts into their religiously mandated prejudices — but all creationists do that. One who certainly knows what he is doing, in anatomy at least, is Jack Cuozzo, who studied some of the original fossils of Neanderthalers; but his 1998 book, *Buried Alive*, is totally off the planet, proposing that Neanderthal skulls like La Ferassie and La Chapelle are the characters who lived shortly after Noah's Flood and, as we are assured by the Book of Genesis, lived to over 300 years — and that their great age explains their peculiar anatomy.

Ah, Noah's Flood. That mainstay of creationists who assume the inerrancy of the Bible. The Garden of Eden is important, but Noah's Flood explains everything. See, it explains the entire geological column — all that thickness of sediment deposited by a massive deluge in less than a year. Goodness me. Duane Gish, like Jack Cuozzo, believes that Neanderthalers and their ilk are "descendants of post-Flood man" because the deposits in which they have been found are Pleistocene, "believed to be post-Flood" (by whom? Ah yes, we forgot: by Gish). If you think that is from left-field, try this, from a creationist called Kofahl: "In fact a number of the man fossils may represent peoples which had suffered degeneration as the result of sin."

As for culture, all this Palaeolithic–Neolithic–Metal Age business must be rejected out of hand (whatever the stratigraphic evidence in archaeological excavations!), because Noah, from whom we are all to have descended, had access to "ocean-liner technology" (so said John Mackay in 1984). Peoples who today use (or until recently have been using) a stone technology have degenerated: "The current status of the races . . . is not a result of innocent people searching for improvement. It is a direct consequence of whether the ancestors of any race worshipped the living God or deliberately rejected Him. . . . [Technologically simple peoples are] spiritual degenerates in need of the gospel of the Creator Christ so they can appreciate education and the relevance of technology" (Mackay again).

It is a disgrace that people who claim to be men of God can not only write like this, but can for no cause accuse scientists of fraud and falsify the evidence to fit their own assumptions. It is an even greater disgrace that they wish to alter school curricula to teach kids this falsified fantasy as science. It is astonishing that so many ordinary people eagerly consume this pernicious rubbish. But they are innocent victims; they do not bear the burden of guilt that the purveyors do. The story of human evolution has many different models and hypotheses and takes slightly different forms in the hands of Chris

Stringer, Leslie Aiello, Gunther Bräuer, Milford Wolpoff, Wu Xinzhi, Alan Thorne, Fred Grine, Ron Clarke, Fred Smith, Yoel Rak, Philip Rightmire, and dozens of other competent paleoanthropologists. Still, they all have one thing in common: They know what they are talking about, they are honest brokers, they try to dupe no one. But then, unlike the creationist rabble, they are scientists. Speaking of science . . .

CHAPTER 2

Evolution of the Miocene Great Apes

The small-bodied ape ran across the top of a tree branch, away from the anger of the dominant male of the group. In haste to get away, however, it had underestimated the thickness and strength of the branch, which gave way. The ape fell to the ground, breaking a forelimb; it yelled in pain. Above, the other apes started a commotion and ran across the branches in anxious movements. The yelling would surely bring carnivores to the small patch of tree cover, which was an island of refuge in the surrounding open country. Soon a lone carnivore appeared and saw its opportunity for an easy kill — it struck swiftly and surely, closing its jaws around the neck of the small ape, crushing its windpipe. Anchoring the head between its jaws, it dragged the ape away from the tree cover. The small apes in the trees could still be heard yelling and thrashing around in the small forest patch.

This sort of thing must occur frequently in tree-dwelling primates. Life in trees can be just as dangerous, in terms of injuries, to those that live in more open habitats. In the mid-1980s a Miocene small-bodied ape fossil was discovered in Kenya and allocated to a new genus, *Turkanapithecus*. The specimen, KNM-WT 16950, was remarkable for its state of preservation, with most of the facial anatomy retained in detail. Also preserved was a puncture mark, most likely from a carnivore canine, located just below the midface or snout. The death of this individual was surely the result of an encounter with a carnivore.

It was probably the mass extinctions of the dinosaurs around 65 million years ago that enabled the diurnal mammals to occupy "in mass" the now-numerous vacant niches, which had until now been closed to them. Among these mammals would have been the primitive and small-bodied primates (similar in appearance and size to squirrels), which appear during the Paleocene epoch. These archaic primates can be differentiated from other small mammals by their possession of numerous primate specializations, including more convergent orbits (stereoscopic vision), a postorbital bar (i.e., the orbit is closed off and not exposed to the temporalis muscle), increased grasping extremities, and nails as opposed to claws (see Fleagle, 1999). Cartmill (1992) suggests that these characteristics evolved as a result of the predatory nature of the archaic primates, for stereoscopic vision is usually observed in predator species that rely on vision to detect their prey, though undoubtedly increased stereoscopic vision would be an excellent exaptation for leaping behavior in primates (Crompton, 1995), or perhaps it was the other way round, that is, exaptation for predator behavior. Conversely, Cartmill notes that most arboreal nonprimate species are not defined by orbital convergence and, in addition, are defined by having claws on their digits to assist in climbing. In early primates, however, the increased ability to grasp by their digits (associated with nails as opposed to claws) is a likely requirement for grasping prey or other food objects, as opposed to assisting in climbing (see also Fleagle, 1999).

With the demise of the dinosaurs, the primitive primates were able increasingly to become more diurnal, and eventually many species adapted completely to life in the sun, depending on daylight feeding patterns. It is with the coming of the late Eocene/Oligocene transition, around 40–37 million years ago, that we witness within the fossil record the first true anthropoids. Most fossil anthropoid species from this temporal region have been discovered over the last 40 years or so in the Fayum depression of Egypt, southwest of modern-day Cairo (Figure 2.1). While earlier primate-like species have been documented in Europe, Asia, and even the New World, Africa appears to represent the place of origin for the earliest true primate species. It is at this time that we see a reduction in the snout of the primates, which can be correlated with a reduction in the olfactory apparatus (sense of smell). This would also enable a further enhancement of the three-dimensionality of these primates (see Fleagle, 1999; Kingdon, 2003). With the emergence of diurnal primate

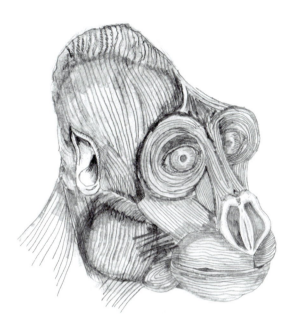

Figure 2.1 ▶   Reconstruction of facial musculature of the Oligocene primate
Aegyptopithecus zeuxis from the Fayum depression of Egypt.

species came an increased ability, in some species at least, for predatory
behavior (e.g., insects), while in others it likely resulted in an adaptive shift
toward larger, often harder fruits; all of which would conceivably result in an
increase in body size/weight (Kingdon, 2003). The large number of archaic
primate species (and genera) that have been identified so far from the Fayum
depression alone supports the extreme range of biological diversity of the
primates (and other nonprimate faunal groups). The primates of the Fayum
represent the likely anatomical condition of the primates that eventually gave
rise to the later hominoids, though whether these primates themselves are the
ancestral population seems doubtful.

The Early Miocene of Africa

During the Eocene, forest vegetation, increased rainfall, and hotter conditions spread from the equator to the poles. The earth's ecology was relatively homogeneous. With the Eocene/Oligocene transition, however,
ecological stability started to break down, and the world was thrown into
the "big chill" (Prothero, 1994). By 25 million years ago, the position and

shape of the continents where beginning to look something like they do today, though North and South America remained separated, the Himalaya and Tibetan plateau had yet to develop, and Africa was still an island continent. At the Oligocene/Miocene transition, however, we see the return of warmer climatic conditions. Even so, from the earliest Miocene, the rainforest belt, which had covered most of Africa, had been breaking up into a number of distinct ecological niches. Instead of the homogeneous tropical cover, we see ever-increasing patches of woodland and grassland interrupting the vast tracts of rainforest. The ongoing continental collisions had reached their zenith during the Miocene, with the major uplift of the Himalayas, the Tibetan Plateau, and the Ethiopian highlands, as continental plates crashed against each other, twisting and thrusting upward from the external land surface. Ecological instability resulted in rapidly fluctuating climatic conditions. This was not only a worldwide pattern but also occurred at a much finer scale, resulting in a patchwork of differing ecological niches within relatively small areas, which were always prone to rapid change or ecological extinction. Ecological change in earlier periods of the Cenozoic had settled down into long periods of stability, but this ceased at the onset of the Miocene (Isaac, 1976; Kennett, 1995; Partridge *et al.*, 1995; Potts, 1996; Denton, 1999; Andrews & Humphrey, 1999).

The earliest Miocene also saw the genesis of the great African rift valleys, as a result of the formation of the Ethiopian highlands. There was massive faulting as the external land surface broke and slipped away, forming the fractured and broken valley floors and walls. This splitting of East Africa's land surface produced a vast increase in volcanic activity within this region (Isaac, 1976; Feibel, 1999). The uplift of the Ethiopian highlands was directly responsible for the formation of a rain-shadow zone because these highlands intercepted the eastward flow of precipitation across the continent. Thus, while western and central Africa continued to receive abundant rainfall, the rift valley systems and East Africa in general (which lay beyond the highlands) were marked by a significant rainfall reduction (Isaac, 1976; Potts, 1996; Andrews & Humphrey, 1999). Indeed, this was soon further exaggerated by the uplift of the Himalayas and the Tibetan plateau, which caused the air to rise and fall in the surrounding region and produced summer heating and winter cooling, and in turn increased the intensity of summer drying in East Africa (Quade *et al.*, 1989; Cerling, 1992; Potts, 1996).

The East African early Miocene families Proconsulidae and Afropithecidae together represent a major biological radiation of apelike primates, arguably the earliest apes, though it is possible that the Proconsulidae at least may represent stem catarrhines (the population that gave birth to the Old World monkeys and apes) of modern aspect (Andrews, 1985, 1992; Harrison, 1987, 1988, 1993, 2002; Groves, 1989a; Begun *et al.*, 1997; Harrison & Rook, 1997; Fleagle, 1999). Representatives of the Proconsulidae include the genera *Proconsul, Rangwapithecus*, and *Turkanapithecus* and have a temporal span from 23 to 15 million years ago, while the larger-bodied apes, allocated to the Afropithecidae, include the genera *Afropithecus, Morotopithecus*, and *Heliopithecus*, which date from between 18 and 15 million years ago (see Andrews, 1992; Harrison, 1992, 2002; Cameron, in press a). (See Figures 2.2 and 2.3.) There is little to support a particularly close phylogenetic relationship between these early Miocene apes and the earlier Eocene and Oligocene primates from the Fayum depression of Egypt. This should not be particularly surprising, because they are separated in time by 10 million years and by almost 4,000 km, though some similarities between the Oligocene primate *Aegyptopithecus* and members of the Afropithecidae have been noted (R.E.F. Leakey *et al.*, 1991).

Members of the Proconsulidae had been living in the forests of Africa long before the appearance of the Afropithecidae, and it is likely that the evolutionary divergence between these two groups began when some groups left the forest and moved out into more open areas. Species of *Proconsul* tended to occupy the more "closed forest" habitat and appear to

(a)   (b)

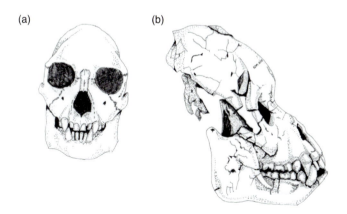

Figure 2.2 ▶   (a) Reconstruction of *Proconsul heseloni* female specimen KNM-RU 7290 (adapted from Walker *et al.*, 1983). (b) Unreconstructed sideview of the same specimen.

Taken from Cameron (in press a).

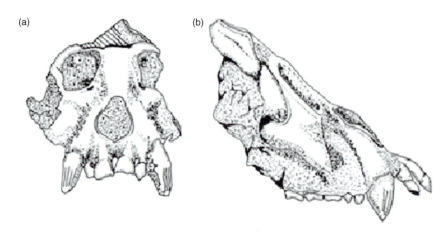

Figure 2.3 ▶   (a) Frontal view of *Afropithecus turkanensis* male specimen KNM-WK 16999. (b) Side view of same specimen.

Taken from Cameron (in press a).

have spent most of their time in the trees. Their postcranial anatomy indicates an above-branch form of locomotion. They were also smaller than the Afropithecidae, and their facial and dental anatomy suggests that they focused on eating soft fruits, for which limited food preparation was required. The ongoing divergence of this group would be emphasized over time as habitat distinctions intensified (see Andrews *et al.*, 1997; Walker, 1997; M.G. Leakey & Walker, 1997; Cameron, in press a).

As the Afropithecidae increasingly occupied more open woodlands and grasslands, they developed a dietary preference for hard and tough food items, which required extensive food preparation prior to digestion. This is evidenced by their strong and robust facial architecture and their large premolars and molars. Members of the Afropithecidae (including emerging daughter species) appear to have extended their behavioral and dietary preference for a more arid habitat and its associated dietary regime. The Afropithecidae adopted a primitive form of terrestrial or semiterrestrial locomotion, associated with increased body size. While a number of unique characters define each of the species within this family, they are united by a general adaptive trend, which is emphasized by selection pressures that are reinforcing these evolutionary-adaptive pathways of coarse and hard food object feeding in a more "marginalized" paleohabitat.

The divergence of these two distinct families can be seen as a direct result of ongoing climatic change, which resulted in a fragmentation of the previous relatively homogeneous ecological conditions expressed in Africa. This ongoing fragmentation starts to impact directly on the early Miocene landscape, but it has a history that extends back even earlier in time.

Later Hominid Phylogenies and Paleobiogeography

During the later parts of the middle Miocene, there was an expansion of continental ice sheets and increasing ice buildup in Antarctica, ultimately the result of the earlier separation of Australia from Antarctica, which had so significantly changed ocean currents and their circulation that it lead in time to a direct effect on world climate, producing global cooling (Williams *et al.*, 1998; Denton, 1999). Significant cooling in the northern high latitudes is also indicated by seasonal ice-rafting of debris into the North Atlantic (Rosen, 1999). As Africa became drier and cooler, it is around this time (the Middle and Late Miocene transition) that we start to witness the biological radiation of Eurasian hominids, whereas hominids in Africa were apparently becoming increasingly rare, though whether this is a real biological phenomenon or the result of fossil sample bias remains unknown.

Whether representatives of the Proconsulidae or the Afropithecidae can be considered the basal population from which the earliest hominids originate remains problematic. The earliest representatives of our own family, the Hominidae, appear around 16 million years ago and are allocated to the subfamily Kenyapithecinae, consisting of the genera *Griphopithecus* and *Kenyapithecus* (Figure 2.4). Most believe that this is the group that represents the basal population from which the hominids emerged, ultimately including the extant hominids, that is, the orangutans, gorillas, chimpanzees, and, of course, humans. The Kenyapithecinae probably gave rise to the later European hominid subfamily Dryopithecinae, *Dryopithecus* (around 12–10 million years ago) and *Oreopithecus* (around 8.5 million years ago) (see Andrews, 1992; Harrison & Rook, 1997; Cameron, in press a). Another fossil ape, *Graecopithecus*, from Greece, dating from between 11 and 10 million years ago, may have been part of a Eurasian dispersal back into Africa and may have shared a close phylogenetic relationship with the late Miocene *Samburupithecus* and the extant *Gorilla* (Andrews, 1992; Cameron, in press a).

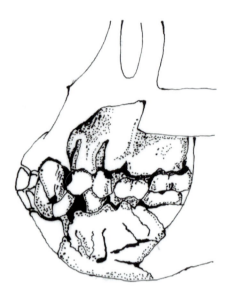

Figure 2.4 ▶   Partial reconstruction of *Kenyapithecus wickeri* from Fort Ternan, Kenya. Adapted from Andrews and Walker (1973).
Taken from Cameron (in press a).

It is suggested that the European subfamily Dryopithecinae originate from the thick, enameled *Griphopithecus* sometime during the middle Miocene between 17 and 14 Ma. *Griphopithecus* is commonly associated with forests in drier and more strongly seasonal conditions, with summer rainfall and prolonged dry seasons (Andrews & Humphrey, 1999); its retention of thick molar enamel can be associated with a dietary regime of small, tough food items. The later Dryopithecinae, which are associated with tropical to subtropical conditions and closed forest habitat, are associated with a number of adaptations, including the evolution of thinner molar enamel and a suspensory form of locomotion to help move through the forest (Figure 2.5) (see Andrews & Humphrey, 1999; Agusti *et al.*, 2001). Thin molar enamel helps further to maintain sharp molar ridges and cusps, which would be beneficial in its dietary preference for soft fruits and vegetative material, that is, it provides increased shearing action. *Oreopithecus*, from 8.5 million-year-old sites in southern Europe, may have evolved from a thin, enamelled dryopithecine or from an as-yet-unidentified ancestral group, though its well-defined pointed molar cusps and sharpened enamel ridges are the ultimate of the thin-enamelled, high-cusped condition (Figure 2.6).

Figure 2.5 ▶ Reconstruction of *Dryopithecus brancoi* from Hungary.
Taken from Cameron (in press a).

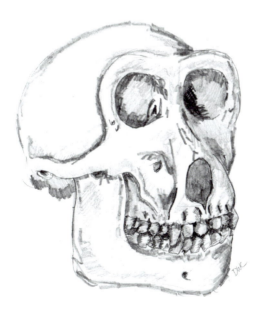

Figure 2.6 ▶ Partial reconstruction of *Oreopithecus bambolii* from Tuscany, Italy.
Taken from Cameron (in press a).

Moyà-Solà and Köhler (1993, 1995, 1996), Agusti *et al.* (1996), and Köhler *et al.* (2001) consider the European later Miocene hominids *Dryopithecus* and *Graecopithecus* as being closely related to the Asian apes, while Begun (1992a, 1994a, 2001, 2002), Begun and Kordos (1997), and Begun *et al.* (1997) consider these same taxa as basal "African" apes. Whether *Dryopithecus* and *Graecopithecus* are "African" or "Asian" hominids has important consequences for interpretations of hominid paleobiogeography. If we accept that they represent basal African hominids, then a Eurasian origin for the extant African hominids is possible, perhaps even likely. This is because many experts are still debating the hominid status of *Kenyapithecus, Otavipithecus,* and *Samburupithecus,* suggesting that they may actually be a phylogenetic sister group to the early Miocene Afropithecidae and, as such, not closely related to the extant African hominids at all (see review in S. Ward & Duren, 2002; also see Begun, 1994b, and Singleton, 2000). If these African fossil taxa are more distantly related, then we currently have no African fossils from the late Miocene, which can be considered immediately ancestral to *Gorilla* or *Pan*. The European hominids *Dryopithecus* and *Graecopithecus*, however, have both at times been argued to represent immediate hominid ancestors (Andrews, 1992; Begun, 1992a; D. Dean & Delson, 1992; Cameron, 1997a; see also de Bonis & Koufos, 1994, 2001; Moyà-Solà & Köhler, 1993, 1995, 1996). If we accept that *Dryopithecus* and *Graecopithecus* represent "Asian" hominids, then the African origins for the extant African hominids at least is still not refuted; that is, we have no Eurasian ancestors for the African hominids.

Moyà-Solà and Köhler (1993, 1995, 1996), Agusti *et al.* (1996), and Köhler *et al.* (2001), conclude that the facial skeleton of the Spanish fossil hominid CLl-18000 shares a number of derived features with the *Sivapithecus–Pongo* clade. This includes a flat zygomatic and the zygomatic foramina position, and its supraorbital region is marked by an orbital rimlike structure. It is also marked by a low glenoid fossa and a reduction in its frontal sinus system, with this latter feature eventually to be lost completely in *Sivapithecus–Pongo* (Figure 2.7) (see also Cameron, 1997a). In its postcranial anatomy, they also conclude that this specimen is closer to *Pongo* in terms of its body proportions, with increased length of its forelimbs, suggesting an adaptation to more frequent climbing and suspensory activities than that observed in the extant African hominids, though, as they partially recognize, this may represent the primitive hominid condition. They conclude that *Dryopithecus* and *Graecopithecus*

Figure 2.7 ▶ Reconstruction of the Greek Late Miocene hominid, *Graecopithecus macedoniensis*.
Taken from Cameron (in press a).

are linked to the *Pongo* clade rather than being a sister group to all extant hominids (see also Andrews & Bernor, 1999). Clearly, from the available evidence, *Dryopithecus* cannot be considered an ancestor to the Asian clade because the European species appear later in the fossil record; that is, *Sivapithecus* appears in Pakistan around 12.5 Ma, while the *Dryopithecus* from Spain (*D. laietanus* and *D. crusafonti*) appears around 9.6 Ma (Agusti *et al.*, 1996). As such, they argue, at least these species of *Dryopithecus* combine the position of being among the more recent but also the more primitive hominid species from Eurasia.

The presence in the Dryopithecinae and the extant African hominids of derived features such as suspensory locomotion and thin molar enamel suggests that *Dryopithecus* is more closely related to the ancestry of the extant African hominids (*Gorilla* and *Pan*), as opposed to the Asian hominid (*Pongo*). The Asian extant and fossil hominids are allocated to the subfamily Ponginae, which have distinct facial and dental features that the African extant hominids and members of the Dryopithecinae lack. The Dryopithecinae, like much of the Middle Miocene European fauna, apparently became extinct with the Vallesian crisis of Europe around 9.5 Ma; if so, the derived form of locomotion shared between the Dryopithecinae and the African hominids must have evolved independently. At any rate, it did apparently evolve separately in the orangutan (whose presumed ancestor, *Sivapithecus*, displays the primitive above-branch locomotion), so there is no reason why it could not evolve independently in the later extant African hominids (see partly Pilbeam, 1996, 1997, 2002).

There is currently no evidence to support the persistence of *Dryopithecus* species after 9.5 million years ago. Nor are they found in the Greek-Iranian Province, which was occupied by the larger apes *Graecopithecus* and *Ankarapithecus*. *Graecopithecus* appears to have anatomical affiliations with an "African ape" condition, but *Ankarapithecus* is considered by almost all to belong to the Asian ape clade, the subfamily Ponginae, which includes the orangutan (Figure 2.8). The earliest fossil representatives of the Ponginae, allocated to *Sivapithecus* (Figure 2.9), had by 12.5 million years ago established themselves in the forests of present-day northern India and Pakistan. The biogeographical divide of the "African-like" *Dryopithecus* from *Graecopithecus* is most likely the result of the separation of the Greek and Albanian landmass from central Europe by the connection of the Aegean with the Paratethys (see R.M. Jones, 1999; Rögl, 1999); thus

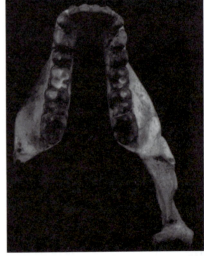

Figure 2.8 ▶ *Ankarapithecus meteai* male specimen AS 95-500.

Photographs kindly supplied to DWC by Dr. Peter Andrews, Natural History Museum, London.

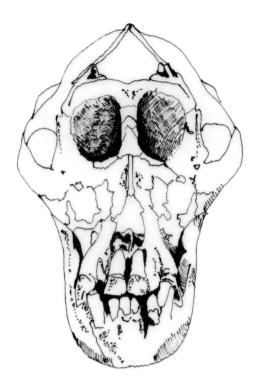

Figure 2.9 ▶   *Sivapithecus indicus* specimen GSP-15000 from the Siwalik foothills, Pakistan; thought by most to be an ancestor to the extant orangutan. Taken from Cameron (in press a).

dispersal into Eurasia from western and northern Europe was not necessarily a "straightforward" proposition.

The Original Miocene "Out of Africa"

The paleobiogeography of the earliest representatives of our own family, the Hominidae, during the Middle and Late Miocene has recently been discussed by a number of researchers (Andrews, 1992; Ciochon & Etler, 1994; de Bonis & Koufos, 1994; Andrews *et al.*, 1996; Pilbeam, 1996, 2002; Begun *et al.*, 1997; Begun & Gülec, 1998; Stewart & Disotell, 1998; Andrews & Bernor, 1999; Agusti *et al.*, 1996, 2001; Begun, 2001, 2002; Heizmann & Begun, 2001; Köhler *et al.*, 2001; Kelley, 2002; Cameron, in press a). Most paleoanthropologists who have discussed an

African or Eurasian origin for the hominids tend to agree with the second scenario; that is, the immediate ancestor to the hominids probably emerged from Eurasia around 14–15 million years ago (Ciochon & Etler, 1994; Begun *et al.*, 1997; Begun & Gülec, 1998; Stewart & Disotell, 1998; Agusti *et al.*, 1996, 2001; Begun, 2001, 2002; Heizmann & Begun, 2001; partly Andrews & Bernor, 1999), the clear implication being that the three hominid genera *Pongo, Gorilla*, and *Pan* are not endemic to their respective extant geographical distributions.

Much of the original evidence used to support a Eurasian origin for the hominids is based not so much on the hominid fossil record but on the well-documented dispersal into Africa of a number of faunal groups, including the Eurasian hipparionine horses and murid rodents that first appear in the African continent around 10–12 Ma (Bernor *et al.*, 1987; de Bruijn, 1986; Flynn & Sabatier, 1984; Ciochon & Etler, 1994; Bernor *et al.*, 1996; Gentry & Heizmann, 1996; Woodburne *et al.*, 1996). Indeed, many "African" mammal faunal groups (e.g., proboscideans, giraffoids, bovids through to rodents) are now recognized as not having any African early or middle Miocene ancestors; rather, their origins appear to be from outside of Africa (Barry *et al.*, 1985; Thomas, 1985; de Bonis *et al.*, 1992; de Bonis & Koufos, 1994; Bernor *et al.*, 1996; Gentry & Heizmann, 1996; Woodburne *et al.*, 1996; Begun, 2001). Solounias *et al.* (1999) have recently argued that much of the extant African savanna fauna migrated into Africa from more northerly latitudes, including Greece and Iran. These migrant fauna are suggested to have replaced much of the African endemic fauna because these new immigrant species were already adapted to the new conditions prevailing in Africa, as a result of global cooling (see later). So much of the previously argued "endemic African savanna fauna" now appears to have its origins within Eurasia (Solounias *et al.*, 1999; see also de Bonis *et al.*, 1988; de Bonis & Koufos, 1994).

Stewart and Disotell (1998) have recently reinterpreted hominoid biographical origins based on extant and fossil catarrhine molecular and fossil evidence (Figure 2.10). When first examining the extant ape molecular data and their geographical distributions alone (i.e., not taking into account the fossil evidence), they recognized two major evolutionary scenarios. The first was that the lineage leading to the extant gibbons dispersed out of Africa to Eurasia around 18–20 Ma, leaving the last common ancestor of the orangutan, gorilla, and chimpanzee in Africa. Later the lineage leading to the orangutan dispersed into Eurasia, while the last common ancestor to *Gorilla* and *Pan* remained in Africa. The second scenario

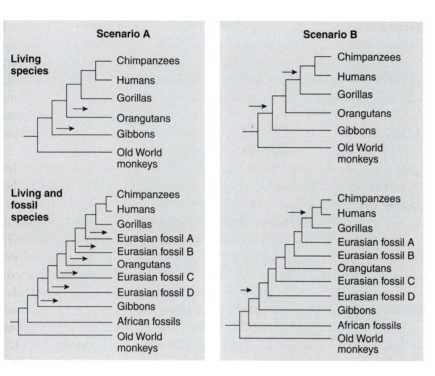

Figure 2.10 ▶ Two dispersal patterns for the hominoids, both assuming an African origin for the Old World primates (catarrhines). Arrows indicate the intercontinental dispersal events required to explain the distribution of the living and fossil species. *Scenario A*: Separate dispersal events from Africa to Eurasia for each of the Eurasian hominoid lineages. *Scenario B*: The common ancestor of all living hominoids dispersed out of Africa, and later the common ancestor of the extant African hominids (gorillas, chimpanzees, and humans) dispersed back into Africa. For just the extant species (top cladograms), these two scenarios are equally parsimonious. When the fossil clades are included (bottom cladograms), scenario B is favored, because scenario A now requires a minimum of six independent dispersal events, while scenario B requires just two (see text for more details).

From Stewart and Disotell (1998), p. 585.

is defined by the dispersing of the last common ancestor of the extant hominoids out of Africa, around 18–20 Ma, which then speciated into the lesser and great ape lineages while in Eurasia. Later, one of these populations moved back into Africa, giving rise to *Gorilla* and *Pan*. Both of these scenarios are equally parsimonious, in the sense that each requires two migratory events. When they included the available fossil hominid material, however, they determined that the first scenario (e.g., African origins) is far less parsimonious, for it required at least six dispersals out of Africa, while the Eurasian origin still only required two such events.

The pattern for African hominid migrations as proposed by Stewart and Disotell, however, is rather narrow in its interpretation, in the sense that they believe it requires every hominid to have its origins in Africa. That is, each is marked by its own migration out of the African continent. For example, they propose that *Griphopithecus* must have moved out of Africa into Eurasia, which was then followed by another separate migration out of Africa by *Dryopithecus*, which was then followed by yet another separate migration out of Africa by *Graecopithecus*. It is doubtful that anyone supporting an African origin for the hominids would agree with such a narrow reading of the "Out of Africa" scenario. Indeed, an African origin can be interpreted in any number of ways. Perhaps the least complicated scenario is that the Asian apes dispersed out of Africa around 18 Ma, followed by the Eurasian hominid ancestor around 16–17 Ma. This Eurasian ancestor then gave rise to the Eurasian hominids, including *Dryopithecus, Griphopithecus*, and *Graecopithecus*, while the ancestor to the African apes, *Gorilla* and *Pan*, remained in Africa. There are three possible middle to late Miocene African hominids, *Kenyapithecus, Otavipithecus*, and *Samburupithecus*, which have been considered by some experts to represent an African hominid ancestor (see Ishida *et al.*, 1984; Conroy *et al.*, 1992, 1993; Hill, 1994; Ishida & Pickford, 1997; Nakatsukasa *et al.*, 1998; Ishida *et al.*, 1999; Cameron, in press a). As such, the extant hominids have been considered endemic to their respective regions, while the Eurasian fossil hominids are thought to have ultimately originated from an African ancestor but are largely endemic to Eurasia. This scheme, like the Eurasian origin for extant and fossil hominids, requires only two dispersal events.

Agusti *et al.* (1996, 2001), however, believe that the pattern of Miocene hominid radiation can be explained as a result of the evolution of the Alpine belt during the Neogene (Figure 2.11). They suggest that around 16 Ma, with the collision of the African and Eurasian plates, an ancestral form of *Dryopithecus* moved into Europe. This was followed by the divergences of *Dryopithecus* from the *Pongo* clade. The rising of the Alpine belt and the expansion of the central European inland sea (the Paratethys) led to the diversification of the later and more specialized southern European hominid *Graecopithecus* (and eventually *Sivapithecus*) from the western European dryopithecine populations. Further, the uplift of the Turkish Plate, producing the Zagros mountain chain, and the continued uplift of the Himalayas and Tibetan Plateau resulted in the independent

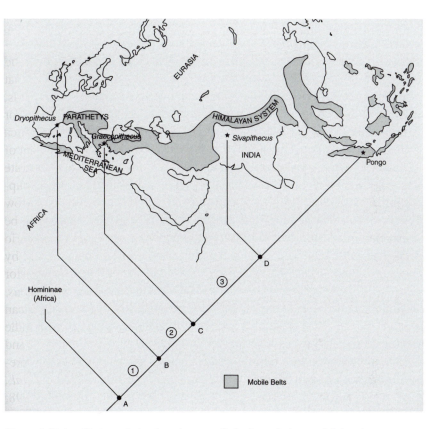

Figure 2.11 ▶   Pattern of vicarious (range-splitting) speciation explaining the independent evolution of the fossil hominids. After the hominids had dispersed into Europe, morphological stasis and persistence of primitive populations took place in western Europe. Isolation by the rising of the Alpine belt, however, may have led to the diversification of more specialized taxa (e.g., *Graecopithecus, Sivapithecus*). The Vallesian extinction event affected all these apes, although some descendants of the *Sivapithecus–Pongo* lineage could persist in refugial areas. Zoogeographic barriers: (1) Tethys–Mediterranean realm, (2) Paratethys realm, (3) Himalayan system (see text for more details).

Agusti *et al.* (1996), p. 153.

evolution of *Sivapithecus* and *Pongo* as a consequence of their long-term isolation. This scenario requires the hominids to be established in both Europe and Asia within a 2-million-year-period, that is, appearing in Europe around 16 Ma and in Pakistan around 14 Ma (this date is based on the requirement to be present in Asia before the major geological barriers). Finally, they suggest that *Dryopithecus* and *Graecopithecus* became extinct in Europe around 9 Ma, while *Sivapithecus* survived to give rise to the Asian extant hominid, *Pongo*. The African extant apes they consider to have evolved from an unknown African ancestor, thus being endemic to this continent (see Table 2.1).

Begun (1992a, 1994a, 2001, 2002) and Begun and Gülec (1998) have recently published a number of papers supporting a Eurasian origin for the African and Asian extant hominids. As discussed earlier, Begun has argued that *Dryopithecus*, defined mostly by the Hungarian Rudabánya specimens, is the basal hominid that gave rise to the African hominids (though Begun [1994a, 2002] does incorporate a discussion of the Spanish specimen CLl-18000 in his hypodigm of *Dryopithecus*). Begun (1992a, 1992b, 1994a, 2001, 2002) and Begun and Gülec (1998) argue that the Asian hominids diverged before the emergence of *Dryopithecus* and, as such, had nothing to do with the origins of the Asian clade. The basal ancestor of the Asian hominid clade (*Sivapithecus, Lufengpithecus*, and *Pongo*), they suggest, may be represented by the late Miocene Turkish hominid *Ankarapithecus* or a hominid very much like it (see Kappelman *et al.*, 2003). The paleobiogeographical interpretation provided by Begun (2001, 2002; see also Begun & Gülec, 1998) is for an African hominoid migration into Europe by 16.5 Ma, as shown by the Engelswies hominoid molar from Germany (Heizmann & Begun, 2001). It is still unknown how closely related this thick molar enamelled hominoid is to

TABLE 2.1 ▶ Spatial and Temporal Information for the Miocene Hominids Considered Here

	Region	Date	References
Engelswies hominid	Western Europe	16.5 Ma	Heizmann & Begun, 2001
Griphopithecus	SW Asia	16.5–15.0 Ma	Andrews *et al.*, 1996
	Western Europe	15.0 Ma	Andrews *et al.*, 1996
Kenyapithecus	Africa	14.0 Ma	S. Ward & Duren, 2002
Otavipithecus	Africa	13.0 Ma	Conroy *et al.*, 1992
Dryopithecus	Western–Central Europe	12.5–9.5 Ma	Andrews *et al.*, 1996; Begun, 2002
Sivapithecus	Asia	12.5–7.5 Ma	Kelley, 2002
Ankarapithecus	SW Asia	11.0–10.0 Ma	Alpagut *et al.*, 1996; Begun, 2002
Samburupithecus	Africa	9.5 Ma	Ishida *et al.*, 1984, 1999
Graecopithecus	SE Europe	9.0 Ma	de Bonis *et al.*, 1988; de Bonis & Koufos, 1999
Oreopithecus	Western Europe	8.5–8.0 Ma	Harrison & Rook, 1989
Lufengpithecus	Asia	8.0–7.0 Ma	Ciochon & Etler, 1994

the other Eurasian hominid *Griphopithecus* (also with thick molar enamel), which first appears in Turkey around this time (Heizmann & Begun, 2001; Andrews *et al.*, 1996). *Dryopithecus*, however, does not appear in the fossil record of Europe until around 12.5 Ma (Andrews *et al.*, 1996; Andrews & Bernor, 1999; Begun, 2001). Begun (2001) suggests that either *Kenyapithecus* from Fort Ternan (Kenya) or *Griphopithecus* from Turkey represent the likely ancestral ape that gave rise to *Dryopithecus*, and he concludes that the African hominid clade evolved in Europe from a *Dryopithecus*-like ancestor. He argues that in terms of its morphological condition, *Kenyapithecus* cannot be linked to the extant African hominid clade (to the exclusion of *Dryopithecus*) because it shares only primitive features with them (see also Begun, 1992b). He believes that the immediate ancestor to the African hominids, while far from resolved, is likely to be *Dryopithecus* or a *Graecopithecus*-like hominid; either way the ancestral "African" hominid is said to be from Eurasia, migrating back into Africa around 9 Ma (see Table 2.1).

Any paleobiogeographical interpretation of hominid origins is bound to be accused of speculation. The usual comment is that preservation bias will give a slanted view and overemphasize the importance of species known as opposed to the great majority of species that we will never know about. This, however, is true for any paleontological interpretation and is not a strong argument for dismissing the available evidence. If this were the case, then we should not attempt to reconstruct phylogenies based on the paleontological record, because they obviously fall "victim" to the same "logic." While certainly not dismissing the potential affects of preservation bias, absence of evidence is not evidence of absence; any hypothesis has to be based on the evidence currently available (Begun, 2001). While there are significant differences in the details of the schemes just reviewed, the overall conclusion is that Eurasia has played a major role in the paleobiogeographical origins/dispersals of the hominids.

   The available evidence suggests that Africa during the middle and later Miocene was impoverished in terms of hominid species, compared to the explosion seen in Eurasia (e.g., see Table 2.1). This cannot be convincingly argued to be the result of a lack of geological exposures from this time period (see discussion in Begun, 2001), which further suggests that the origin of the extant hominids is from Eurasia and not from an endemic African ape population. As already discussed, most experts accept that the nonhominid fossil samples (which also suffer from preservation bias) are strong evidence for an influx of nonendemic fauna into Africa during the later Miocene (Barry et al., 1985; Thomas, 1985; Tong & Jaeger, 1993; Bernor et al., 1996; Gentry & Heizmann, 1996; Solounias et al., 1999; Begun, 2001). Given that the available hominid fossil evidence also points to this conclusion, the most parsimonious scheme is to accept that the basal hominid also originated from outside of Africa.

The differential distribution of the hominids can be explained by a change in worldwide climatic conditions during the middle Miocene. Around 15 Ma there was increasing expansion of continental ice in high latitudes, resulting in the establishment of a large ice cap in the region of Antarctica. Significant cooling in the northern high latitudes is indicated by seasonal ice rafting of debris into the North Atlantic. This cooling affected Africa, because it became drier and cooler. Climate change in Africa was further emphasized by the major continental uplift that affected both eastern and southern Africa. This rise of interior plateaus, especially the Ethiopian Plateau, blocked moisture from the Atlantic Ocean in the west, leading to a rain-shadow effect over much of eastern Africa, with increased seasonality in precipitation accentuated (Owen-Smith, 1999; see also Kingston, 1999). Finally, with the main uplift phase of the Himalayas and the Tibetan Plateau by the late Miocene, monsoonal climates brought on further climatic and paleoenvironmental changes (Cerling et al., 1997; 1998; R.C.I. Wilson et al., 2000).

The previously held "consensus" view that the late Miocene and early Pliocene epochs of northern and eastern Africa were marked by a rapid "drying out" and that the endemic fauna were characterized by an adaptive shift to cope with the changed circumstance (Andrews & Van Couvering, 1975; Coppens, 1994, 1999) may be an over simplification of events. As argued by Solounias et al. (1999), there is evidence that much of the extant African savanna fauna do not have origins in Africa but, rather, migrated into Africa from more northerly latitudes, including Greece and Iran. These migrant fauna will have replaced much of the African endemic fauna because these new immigrant species were already adapted to the new conditions prevailing in Africa as a result of global cooling. As such, what was previously designated as "African fauna" associated with Graecopithecus (de Bonis & Koufos, 1994, 1999, 2001; de Bonis et al., 1992) may actually have been endemic to Eurasia.

During the Early Miocene, much of Eurasia was covered in evergreen woodlands associated with increasing seasonality and aridity. During the early/middle Miocene, African mammal species, including primates, expanded their range from the African tropical forests and woodlands to the Eurasian woodlands, which were probably much more open in structure (Andrews & Bernor, 1999). Finally, in Eurasia, the African elements further adapted to the Eurasian conditions, and eventually their adaptations became exaptations for the changing conditions in Africa, which may have enabled them to occupy successfully and outcompete much of the endemic fauna. Indeed, the hominoid fossil record supports this interpretation because the Eurasian hominids (e.g., *Graecopithecus*) bear a closer relationship to the extant *Gorilla* and *Pan* than do any of the documented "endemic" Miocene hominoids (Andrews, 1992; D. Dean & Delson, 1992; Cameron, 1997a; partly Begun, 1994b; Singleton, 2000; S. Ward & Duren, 2002; Cameron, in press a).

It is probable, given the available evidence, that the later hominids of Eurasia originated from the thick-enameled *Griphopithecus* sometime during the middle Miocene, between 16 and 14 Ma. *Griphopithecus* is commonly associated with forests in drier and more strongly seasonal conditions, with summer rainfall and prolonged dry seasons (Andrews & Humphrey, 1999). This may have led to the evolution in western Europe of the thin-enameled *Dryopithecus*, which is associated with tropical to subtropical conditions, while in Eurasia this resulted in the appearance of the hyperthick molar enameled *Graecopithecus*, which occupied conditions not too dissimilar to its presumed *Griphopithecus*-like ancestor. Thus, the Eurasian hominids retained the primitive condition of thick molar enamel, while *Dryopithecus*, which was occupying a new environmental niche (requiring a more frugivorous diet), evolved thinner molar enamel (to help further define its postcanine occlusal cristae) as well as suspensory locomotion to help move through the closed forest habitat (see Andrews & Humphrey, 1999; Agusti *et al.*, 2001).

The primitive hominid nature of *Dryopithecus*, compared to the derived status of *Graecopithecus*, may be a partial result of its inferred isolation from the southeastern Mediterranean hominids, caused by the separation of the Greek and Albanian landmass from central Europe by the Aegean connection with the Paratethys (see R.M. Jones, 1999; Rögl, 1999). *Dryopithecus* species and fossil small-bodied apes (e.g., *Anapithecus, Pliopithecus*) of western Europe appear to have adapted successfully to the more wet subtropical forests of northern Europe, as evidenced by their dental morphology and suspensory locomotion (Andrews & Bernor, 1999). With the encroaching cooler and drier conditions, however, they appear to have failed to adapt to the more open habitat, thus "slowly" being driven to extinction. For example, during the Middle Miocene, deciduous plant communities invaded Europe gradually, replacing the tropical plant communities (Andrews & Bernor, 1999). This climate change is further indicated by the retreat of the z-coral communities (which today thrive in the Caribbean and Indopacific) from the then-flooded Vienna Basin, southward to the northern coast of the Mediterranean during the Late Miocene (Rosen, 1999; Andrews & Bernor, 1999). The demise of *Dryopithecus* and the small-bodied apes of Europe can be equated with processes resulting in numerous "natural" extinctions observed today. These include: changed paleohabitat; resource competition by other nonprimate species as they occupied the new, changing environmental conditions; increased predation by new, large-bodied carnivores; and/or subsequent genetic isolation, finally pushing them into extinction.

   Graecopithecus is likely to have been less affected by climate and paleohabitat change, because of its preexisting conditions and its ability to move between Eurasia and Africa. Indeed, its hyperthick molar enamel and inferred terrestrial locomotion suggest that it was well adapted to the drier conditions of the eastern Mediterranean, with a broad dietary regime (Andrews & Bernor, 1999). Indeed, *Dryopithecus* from western Europe disappears from the fossil record 1 million years earlier than the southern hominids; although its relative *Oreopithecus* persists for an additional 2 million years. While *Graecopithecus* disappeared from the Eurasian fossil record by 10 Ma, it is possible that southern populations of it (or other closely related taxa) may have further increased their range to include northern Africa, by fragmentation or splitting of territorial ranges, e.g., vicariance (see Andrews & Bernor, 1999; Strait & Wood, 1999). It is also possible that surviving populations of these hominids in Eurasia migrated (along with other fauna) back into Africa in order to take advantage of the changing conditions within this continent while the remaining hominid populations in Eurasia eventually succumbed to competition from other, nonprimate species and/or ever-increasing climate and habitat change.

   It is only after the original "Out of Africa" by early–middle Miocene African hominoid taxa into Eurasia around 16 million years ago that we see the explosion in hominid species and genera, so the modern African hominid lineage is likely to have its origins in Eurasia (Figure 2.12). This ancestor gave rise to the last common ancestor of the extant African and Asian hominid lineages. By 15 million years ago the basal hominid was present in Eurasia, the best candidates being either *Griphopithecus* or a proto-*Dryopithecus* population. From this ancestor, the extant apes ultimately emerged. The Asian hominid lineage was first established with

Figure 2.12 ▶   Miocene hominid dispersals out of and into Africa. (1) Kenyapithecinae (*Griphopithecus*) dispersal into Eurasia around 17 million years ago, while Kenyapithecus is endemic to Africa. (2) Evolution of the Dryopithecinae (*Dryopithecus* and *Oreopithecus*) in Europe between 12 and 8 million years ago. (3) Dispersal into Asia around 13 million years ago from Eurasia of the Miocene Ponginae (*Ankarapithecus, Sivapithecus*, and *Lufengpithecus*). (4) During the Late Miocene *Graecopithecus* disperses back into Africa, following other large and small mammal dispersals from Eurasia into the African continent due to environmental change, replacing many previous Miocene endemic fauna.

species occupying present-day Turkey (*Ankarapithecus*), Pakistan and India (*Sivapithecus*), and China (*Lufengpithecus*), while slightly later the African lineage was established in Eurasia with the appearance of *Graecopithecus*. With the changing climatic conditions of the later Miocene, *Graecopithecus* or a closely related species may have increased its range from present-day Greece–Albania into Eurasia proper as well as into Africa. This migration was likely part of a general faunal migration into Africa from the surrounding regions. If this scenario is correct, then the early African hominids, which gave rise to the extant African hominids, were not endemic to Africa.

Until very recently, the available evidence suggested that Africa during the middle and later Miocene was impoverished in terms of hominid species, compared to the explosion seen in Eurasia. Most experts now accept that the nonhominid fossil record provides strong evidence for an influx of nonendemic fauna into Africa, during the later Miocene; the recent discoveries of *Orrorin* and *Sahelanthropus* and the earlier discovery of the specimen from Lothagam (all discussed in the next chapter) suggest that the Late Miocene of Africa was not as impoverished as we once may have thought (see also S. Ward & Duren, 2002). There is little evidence to support a close phylogenetic connection between the extant African hominids and the dryopithecines from Europe, though such a relationship with the Eurasian hominid *Graecopithecus* cannot be ruled out (see Andrews, 1992; D. Dean & Delson, 1992; Cameron, 1997a, in press a). While the explosion of the Eurasian hominids during the middle and later Miocene resulted, in most cases, simply in extinction, some unknown hominid genera, possibly descendants of a *Kenyapithecus*-like or *Graecopithecus*-like hominid, survived to give rise to the proto-gorilla and proto-chimpanzee and, of course, ultimately proto-humans.

CHAPTER 3

The Later Miocene and Early Pliocene Hominids

The thunderstorm could be seen in the distance. The rain had yet to reach the group, but it would soon be upon them. They had started out earlier that morning from the forests near the lake in search of tubers that they knew to exist within the grass plain, beyond the relative safety of the woodlands. They had moved farther from the forest than expected, and it would soon be night. Clearly the group would have to spend the night in the savanna, and they were now looking for a place of refuge away from the roaming carnivores. The tallest member of the group was the dominant male, who stood around 4½ feet tall. While most individuals walked erect, occasionally some would fall back into a knuckle-walking stance, especially when waiting for others to catch up. Their chimpanzee-like faces were expressive, and occasionally a loud yell would penetrate the stillness of the savanna. The dominant male was impatient with the slowness of the mothers and their infants, who tended to fall behind.

Soon they came across a small ravine slicing through the plain. There were no trees into which they could climb and spend the night, so this would have to do. They could hide from the bands of roaming hyenas and the large predatory big cats. It would also be a relatively comfortable place, away from the winds that were now starting to pick up in strength. Early tomorrow they would be safe back in the forest, but tonight they would all have to risk sleeping in the small ravine. The group started to settle for the night. The mothers and their infants slept in the middle of the group, while the males and older females settled around the periphery. They made little noise, all knowing that their safety depended on not being heard or spotted by the large carnivores that would be patrolling close by. Most went to sleep with the distant sound of rolling thunder.

Some awoke just moments before the catastrophic event, hearing an increasing roar, while around them the walls and floor of the ravine gently shook. Others remained asleep, oblivious to their fate. It had only been a few hours since they had settled for the night, but now a huge and concentrated wall of water was rushing through the ravine, carrying all in its wake. Almost as soon as the flash flood had come, it was gone, and the dry wadi was now a soaking, mud-filled channel. All evidence of the group had been swept away; they all lay washed farther down the ravine and covered deep in meters of mud. The savanna did not notice their passing.

In the mid-1970s, shortly after the discovery of the now famous "*Australopithecus*" *afarensis* skeleton, commonly called Lucy, Johanson's team also found a rich fossil bed containing a large number of *afarensis* fossil remains. They named this fossil site Locality 333. The preserved bones indicated that 13 individuals were present: males, females, and at least four infants. This is the group of fossil specimens that has commonly been referred to as the "first family." The peculiar thing about this fossil bed is that the only fossils found were of the hominids; there was almost no "background noise" of other fossil animals (see Johanson & Edey, 1981). What befell this "family" group to, presumably, kill them all in one catastrophic event and leave their remains to be discovered almost 3.2 million years later at Locality 333? There was no evidence of carnivores that might have eaten them, and no remains were found of the large grazing animals that shared the plains with them. Johanson's explanation seems the most logical: They had been caught in a flash flood while sleeping or resting in a wadi, or dry riverbed. Geological studies of this locality tend to support this hypothesis: The remains are associated with thin clay sediments, suggestive of a sudden event like a flood (Johanson & Edey, 1981). Wadis are notorious for flash flooding; in the badlands, rain from many miles away can suddenly appear from nowhere, hauling down a wadi like a confined mini-tidal wave, carrying away everything in its path. The finding of this group is, of course, significant in the interpretation of the social dynamics of these hominids—if they really were all one social group (which appears likely), then they must have lived in relatively large groups of mixed sexes and ages. What it does not tell us is whether this was a permanent social group, like that of gorillas today, or whether it was a subgroup of a larger community, like those of present-day chimpanzees . . . or humans.

Before commencing our examination of the hominids and hominins, it is necessary to clarify what we mean by these terms. When we refer to a group of fossils as *hominids*, this means that their phylogenetic status remains obscure, relative to those considered part of the human lineage. That is, they could represent ancestors to orangutans, gorillas, chimpanzees, humans, or none of the above. In other words, they might have their own distinct evolutionary history, not closely tied to the emergence of humans and their immediate ancestors. *Hominid* is the vernacular from the family Hominidae, which nowadays incorporates all of the living great apes and humans.

The term *hominin* is used to describe those groups considered to be closely associated with the emergence of the human lineage. This does not necessarily mean they need to be ancestral to humans. Rather, they share a number of derived features with humans; they share an immediate common ancestor to the exclusion of the other hominids. For example, while the species within *Paranthropus* are not ancestral to *Homo*, they are nonetheless "hominins." The word is derived from the Hominini, the name of the tribe that includes those forms that are closer to humans than to chimpanzees.

The hominid or hominin status of the taxa to be discussed in the next few chapters will be addressed in Chapter 5, which will present a phylogenetic analysis (evolutionary history) of the late Miocene and Plio-/Pleistocene taxa. For convenience, in the next two chapters we will often refer to species as hominids, which encompasses the subcategory hominin.

The Emergence of *Sahelanthropus, Orrorin,* and the Lothagam Hominids

The recent significant discovery and description of *Sahelanthropus tchadensis* from Chad by a joint French and Chadian paleoanthropological team, dating to between 6 and 7 million years ago (Brunet *et al.*, 2002; Vignaud *et al.*, 2002; see also B.A. Wood, 2002), has done much to refocus our attention on the divergences of the hominins from other hominids. One of the key differences of the hominins from most other hominids is the development of a primitive form of bipedal locomotion, which would likely postdate the development of increased ability for upright posture (see Wood, 2002). Around 5–6 million years ago, the common human–chimpanzee ancestor had what we now think is the primitive hominid locomotion pattern, a quadrupedal, knuckle-walking ability (Richmond & Strait, 2000). Shortly after the split from the last common ancestor with the chimpanzee, the hominins must have developed a more upright gait, with some form of primitive bipedal walking. If *Sahelanthropus* was bipedal, as suggested by Brunet *et al.* (2002), either it was a hominin or, if it represents a common ancestor to the chimpanzee–human lineages, knuckle-walking must have developed independently in both *Pan* and *Gorilla*. For reasons to be discussed presently, we think that the latter is most unlikely.

Associated with the development of these patterns of locomotion and positional behavior would be the development of numerous features of the skull and teeth. In the proto-chimpanzee, we would see much the same condition observed in the last common ancestor with the proto-gorilla, which presumably, like extant *Gorilla* and *Pan*, would include the following: large male canines, a moderately small premolar and molar complex, lower third premolar unicuspid, thin molar enamel, prognathic premaxilla (snout), marked postorbital constriction, developed supraorbital torus, a braincase pushed well back from the face, a low frontal, large masticatory muscles and corresponding skeletal attachment sites for these muscles, and, finally, a relatively small brain (Cameron, in press a). Some of these features were of course retained in the earliest hominins, after the split from the chimpanzee, as primitive features.

In the earliest hominins, we see, along with the derived form of bipedal locomotion, some reduction in male canine size, development of a bicuspid lower third premolar, and an increase in molar enamel thickness. Somewhat later, hominins developed a number of uniquely shared features not seen in the other clades: an absolute increase in brain size, a further increase in molar enamel thickness, further reduction in male canine size, reduced supraorbital torus, an absolute reduction in postorbital constriction, development of a relatively high frontal, and a nonprognathic premaxilla. The significance of *Sahelanthropus* is that, while we know very little about its pattern of locomotion or positional behavior, we finally have evidence of the facial and dental anatomy of a species at the point of divergence from the chimpanzee. Does *Sahelanthropus* follow the proto-chimpanzee condition or the basal hominin condition?

Not surprisingly, *Sahelanthropus* retains a number of primitive hominid features, including a small brain, developed postorbital constriction, and a low frontal. All of these features can be considered as part of the primitive condition called neuro-orbital disjunction (the frontal lobes of the brain are placed well back from the face). What might tend to refute a close relationship with the chimpanzee lineage, however, are a number of claimed hominin features, including its proposed bipedal locomotion, reduced canine size, increased molar enamel thickness, and absolutely reduced facial prognathism. A strict cladistic interpretation would thus place *Sahelanthropus* within the hominins, though at this point we are hesitant to do this, for several reasons. First is that the published illustrations (Brunet *et al.*, 2002) (Figure 3.1) show a very long basicranium; though it is distorted, there is no doubt that in this respect, it is outside the hominin range and thus comparable to a gorilla. The claim of bipedalism rests on the interpretation of the basicranium as *short*; but this is absolutely not so! Second, the enamel thickness is uncertain; it actually matters rather little, because the earliest hominin of which most are fairly confident, *Ardipithecus*, had thin enamel (though Cameron doubts the hominin status of *Ardipithecus*). Third, the short face of *Sahelanthropus* is duplicated in some modern gorilla skulls.

One feature of *Sahelanthropus* that is surprising is the very small canines. The describers maintained that the almost complete *Sahelanthropus* skull TM 266-01-060-1 is a male, but there seems no convincing reason for this. If we contemplate the alternative, that it might be a female, then the canine size suddenly becomes comparable to that of the late Miocene hominid *Graecopithecus* from Greece dating to about 9.5 million years ago, which has been argued by some to be a "proto-gorilla" (Dean & Delson, 1992; Cameron, 1997a). The canines of *Graecopithecus*, though still sexually dimorphic, were unexpectedly small. Perhaps in this instance parsimony fails: The common ancestor had rather small canines, and in their subsequent evolution, both gorilla and chimpanzees, independently, evolved large canines.

Farther east of Chad by almost 2,500 km, and dating to somewhat later than *Sahelanthropus*, is *Orrorin tugenensis* from Lukeino in Kenya, dated to the latest Miocene. Senut *et al.* (2001) claim that it represents a direct human ancestor, largely because of certain features of the femur; one commentator

Figure 3.1 ▶   Professor Michael Brunet with the now famous "Toumai" skull representative of the late Miocene hominid *Sahelanthropus tchadensis* from Chad. Adapted from a photograph by Patrick Robert in *Time* Magazine (July 22, 2002).

has strongly criticized the original arguments (Haile-Selassie, 2001), but more detailed arguments have recently been put forward (Pickford *et al.*, 2002). Specifically, in *Orrorin* the shaft of the femur is anteriorly convex and the pilaster is vertical, and it has a distinct intertrochanteric line, an elongated and strongly anteroposteriorly compressed femoral neck, and a developed *obturator externus* groove. In the neck of the femur, the cortex is thick inferiorly and thin superiorly, as in bipeds (but less markedly so than in *Homo*), rather than equally thick all round, as in nonhuman apes. Though Pickford *et al.* (2002) acknowledge that in some respects the bipedal condition of the femur is distinctly less developed than in later hominins, it is clear that the argument that *Orrorin* is a hominin has been much strengthened by this analysis. Less clear is the significance of some features that appear to be even more human-like than in australopithecines, such as the larger size of the femoral head and its anterior (instead of posterior) twist, and the more medially (less posteriorly) oriented lesser trochanter. The authors suggest that the australopithecine type of bipedalism may not be intermediate between *Orrorin* and *Homo*, and either the evolution of human locomotion took an indirect route (a view that was in fact already implicit in the work of such authors as McHenry [1986]), or else, more controversially, all known species of australopithecines are to be excluded from the human line altogether.

Further material of *Orrorin*, especially craniodental, is needed. Already, geological and geophysical work have refined the age of the deposits; early reporting gave a date of "about 6 million" years (Senut *et al.*, 2001); the Lukeino Formation is now dated by K/Ar, backed up by paleomagnetism, to between 5.7 and 6.0 Ma, and the most significant specimens (from Kapsomin) are 5.8–5.9 Ma (Sawada *et al.*, 2002).

If both authors are readier than formerly to accept a basal hominin status for *Orrorin*, the two of us differ in our interpretation of *Sahelanthropus*. Cameron believes that it probably does represent a very early hominin. For example, orthognathy in gorillas, though it does exist, is rare; it is more parsimonious to accept that the last common ancestor of *Gorilla* and *Pan* had not only a prognathic face but also thin molar enamel and that the reduced prognathism and thick molar enamel observed in *Sahelanthropus*, and most hominins, developed after their split from the last common ancestor with the chimpanzee. This view, however, is at odds with the molecular evidence; if we maintain a divergence time of 4.6–6.2 million years ago, as calculated by Chen and Li (2001) in what seems much the most thorough survey to date, then *Sahelanthropus* is too early to be a hominin (or else there is something a great deal wrong with Chen and Li's molecular clock). Likewise, *Ardipithecus*, which has thin molar enamel, would represent the survivor of a lineage more primitive than *Sahelanthropus*. Groves, however, is convinced that the long basicranium of *Sahelanthropus*, the more sagittal orientation of the petrous bone, and the implied steep angle of the nuchal plane exclude it from the Hominini, except perhaps at its very base. In essence, there is no evidence that it has any hominin character states at all. Only more detailed studies and publications on this most important specimen will resolve these and other issues.

An important hominid, which spans the divide between *Sahelanthropus* and *Orrorin* and the australopithecines, is the mandibular fragment from Lothagam in Kenya, dated only as "older than 4.2 Ma and younger than 5.0 Ma" (McDougall & Feibel, 1999). Kramer (1986) and T.D. White

(1986), writing at a time when *Australopithecus afarensis* rated as the earliest member of the human lineage, both argued that the Lothagam mandible has detectable australopithecine affinities, though with plesiomorphic features. Most recently, M.G. Leakey and Walker (2003) have suggested a close relationship to *Australopithecus anamensis*, though they stress that until additional specimens are discovered, it is not possible to allocate it to any known hominin species.

It matters whether *Sahelanthropus* is the earliest hominin or whether this position goes to *Orrorin*, *Ardipithecus kadabba* (see next section), or the Lothagam mandible, because if one or more of them is hominin, then it will give us a minimum date for the separation of the human and chimpanzee lineages. But even if none proves to be hominin, their significance will be undiminished, because we will have a glimpse of the morphology of immediate "presplit" apes.

The Emergence of *Ardipithecus* and Early Australopithecines

By 5 Ma, the hominid populations had split into those that began to walk erect and those that continued to be knuckle-walkers. The adaptation to bipedal walking appears to have occurred over a long period of time, because the earliest hominin fossils with preserved wrist joints, from Ethiopia and Kenya, indicate they retained knuckle-walking adaptations (Richmond & Strait, 2000), like gorillas and chimpanzees. Because gorillas and chimpanzees are both knuckle-walkers, but chimpanzees are more closely related to humans, parsimony dictates that the preferred hypothesis be that knuckle-walking was the locomotion of the last common ancestor of humans and chimpanzees. Thus the presence of leftover knuckle-walking features in the earliest hominins was predictable. Whether this implies that they still, on occasion, actually walked quadrupedally on their knuckles — or that these features are just relics of a knuckle-walking past, not yet lost by the new bipeds — is debatable.

In 1993, a joint American, Japanese, and Ethiopian paleoanthropological team working in Middle Awash, Ethiopia, discovered what they claimed to be the oldest fossil hominin remains thus far known (T.D. White *et al.*, 1994). The species occupied the African landscape around 4.4 Ma and in terms of time are very close to the split between the chimpanzee and earliest hominins, as calculated from molecular clocks (WoldeGabriel *et al.*, 1994; T.D. White, 2002). Though first classified as *Australopithecus ramidus*, these fossils have now been allocated to their own genus, *Ardipithecus*, and very recently an even more primitive form has been described from deposits 5 Ma that is supposed to be a subspecies of the same species, *A. ramidus kadabba* (Haile-Selassie, 2001; T.D. White, 2002). The teeth of *Ardipithecus* are generally similar to those of the chimpanzee (including thin molar enamel), but the canines are shorter, though still sexually dimorphic. In the features of the rather fragmentary cranium, however, it resembles later australopithecines; its cranial base is said to be short, implying that the head is balanced "on top" of the spinal cord and needs less muscle support and that it walked upright. We must admit, however, that this is difficult to judge because no detailed illustrations of the basicranium have yet been published. Whether *Ardipithecus* really shares any derived features with later hominins is difficult to tell. If the description is accurate, then it certainly does; but without detailed figures and descriptions of these important specimens, nothing more can be said.

With all the significance that has been attached to bipedalism, a word of caution is in order. 9–7 million years ago, on the Tyrrhenian Island (as it then was) in Italy, lived a small hominoid of uncertain but definitely non-hominin affinities: *Oreopithecus bambolii*. Many aspects of its anatomy, not least its bizarre dentition, place it outside the range of modern hominids, and whether it can be viewed as a hominid at all has been disputed. But of one thing there is no doubt — the structure (including the microstructure of the cancellous network) of the pelvis shows that it was upright and bipedal (Rook *et al.*, 1999). Bipedalism can and did evolve more than once, and with some, at least, of the same osteological modifications. We need a total morphological pattern, not only bipedalism, to identify a hominin.

The current archeological and paleontological evidence suggest that *Ardipithecus* preferred a forest habitat; the fossils are found with a typical forest fauna (WoldeGabriel *et al.*, 1994; Andrews & Humphrey, 1999; Denys, 1999). This is reinforced by its implied dietary adaptations. For example, the thin molar enamel and small molar teeth suggest that its diet was based more on leaves, possibly soft fruits, and other soft vegetative material. If *Ardipithecus* did walk bipedally, then it was doing so before the climate shift (around 2.5 Ma) that led later hominins to occupy a more open habitat. Thus early bipedalism evolved in the forests, suggesting that it was not an adaptation for occupying the savanna (Andrews & Humphrey, 1999; Feibel, 1999; Foley, 1999).

The early development of bipedalism, then, seems not to have been an adaptation to life in the savanna as such but was perhaps associated with the need to move from one forest patch to another (Rodman & McHenry, 1980; Cameron, 1993a; Cameron & Groves, 1993). While bipedalism is an inefficient system for short-term sprinting, it is very efficient for long-distance travel, though its earliest form, as observed in the australopithecines, is said to be a "cheap way of moving" (Taylor & Rowntree, 1970; Fleagle, 1999). Walking upright will also mean that the body is a smaller target for the sun, and thus the body will remain cooler than if walking on all fours, where much more of the body is exposed to the sun (Wheeler, 1984, 1991, 1993; Foley, 1987; Aiello & Dean, 1990). Richard Leakey has suggested, in lectures, that in open dry forests, a large-bodied ape is forced to be terrestrial yet may still depend on an arboreal diet, such as seed pods. To reach them, it is forced to rear up. This explanation for bipedalism strikes us as a most perspicacious piece of lateral thinking. Nor can we exclude the Aquatic Ape Hypothesis (AAH). Elaine Morgan has long argued that many aspects of human anatomy are best explained as a legacy of a semiaquatic phase in the proto-human trajectory, and this includes upright posture to cope with increased water depth as our ancestors foraged farther and further from the lake or seashore. At first, this idea was simply ignored as grotesque, and perhaps as unworthy of discussion because proposed by an amateur. But Morgan's latest arguments have reached a sophistication that simply demands to be taken seriously (Morgan, 1990, 1997). And although the authors shy away from more speculative reconstructions in favor of phylogenetic scenarios, we insist that the AAH take its place in the battery of possible functional scenarios for hominin divergence.

Almost 300,000 years after the disappearance of *Ardipithecus* from the fossil record, the first of what we call the australopithecines appears, *Australopithecus anamensis*. This species first appears in deposits at Kanapoi, south of Lake Turkana, Kenya, dating to around 4.2 Ma and disappears from the fossil record around 3.8 Ma (M.G. Leakey *et al.*, 1995; C.V. Ward *et al.*, 2001). Certain anatomical features of the jaw joint, as well as the increase in molar size and enamel thickness from that observed in *Ardipithecus*, suggest that it employed a very different pattern of food processing and/or food types. Although the little we know of the skull of this species indicates it is primitive, with rather large canines and narrow, rectangular jaws, the leg bone fragments, which consist of parts of the knee and ankle joints, are more similar to members of our own lineage, *Homo*. Indeed, if only the leg bones had been discovered, it is likely that they would have been allocated to a species of *Homo*. And it is significant that as long ago as 1967, the lower end of a humerus, part of the elbow joint, was discovered at Kanapoi, and for a long time (until excavations at Kanapoi were reopened in the 1990s) it remained rather a mystery — a "curiously modern-looking" elbow joint older than the more primitive australopithecines (see Senut & Tardieu, 1985; M.G. Leakey *et al.*, 1995).

Doubts still remain. Andrews (1995) has suggested that the postcranial bones and the skull parts of this hominid are from two distinct species: that the skull and teeth remains, which were found in geological deposits dating a little earlier than the recovered leg bones, belong to an extinct great ape, while only the leg bones, he argues, should be allocated to this new species. There is little in the preserved anatomy of this species that suggests a close relationship to later hominins. Certainly its cranial morphology is primitive, and even if the leg bones do belong to the same species, we know, or think we know (from *Ardipithecus* and probably *Orrorin*), that bipedal locomotion had already had a long history, of a million years or more. And let us always bear *Oreopithecus* in mind.

As with *Ardipithecus*, *A. anamensis* appears to have occupied a gallery forest, although other areas from this time, which also contained specimens of this species, are known to have been more open, wooded or largely bushland (M.G. Leakey *et al.*, 1995; Andrews & Humphrey, 1999).

Perhaps the best-known Pliocene hominid is the world famous "Lucy" skeleton found by Johanson's group in the early 1970s, dating to between 3.6 and 2.9 Ma (Johanson & Taieb, 1976; Kimbel *et al.*, 1994; T.D. White, 2002). "Lucy" and her kind were originally named *Australopithecus afarensis* (Johanson *et al.*, 1978), but Strait and Grine (2001) have allocated them to the genus *Praeanthropus* (see also Strait *et al.*, 1997) (Figure 3.2), originally erected by Senyurek in 1953 for the Garusi maxilla, which is now catalogued as Laetoli Hominid 1. Although Senyurek called LH1 *Praeanthropus africanus*, the International Commission on Zoological Nomenclature has recently suppressed this usage of the name *africanus* and placed *afarensis* on the Official List of Names in Zoology. We provisionally accept this revision, but will return later to the whole question of taxonomy.

Figure 3.2 ▶   Reconstruction of *Praeanthropus afarensis* from the Pliocene of Hadar, Ethiopia.

Lucy was just over a meter tall (about 3½ feet), with long arms, a short trunk, and short legs. Taking her skeleton together with the more fragmentary remains of other individuals, we can say that the foot bones of *Pr. afarensis* are generally human-like, but the ankle joints are chimpanzee-like in their overall flexibility (see Richmond & Strait, 2000). This has caused great controversy. To some experts, it suggests that tree climbing was still an important ability; and those who argue for tree-climbing point out that the toe bones are curved and strong, the ribcage is funnel-shaped and apelike rather than barrel-shaped and human-like, and the arms are strongly muscled (Susman, 1979; Stern & Susman, 1983; Susman *et al.*, 1984; Heinrich *et al.*, 1993). To others, however, these features are simply primitive traits left over from a more arboreal ancestry. The shape of the pelvis of *Pr. afarensis* indicates very clearly a bipedal gait, and the femur is valgus (meaning that it is orientated inward at the knee, which is typical of humans, whereas in apes it is more vertical); in particular, the big toe is not markedly — or at all — divergent, as one would expect if it were doing much climbing (Lovejoy, 1974, 1981; T.D. White, 1980; T.D. White *et al.*, 1983; see also partly Robinson, 1972). Footprints of *Pr. afarensis* were discovered at Laetoli during the mid-1970s. Although their interpretation is still very controversial, the one thing that all agree upon is that they show a big toe aligned with the other toes, with only a slight gap (M.D. Leakey & Hay, 1979; Day & Wickens, 1980; T.D. White, 1980). A third school of thought accepts that *Pr. afarensis* was basically bipedal but finds it difficult to overlook altogether the survival of these apelike, presumably arboreal, features (B.A. Wood, 1992). Wood suggests that *Pr. afarensis* was neither predominantly arboreal nor fully bipedal (B.A. Wood, 1992). And then we have to come to terms with the recent revelation that studies of the wrist bones also suggest that some form of knuckle-walking may have been involved (Richmond & Strait, 2000) (or were the knuckle-walking adaptations, too, a mere survival from a primitive ancestor?).

One very obvious feature of *Pr. afarensis* is the large difference in body size between males and females. Males average in height 4 feet, 10 inches and in weight 143 lb, while females average in height only 3 feet, 3 inches while only weighing 66 lb (Stringer & McKie, 1996). Thus females average only 80% of male height and only 46% of male weight. This degree of sexual dimorphism, while common in the great apes (orangutans and gorillas but not chimpanzees) and many monkey species, is unusual for hominins. So marked is this size dimorphism that one of us (CPG) had previously been reluctant to accept such an interpretation, preferring to believe in the coexistence of a small and a large species (Groves, 1989a), but now tends, still not without some misgivings, to accept the homogeneity of the species.

There is no evidence of tool making by *Pr. afarensis*, and its brain was little or no bigger than that of a chimpanzee. Unlike *Ardipithecus*, but like members of *A. anamensis*, the molar teeth in *Pr. afarensis* were large in size, and the molar enamel was thick, suggesting that it was eating tougher food types, including hard fruits and possibly nuts.

Around 3.3 Ma some populations of *Pr. afarensis* in east Africa were living near a large lake, surrounded by forests and bushland. Over the next 500,000 years, the region fluctuated in climate and habitat types, the lake receded and then expanded, and the fauna shifted from a more closed forest type to a more open grassland type and then back again to a closed forest type. When *Pr. afarensis* disappeared from the fossil record, around 3.0 Ma, savanna dominated the region. By 2.8 Ma, however, the region had reverted to the bushy, forested conditions that held sway almost

500,000 years earlier (Potts, 1996; see also Andrews & Humphrey, 1999; Feibel, 1999).

Contemporary with *Pr. afarensis* in Ethiopia is the recently discovered australopithecine from Chad, some 2,500 km west of the Rift Valley (Brunet *et al.*, 1995, 1996). The specimens were originally thought to represent a western population of *Pr. afarensis*, but in 1996 they were allocated to a new species of *Australopithecus, A. bahrelghazali*. Very little is known of this species at present, and whether its ascription to *Australopithecus* is justified can only be answered with the discovery of more complete material. Like *A. anamensis* and *Pr. afarensis*, it has thick molar enamel, but its mandible is of a lighter construction and its premolar cusps are less developed than in *Pr. afarensis* (Brunet *et al.*, 1995). The deposits containing *A. bahrelghazali* reflect a lakeside environment with both perennial and permanent streams, and a vegetation mosaic of gallery forests and wooded savanna with open grassy patches (Brunet *et al.*, 1995), rather like that occupied by earlier and contemporary proto-australopithecines of east Africa.

The most recent Pliocene "australopithecine" species discovered, *Australopithecus garhi*, appears in the fossil record of the Middle Awash of Ethiopia around 2.5 mya (Asfaw *et al.*, 1999). Its discoverers suggested that this is the likely direct ancestor to *Homo*. In overall anatomical features (Figure 3.3) it is similar to *Pr. afarensis*, including a similar cranial capacity, around 450 cc, its projecting facial profile, and, if the limb bones are correctly associated, its body proportions with typically short legs and a short trunk but relatively long arms. Yet it has some features that are said to be significantly different from those of the earlier australopithecines. These include the forward positioning of its cheekbone, a more oval premolar shape, and huge premolars and molars. Actually, it shares no unique characters with *Homo*, and we suspect that it represents either a species of *Praeanthropus* (*Pr. garhi*) or perhaps a new genus. It certainly cannot be considered a species of *Australopithecus* (see Strait & Grine, 2001; Cameron, in press b). It has been suggested that it made stone tools (Heinzelin *et al.*, 1999): Animal bones with butchering marks (cut marks) as well as some primitive stone tools have been found close by, though whether these were actually the work of members of the "*garhi* group"

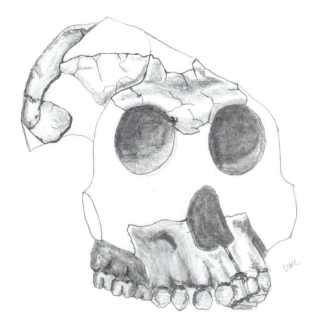

Figure 3.3 ► Reconstruction of *"Australopithecus" garhi* holotype specimen BOU-VP-12/131 from Bouri, Ethiopia.
Adapted from T.D. White (2003).

remains problematic, for early representatives of *Homo* are also found in this region dating to around the same time or just a little later (Kimbel *et al.*, 1997).

Early Hominin Social Dynamics

Three main archaeological interpretations of Plio-/Pleistocene hominid(in) behavior have been proposed. While the models emphasize the behavioral repertoire of early *Homo* — for they focus on stone tool technologies and their distribution over the landscape — they are probably relevant to the behavioral features of the australopithecines too. The first was proposed by the Harvard archaeologist Glynn Isaac, who became famous for his archaeological excavations and interpretations of the localities centered around Koobi Fora and Olorgesailie (see Isaac, 1977). Isaac (1976, 1978, 1986) formulated the *Central Place Foraging model*, which argues that early hominins required a central location to which they could retreat in order to eat in safety and at leisure. The main emphasis is on a safe refuge, with secondary importance given to food sharing and development of a sexual division of labor. In this model, males focused on large animal hunting and the scavenging of animal carcasses, while females and juveniles focused on opportunistic hunting of small animals and on the gathering of edible plant material.

The second model, proposed by the North American archaeologist Louis Binford, is the *Scavenging hypothesis*. Binford (1981) argues that early hominins were not involved in hunting at all but, rather, roamed the African savanna, scavenging the kills from other carnivores, thus obtaining only low-food-utility items, and then, like other scavengers, moved off to a more protected location. His hypothesis is based on a statistical interpretation of patterns observed at known carnivore scavenging sites, which were then compared to early hominin archaeological sites. Binford argued, based on his number crunching, that the early hominin sites were more similar to the carnivore scavenging sites than to known later hominin hunting sites. He believed that the focus on low-food-utility items (those not consumed by large carnivores) enabled the early hominins to carve out a niche for themselves as marginal scavengers.

Finally Richard Potts (1988, 1996), of the Smithsonian Institution, has proposed a third model called the *Stone Caching hypothesis*, which argues that early hominins were not using a central place foraging system but, rather, were bringing hunted and/or scavenged animal remains back to a number of stone caches situated optimally throughout the landscape. These stone caches, he suggests, were *an aggregation of transported stone, including modified and unmodified pieces, which were repeatedly visited to obtain or manufacture tools and to use them in processing food.* The model is based on the presence of rock types (some of which have been made into tools) that are not naturally found at Olduvai, but farther afield. Hominids, he argues, must have brought these rocks from the surrounding regions; whenever they were able to kill, or scavenge from other carnivores, the carcass or parts of it were carried to the nearest stone cache in order to speed up the butchering process and reduce the chance of confrontation between them and other carnivores.

These three models are of course based on the archaeological evidence. In the early 1990s, the authors decided to test these three models to determine which, if any, was the most likely (Cameron, 1993a; Cameron & Groves, 1993), but we used a completely different data set. We examined the literature of ape (*Pongo, Gorilla*, and both species of *Pan*) and human behavior, and defined a number of behavioral features common to *all* species; these features were also likely to have been present in the earliest hominins, by parsimony. We further examined the behaviors of humans and the two species of chimpanzee, and looked for common features shared by all three; again, the earliest hominins would also likely share these more derived features (Figure 3.4, Table 3.1).

Our study strongly indicated that the social system of the immediate common ancestor of chimpanzees and humans was one where both hunting and an incipient division of labor were present, with little evidence for scavenging as a major component of early hominin behavior. This is not to argue that some scavenging did not occur, just that it is not likely to have reflected the primary strategy for early hominin activities. Stone caching is also unlikely because it is a unique condition of just one group of common chimpanzees (from the Tai Forest, Ivory Coast). There is evidence, however, that some form of central place foraging strategy, based on

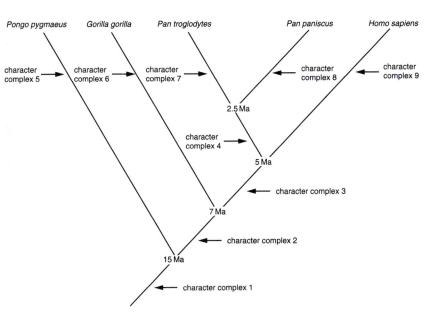

Figure 3.4 ▶ Cladogram of the extant hominids based on the molecular evidence. See text and Table 3.1 for further details.

From Cameron (1993a).

TABLE 3.1 ▶ Behavioral Features Characterizing Each Node (Character Complex) in Figure 3.4

Character Complex 1
  1. Variety in diet
  2. Ranging pattern depended on seasonality of available foods

Character Complex 2
  1. Social

Character Complex 3
  1. Community not necessarily controlled by dominant male
  2. Omnivore
  3. Food-sharing present
  4. Engaged in opportunistic and/or hunting activities to obtain meat

Character Complex 4
  1. Sexual relationships that are polygynous and promiscuous
  2. A medium degree of emphasis placed on a sexual division of labor
  3. Known to move food away from source location
  4. Males retained, females migrate

Character Complex 5
  1. Females that have exclusive core areas, while males have larger overlapping areas
  2. Social systems based on lone individuals, not group structure
  3. Lone male territorial patrol
  4. Polygamous and opportunistic mating strategy
  5. No sexual division of labor
  6. Females and males very rarely form into groups, and when they do it is for a very short period of time
  7. Nests rarely reused

Character Complex 6
  1. Polygamous mating strategy
  2. Weak sexual division of labor

Character Complex 7
  1. A group of males may patrol alone or may be accompanied by estrus females
  2. Weak female–female bonding but strong male–male bonding
  3. Dependent on small food patches
  4. Adult males frequently share meat among themselves and with estrus females
  5. Meat moderately important to the diet
  6. Food occasionally taken back to a nest
  7. Nests infrequently shared
  8. Food may be taken back to optimally placed material caches for processing

Character Complex 8
  1. Increased period of estrus
  2. Weak male–male bonding but strong female–female bonding
  3. Meat of only minor importance to the diet
  4. No pirating of meat
  5. No scavenging of meat
  6. Nests commonly shared

Character Complex 9
  1. Group of males usually patrol alone
  2. Extreme periods of estrus

TABLE 3.1 ► Continued

3. Strong bonding among all classes
4. Mating strategy extremely variable
5. Extreme sexual division of labor
6. Dependent on concentrated food patches
7. Female attitudes to new group female members extremely variable
8. Capture of prey based on complex organization
9. Meat important to the diet
10. Food usually removed to another location
11. Food usually taken back to a nest
12. Nests almost always shared
13. Same nest may be reused for a prolonged period of time
14. Food usually processed at camp, not at optimally dispersed material caches.

Synapomorphies of Pan troglodytes and Homo
1. Both based on unstable ranging pattern
2. Both marked by intercommunity relations that are based on physical agression
3. Males the predominant hunters, and they tend to hunt in male groups
4. Both use stone and plant tools in food processing
5. Both known to pirate meat
6. Scavenging infrequent, although it does occur

Synapomorphies of Pan paniscus and Homo
1. Meat usually shared with little distinction between the sexes.

From Cameron (1993a).

shared derived behavioral conditions of the chimpanzee and human lineages, applied. For example, common chimpanzees have been observed to take meat back to a night nest, although it is not reported that they share the meat there. This is close to central place foraging and differs only in degree, not kind. In both common and pygmy chimpanzee groups, food sharing is usual, so this is a common derived feature of the human and chimpanzee lineages, and it is clear that a sexual division of labor is incipiently present in all African apes (gorillas, chimpanzees, and humans). In addition to these features, the earliest hominin social systems were probably characterized by the sharing of meat, with little distinction between the sexes. Hunting of small mammals as well as scavenging of small and large mammals took place; infrequent pirating of meat is also likely to have occurred. The hunting parties were cooperative and consisted predominantly of male members. These Pliocene hominid groups were probably based on unstable ranging patterns, which were dependent on seasonality of available foods; intercommunity relationships would have been based on physical aggression. Some stone and plant tools were also probably used in food processing. Sexual relationships were probably based on promiscuous behavior and occasional consortships.

Recently, Wrangham (2001) has proposed a behavioral model similar to the one we propose, but it has the addition of a few derived features to help explain the original emergence of the proto-human from the proto-chimpanzee. He suggests that the earliest proto-humans, like extant apes, originally occupied a forest habitat (with many of the behavioral features we have just discussed), but at least one of these populations had acquired the ability to search for and obtain underground storage organs (USOs), for example, tubers, corms, rhizomes, and other roots. With the depletion of the forests around the time of the Mio-/Pliocene transition, most proto-human populations must have died out. But some that had already adapted to these foods would be able to survive in the increasingly drier habitats. These foods would be an important food source, because they are not readily available to other animals (they are hard to dig up and often tough or rich in toxins). Numerous "digging tools" have been found in association with fossil hominin specimens. Indeed, the increased thickness of molar enamel and increased molar size in later fossil hominins may be associated with such an adaptation. As such, these foods can be seen as a fallback position to preferred foods that may have become increasingly difficult to obtain. As suggested by Wrangham (2001:128) "This was a critical adaptation enabling the woodland apes to survive when natural selection was at its most intense."

We suggest, then, that the evolution of the Pliocene hominins emphasized a number of behavioral features present in the last common ancestor of the chimpanzee and human lineage. The unique derived features of living human groups cannot alone be used to explain the behavior of the earliest hominins, just as the behavior of living chimpanzee groups cannot alone be used as an analogy for early hominin behavior, though we do accept that Wrangham's USO hypothesis is an intriguing model for the initial split between the proto-chimpanzee and proto-humans around 5 million years ago. In examining the more primitive and shared derived features of the apes, we can say that of the three archaeological interpretations proposed so far for early hominin behavior, Isaac's Central Place Foraging model is the most likely.

INTERLUDE 2
The Importance of Being an Ape

We are animals, of course. (A creationist might dispute that. What are we, then — plants? fungi?) We are vertebrates — animals with backbones. We are mammals — vertebrates with body hair, and "we" secrete milk for the newborn ("Man is an animal that suckles his young," in prenonsexist parlance). We are primates — mammals with grasping hands and feet, Meissner's corpuscles to enhance the sense of touch, and stereoscopic vision, and the males have dangling penises. We are haplorrhines — primates with a dry nose, a macula in the retina of the eye, and born of a vascular placenta of the type called haemochorial. We are catarrhines — haplorrhines with an auditory tube conducting sound from the external ear to the middle ear, and only two premolar teeth (in each half of each jaw). And we are apes — catarrhines that sit upright and sometimes stand upright and have a shortened lumbar spine, lowering the center of gravity for this purpose; that have shoulder joints with an all-round hemisphere of rotation; and that have loose, flexible wrists and no tail; that have simple molar patterns and shortened canine teeth; and that have an appendix in the gut. The other apes are gibbons (called lesser apes), orangutans, gorillas, and chimpanzees (called great apes).

Until the 1960s, textbooks placed humans in one family, Hominidae, and "the apes" in another, Pongidae. Or the lesser apes might be classified in a third family, Hylobatidae. The evolutionary diagrams had the human stem separating from the ape stem way back, perhaps in the Oligocene. Work in the 1960s on a mid-Miocene ape, *Ramapithecus*, apparently confirming it as a human ancestor, seemed to corroborate this.

Then came the work of Morris Goodman. In a classic paper in a 1963 book (*Classification and Human Evolution*, edited by Sherwood L. Washburn), Goodman illustrated the results of his comparisons of the serum proteins of humans and great apes. According to the "traditional" model, the serum proteins of humans should have been the most distinct; instead, those of the orangutan were the most distinct. Next most distinct were those of the gorilla. Those of humans and chimpanzees were very alike indeed. Goodman proposed that the gorilla and chimpanzee should be taken out of the Pongidae and placed in the Hominidae.

Goodman was way ahead of his time. What price serum proteins, when we had all that anatomy telling us that man was so very different from the great apes: the habitually bipedal locomotion, the hairless body, the noble brow, the huge brain, the ability to make tools and television sets.

Another biologist, meanwhile, was white-anting the establishment in a different way. Jane Goodall had begun her successful long-term field study of chimpanzees in Gombe National Park (as it is called today), in Tanzania. In her paper in a 1965 book (*Primate Behavior*, edited by Irven DeVore), she revealed that chimpanzees make tools — not television sets, but simple tools of twigs and grass stems, to get termites and ants out of their nests to eat.

The collapse of the traditional model probably started about 1970, as other work in both genetics and psychology showed how close humans and great apes really are, and the genetic work in addition confirmed Goodman's conclusion that chimpanzees are our closest relatives. In 1980 the final stroke that brought down the old edifice was the demonstration, in a classic paper by Peter Andrews and Jack Cronin, that our mid-Miocene supposed ancestor, *Ramapithecus*, had been outrageously misrepresented and that the available evidence showed it as barely, or not at all, different from *Sivapithecus*, which is, in its turn, a member of the orangutan lineage.

All primates are intelligent, as mammals go, and monkeys are intelligent as primates go. But with the apes, it is not just a matter of intelligence. The great apes have a theory of mind: They know how their own and others' minds work, so they can anticipate others' reactions, empathize with them, and anticipate their own reactions. They can imitate others in a way that monkeys cannot; and, being self-aware, they can learn to recognize themselves in mirrors. A monkey will continue for hours on end threatening that other monkey it sees in the mirror, or looking behind the mirror for it; most chimpanzees will take a matter of days, at the most, to realize who they are looking at, and most orangutans, too, though only about 25% of gorillas (so far).

Having an insight into ones own and others' mental processes has other implications. Great apes can learn from each other and take turns at doing some complex actions. One chimpanzee works the lever, and the other operates the food tray; then they change places — the one who was operating the food tray wants a turn at the lever. Kanzi, the bonobo (so-called "pygmy chimpanzee") who lives in the Language Research Center in Atlanta, Georgia, strikes a flake off a stone core, tests it for sharpness, and uses it to cut a rope and thereby release the door of a food box. Orangutans are more patient and methodical than chimpanzees. One used to escape from his cage in the London Zoo in the early 20th century by fashioning a key to the cage door out of wood, hiding the half-finished key under the straw whenever a keeper came into view.

"Ape language" has always been controversial — needlessly so, in fact. When, in the early 1970s, the first chimpanzees were taught "sign language" (using hand signs as symbols for objects and actions), there was ill-tempered polemic over whether they "had language" or not. This sort of argument is ultimately sterile; it just depends where you want to put the barrier between language and not-language. A much more productive argument would have been to ask what language-like features this chimpanzee hand-signing has and whether the symboling abilities are homologous with human linguistic abilities or whether they arise from some other aspect of the complexity of chimpanzee cognition. Because people asked the wrong questions, ape-language trainers became incautious and claimed more and more for the apes (mostly chimpanzees) in their care. So when, in 1981, Herb Terrace and his colleagues showed that the signs the chimpanzees were making were not sentence-like and that they were not taking turns in their conversations with their trainers but that the trainers were often inadvertently cuing them, it was too easy for human chauvinists to breathe a sigh of relief and say, "I knew it — apes ain't got language."

It is due largely to the work of Sue Savage-Rumbaugh and her indefatigable colleagues in the Language Research Center that ape-language studies have been revived and brought back on track, first using common chimpanzees (*Pan troglodytes*) and then turning to bonobos (*Pan paniscus*). The work at the Language Research Center with Kanzi, Panbanisha, and other bonobos has finally allowed nonhuman great apes to approach us human great apes in the linguistic sphere as they have done in other cognitive aspects. The way this has been done is really so simple. Consider this amazing fact: Children learn language not by making the sounds and then associating them with objects and actions, but by understanding first. So did Kanzi: He watched over his mother's shoulder while she was being taught, unavailingly. Just like a child, he understood first — only then did he utter.

The Great Ape Project, first proposed in 1994 by Peter Singer and Paola Cavalieri in the book of that name, promotes the idea that nonhuman great apes deserve a version of human rights. With such impressive psychological support, it is no wonder that the progress of this proposal has been accelerating. It would be a foolhardy biomedical researcher nowadays who admitted to performing disabling or distressing research on chimpanzees. On October 7, 1999, the New Zealand Parliament passed an amendment to its Animal Welfare Act that "nonhuman hominoids . . . must not be used in research, testing or education, unless the government official responsible for animal welfare is satisfied that such use is for their benefit, either as individuals or as a species." Some other countries have legislation that is similar in tenor, if weaker; thus the British government in November 1997 stated that it would "not issue any licenses to use great apes in scientific procedures."

Why great apes have these cognitive skills is obscure. The best that anyone can suggest is that it is something to do with their large size. But whether it is somehow selected for because they are large or is a simple epiphenomenon of it is still argued.

Other animals? Whales and dolphins seem to have complex cognition but are harder to test than apes, for obvious reasons. Elephants too.

If great apes, maybe gibbons? Maria Ujhelyi has found mirror recognition in the large Siamang gibbon (*Symphalangus syndactylus*) and in the smaller white-cheeked gibbon (*Nomascus leucogenys*), but she has failed to elicit mirror recognition in the still smaller white-handed gibbon (*Hylobates lar*). Does this support the body-size hypothesis? It is too early to judge. But it seems to make sense to us that, rather than a cognitive divide between great apes and everybody else, we have a sort of gradation, though the nature of this gradation remains to be explored.

CHAPTER 4

Our Kind of Hominins

A group of robust and heavyset individuals was moving through the valley in search of food and shelter. It was the changing of the seasons and they had moved into the valley to take advantage of the warmer conditions. While they had been in this valley a few times previously, they had never before needed to spend the night here. Just before the setting of the sun, the group moved farther up from the valley floor in order to reach a cave entrance. They had noticed the cave earlier that day and had decided that it would be a good place to spend the night, after briefly checking it out. The day had been productive with tubers found in abundance; indeed, they were even able to bring some with them back to their intended sleeping place. Unfortunately, they had come across no opportunities for obtaining meat. One individual cautiously entered and moved into the first large cavern, close to the entrance. There was a strong smell of animal droppings, but there still appeared to be no other animals in the cave. He moved back out of the cave and motioned the others to join him. They settled in for the night after finishing off the remaining tubers. The lone mother and child settled into the rear of the first chamber, away from the cave entrance.

All was not as it seemed. Within the deeper recesses of the cave, a leopard, upon smelling and hearing prey, had crept slowly forward toward the entrance. This was not only the sleeping place of the big cat, but in the deeper recesses of the cave, the cat had established

its primary feeding site, where, safely tucked away from other carni-
vores, it could consume its prey without interruption. This was not its
first ambush from within the cave; the bones of baboons and hominins
were scattered around the cave floor, well away from the entrance,
a testimonial to the success of this site as a source of fresh meat.

The leopard chose its own time to strike; it also chose its prey from the
group. Conveniently for the leopard, the mother and child were closest.
And when hominins were in a group, a child or elderly individual was
the preferred prey. The mother cuddled up to her child as they settled for
the night. It was at this point that the leopard pounced. From a small ter-
race within the cave that was slightly above the mother and child, the
leopard sprang into life. Its powerful jaws clenched the lower leg of the
child and dragged it from its mother. Both mother and child screamed in
fear and terror. The big cat was now positioned between the mother and
child, flashing its large canines, growling. It reclined back in a position
to bounce if attacked. The mother yelled at the cat, flinging her arms,
while the other members of the group ran to her assistance, yelling and
screaming. The cat leaped at the mother, slashing through a thigh mus-
cle. The other hominins jumped back, grabbing her by the arm, forcing
her out of the cave. Nothing could be done. The child could be heard in
the cave for a short time, yelling and crying; then there was silence. The
big cat had clenched the child's neck in its jaws and cut into its wind-
pipe. The child was soon dead. The leopard dragged the body by its
head, moving back to its feeding area in the darkest recesses of the cave.

The classic excavations at the Swartkrans cave locality, South Africa,
have yielded a large number of hominin specimens, attributed mostly
to *Paranthropus robustus*. One of these, SK 54, is a young child with two
very distinct puncture marks on its cranium. It was Brain (1970) who originally suggested that a leopard had killed the child. He demonstrated that
the puncture marks were spaced so as to match precisely the canine spacing
of a primitive leopard's jaw, specimen SK 349, from the same deposits. The
child, he suggested, having been killed, was taken by the leopard into a
tree, where it was consumed, and pieces of the body had fallen from the
tree into a nearby sinkhole. Over time the surrounding surface was eroded
and the former sinkhole became a cave. Brain has since revised aspects
of his model; while a leopard attack was still the likely cause of death for
SK 54, the child was probably taken from a badly chosen sleeping site

Brain, 1981). Leopards have been seen to kill baboons in a similar fashion. Brain suggests that if leopards were involved, which seems almost certainly the case, then the overwhelming number of baboon and hominin remains suggests that a sleeping site was being exploited — it was not leopards hunting normally in the open country.

The Emergence of *Kenyanthropus* and *Australopithecus*

Kenyanthropus platyops originates from geological deposits dating to around 3.5 million years ago (M.G. Leakey *et al.*, 2001). Its discovery so recently, with its unique anatomical features so unexpected, sent shockwaves through the anthropological world and started a flurry of speculation. Indeed, recently T.D. White (2003) has suggested that the cranium of the type specimen of *Kenyanthropus platyops* (specimen KNM-WT 40000) may actually represent a specimen of *Praeanthropus*, for he suggests that the distortion of the specimen has resulted in a misdiagnosis. While this suggestion deserves serious consideration, some derived features of *Kenyanthropus*, which are unlikely to be overtly influenced by the type of distortion, suggest otherwise. For example, *K. platyops*, unlike *Pr. afarensis*, does not have an occipitomarginal sinus or a compound temporonuchal crest; it has reduced incisor heteromorphy, the upper molars are also significantly reduced in size relative to *Praeanthropus*, and finally its enamel thickness is reduced (see M.G. Leakey *et al.*, 2001). Therefore, the present authors recognize *Kenyanthropus* as a distinct taxon.

It has recently been suggested by D.E. Lieberman (2001) that *Kenyanthropus* probably evolved from members of the "*anamensis* group" or a species very much like it, and that *Kenyanthropus* is the immediate ancestor to the species called *Australopithecus rudolfensis* by some (B.A. Wood & Richmond, 2000; B.A. Wood, 2002; Walker & Shipman, 1996; partly Walker, 1976) and *Homo rudolfensis* by others (D.E. Lieberman *et al.*, 1996). If so, the species *rudolfensis* should be placed within *Kenyanthropus* as *K. rudolfensis* (D.E. Lieberman, 2001; see also partly M.G. Leakey *et al.*, 2001). This scheme would go some way toward helping explain the "problematic" fossils currently allocated to *rudolfensis*.

As discussed by M.G. Leakey *et al.* (2001) and D.E. Lieberman (2001), *Kenyanthropus* is distinct from the proto-australopithecines and more like later hominins in several features, notably its facial skeleton (Figure 4.1).

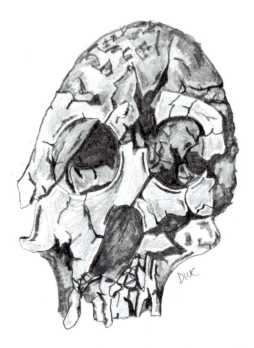

Figure 4.1 ▶   Frontal view of distorted *Kenyanthropus platyops* specimen KNM-WT 40000.

Adapted from M.G. Leakey *et al.* (2001).

The anterior insertion of its cheekbone gives its face a less prognathic appearance, further emphasized by the relatively flat subnasal region, the area in which the upper incisor roots are implanted; it also has a tall cheekbone, like later hominins. The supraorbital torus (browridge) is also less developed and more gracile, as in later hominins. And while it has a small cranial capacity (around 350 cc), its frontal is set high above the face, with no frontal sulcus. Indeed, D.E. Lieberman (2001) suggests that except for its small cranial size, *Kenyanthropus* is strikingly similar to the famous facial skeleton KNM-ER 1470, which largely defines the anatomical condition of the "*rudolfensis* group." This would be remarkable because members of the "*rudolfensis* group" do not appear in the fossil record until almost 1.5 million years later. *Kenyanthropus* does share some features with some proto-australopithecines, including a small auditory meatus, thick molar enamel, a small brain, and a flat inferior nasal margin (D.E. Lieberman, 2001); though these are primitive features, all of which have been retained from its evolutionary past and are of no significance in terms of working out phylogenetic relationships.

Given the unique combination of anatomical features, primitive and derived, we agree with M.G. Leakey *et al.* (2001) and D.E. Lieberman (2001) that if we continue to accept the validity of *Australopithecus* and *Paranthropus*, then the description of the new genus *Kenyanthropus* is justified. A genus, unlike a species, is recognized not on the basis of anatomical variability but on its implied evolutionary relationship with existing taxa (see Chapter 1). If a taxon does not share a close sister-group relationship or a recent common ancestor with any previously known genera, then it is a candidate for its own genus. *Kenyanthropus* cannot easily be accommodated within either *Australopithecus, Paranthropus*, or *Homo*, as Cameron (in press b) has shown in his phylogenetic analysis of early hominins (see also the next chapter); in some ways it appears to represent a good ancestral type for both *Australopithecus* and early *Homo*.

Recent paleontological and paleobotanical studies at Lake Turkana (Kenya) and within the Omo region (Ethiopia) have examined deposits from the time of the emergence of *Kenyanthropus*. They indicate that a complex mosaic of woodlands and grasslands stretched over the region, interrupted by gallery forests that occupied and contracted over time. From 4 to 1 Ma, the gallery forests underwent a rapid and dramatic floral change. In addition, at Lake Turkana, around 3.4 Ma, fossil snails typically associated with tropical rainforests have also been found (Potts, 1996; see also Isaac & Behrensmeyer, 1997).

Just when *Praeanthropus afarensis* and the "*bahrelghazali* group" were disappearing from the fossil record at the other end of the continent, in South Africa *Australopithecus africanus* appeared — the first African Pliocene hominin ever discovered and described. The earliest appearance date for this species is around 3.5 Ma (Strait & Wood, 1999; B.A. Wood & Richmond, 2000), which is similar to the proposed dates for *Kenyanthropus*. As will be discussed in the next chapter, this is the only species that really belongs to the genus *Australopithecus*. We also believe that, unlike the protoaustralopithecines, *Australopithecus* and *Kenyanthropus* represent the earliest unchallengeable hominins to date.

In 1925 the Australian Raymond Dart, who held the Chair in Anatomy at the University of Witwatersrand, South Africa, was digging around in a box of fossil specimens supplied to him by the owner of the Rand Mines, when he identified among the numerous baboon skulls, the skull and associated fossilized brain of what became known as the Taung child (Figure 4.2). Realizing their significance, he quickly drafted a paper to the prestigious journal *Nature*, announcing the discovery of a

> manlike ape . . . and that the first known species of this group be designated *Australopithecus africanus*, in commemoration, first, of the extreme southern and unexpected horizon of its discovery, and secondly, of the continent in which so many new and important discoveries connected with the early history of man have recently been made, thus vindicating the Darwinian claim that Africa would prove to be the cradle of mankind (*Dart, 1925:198*).

Dart recognized that this apelike individual walked erect because the foramen magnum, through which the spinal cord connects with the brain, was located directly underneath the cranial base. This means that the head was

Figure 4.2 ▶   The Taung Child from South Africa, the type specimen of *Australopithecus africanus*.

Adapted from Dart (1925).

balanced on top of the vertebral column, unlike other apes, where the fora-men magnum is located toward the back of the skull. Unfortunately for Dart, the first decade of the new century had focused paleoanthropological atten-tion onto the English Piltdown specimen, a bizarre forgery, a composite of an orangutan's jaw and a modern human cranium (see F. Spencer, 1990), and this indicated that early hominins had a large brain and apelike teeth, pre-cisely what most researchers at the time predicted. The Taung child, how-ever, had the reverse condition, a small brain and human teeth; so many people considered it a primitive chimpanzee, having little or anything to do with human evolution, and it was argued that the unusual location of the foramen magnum was because it was a young child and by adolescence the foramen magnum would have moved more posteriorly. It would be almost another 30 years before Piltdown was finally shown to be a scientific forgery (Oakley & Hoskins, 1950; F. Spencer, 1990) and thus room made for the australopithecines within the human evolutionary tree. And in the meantime, more and more of them have been discovered in sites in South Africa.

Like most proto-australopithecines, *A. africanus* was probably not a stone toolmaker, though it was possibly a tool user — after all, chimpanzees use both wood and stone tools and even make some tools of flexible, perishable materials (though they do not make tools of stone); so presum-ably the australopithecines could do no less. In the 1950s, however, Dart found what seemed to him an unusual sample of fossil antelope and pig bones in the cave site of Makapansgat, which contained specimens of *A. africanus*. The way in which they had been broken suggested to him that they had been selected and modified to use as tools. *Australopithecus africanus*, he argued, had chosen suitable bones from carcasses and kept them handy for use in butchering animals they had killed, and some of them perhaps were even weapons for prehistoric warfare, including cannibalism (Dart, 1957, 1959, 1960). Large mammal long bones were clubs, animal jaws with teeth were tools used to help cut material, while other animal parts were the remains of prehistoric feasts. This whole toolkit was given the name *osteodontokeratic culture* (*osteo-* meaning "bone," *odonto-* meaning "teeth," and *kerat-* meaning "horn"). Dart believed that the australopithecines were marauding killer apes, with a social system based on brutality. It is likely that Dart's experience as an Australian med-ical officer in the fields of Flanders during the First World War influenced this very pessimistic outlook of human behavioral evolution. We now know, from careful analysis of the way that modern skeletons disintegrate and are taken apart by scavengers, that it was a perfectly natural assemblage;

indeed, rather than doing the eating at Makapansgat, the australopithecines were the prey! The evidence suggests that these bone accumulations at australopithecine sites represented hyena or sabertooth dens and/or leopard kills (Brain, 1981).

Australopithecus africanus (Figure 4.3) was similar in body type and brain size to its often-presumed ancestor *Pr. afarensis*, though there are many differences between the two: Canine teeth of *A. africanus* were smaller, and its molars were larger, with even thicker enamel; and its facial skeleton was buttressed with thickened bone alongside the nasal aperture in what Rak (1983) has called "the nasoalveolar triangular frame." Surviving postcranial skeletal parts of *A. africanus* suggest that, like *Pr. afarensis*, it was adapted to some form of bipedal locomotion. However, the muscles of the lower limbs may have been arranged in a way unlike those of modern apes and humans (B.A. Wood, 1992), and the arms may have been longer and more powerful — but this is very controversial.

With the arrival of drier climates and the expansion of the savannah around 2.5 Ma, *A. africanus* disappears from the fossil record. The habitat of *A. africanus* ranged from a wet forest with high rainfall at Makapansgat

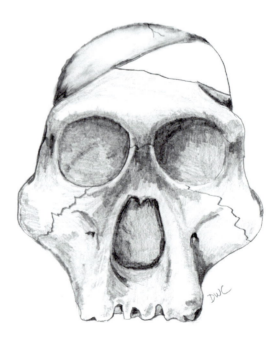

Figure 4.3 ▶   South African Sterkfontein *Australopithecus africanus* specimen Sts 5.

to a dry partially wooded savanna at the later sites of Sterkfontein and Taung (Andrews & Humphrey, 1999; McKee, 1999). The drier and more patchy forest cover suggested by the later localities fits in well with the evidence that this is the time of the earliest northern hemisphere glaciation, which started around 2.5 mya and resulted in more arid conditions in Africa (Vrba, 1999). So *A. africanus* — like *Pr. afarensis* — experienced a diverse succession of environmental conditions over time (Potts, 1996).

Another genus, *Paranthropus*, appears in the fossil record at about the same time that *A. africanus* vanishes and that the "*rudolfensis* group" and early *Homo* are first emerging (see Delson, 1988; Grine, 1988; Suwa *et al.*, 1997; Deacon & Deacon, 1999; Dunsworth & Walker, 2002; T.D. White, 2002). In terms of competition and access to available food resources and territorial ranges, *Paranthropus* are likely to have played some indirect role in the adaptive trends adopted by the earliest humans.

The Emergence of *Paranthropus*

With the discovery by Alan Walker of KNM-WT 17000 (nicknamed the "Black Skull" due to its color; see Figure 4.4) in Kenya during the mid-1980s, another major problem in palaeoanthropology was largely settled, the origins of the unique robust hominins of the genus *Paranthropus* (see Walker *et al.*, 1986; Walker & Leakey, 1988). Until the discovery of this specimen, dated to 2.5 Ma, there was no fossil evidence for their origins: in East Africa, *P. boisei* appears around 2.3 Ma and disappears around 1.4 Ma; in South Africa, *P. robustus* appears around 1.8 Ma and disappears around 1.6 Ma (Keyser, 2000; T.D. White, 2002). The Black Skull is anatomically intermediate between *Pr. afarensis* and the later species of *Paranthropus*. The species that traditionally make up species of *Paranthropus* are monophyletic, and thus their generic distinction from other hominins is clearly justified (see Groves, 1989a; Strait *et al.*, 1997; Strait & Grine, 2000; Cameron, in press b; and the appendix); as long as we are going to split them up into different genera at all, of course.

To what species should we allocate the Black Skull? Some authorities see it as a simple lineal ancestor of *Paranthropus boisei* and so place it in that species. Yet it does fall outside the range of *P. boisei*, with its smaller cranial capacity, its larger front teeth (as suggested by its surviving incisor roots, which are well developed), and its very prominent, prognathic lower face. Besides, it is probably ancestral to the South African species

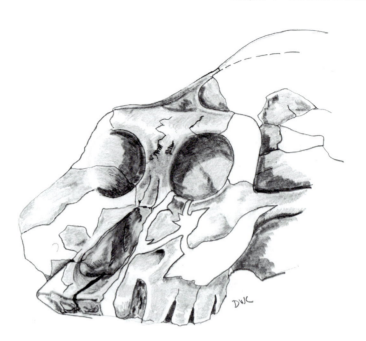

Figure 4.4 ▶ Cranial specimen of *Paranthropus walkeri* (KNM-WT 17000), found by
Alan Walker in West Turkana, Kenya, dated to around 2.5 million years ago.

P. robustus as well, though not all specialists agree about this (see Walker
et al., 1986; Clarke, 1988; Kimbel *et al.*, 1988; Walker & Leakey, 1988;
B.A. Wood, 1988; B.A. Wood & Richmond, 2000).

In 1968, Arambourg and Coppens discovered a large and strange-looking
jaw (Omo-18) in 2.6-million-year-old deposits in the Omo valley and placed
it in a new genus and species, *Paraustralopithecus aethiopicus* (see
Arambourg & Coppens, 1968; Howell & Coppens, 1976). Walker *et al.*
(1986), when describing the Black Skull, recommended that studies should
be made to determine whether it might represent the same species as the
Omo-18 jaw and, if so, it should be called *Paranthropus aethiopicus*. No
one seems to have taken up this sensible suggestion; instead, people have
rushed right in and ascribed the Black Skull to *Paranthropus* (or even
Australopithecus) *aethiopicus* without any special study at all. We believe
that it is best to avoid this designation, for two main reasons: (1) because
Omo-18 and the Black Skull might not, after all, be the same animal (and
we now know of another big-toothed species that was living at the same
time as the individual defined by the "Black Skull," the so-called *A. garhi*);

(2) because the name *Australopithecus aethiopicus* had previously been used by a few anthropologists for "Lucy" (see Tobias, 1980) and, the rules of nomenclature being very tortuous, this would restrict the use of the name *aethiopicus*. Ferguson (1989) gave the Black Skull its own species name, *walkeri*, and all in all we think it safest to call the species represented by the Black Skull *Paranthropus walkeri* (Groves, 1999), until it is demonstrated that the large robust Omo mandible and the Black Skull do indeed belong to the same species.

Recent archaeological interpretations at Swartkrans in South Africa suggest that *P. robustus* may have used animal long bone shafts as digging "sticks." There is also the suggestion of burned bone, but the geological deposits containing these fossils also overlap with the presence of early *Homo* fossils (B.A. Wood, 1992). Thus, we cannot tell whether it was *Paranthropus* or early *Homo* who burnt these bones (Pickering, 2001), or whether it was either of them, because the burns could even have been the result of natural processes such as bushfires.

Later *Paranthropus* species have a large flat face, and the brain was relatively small — somewhat bigger than a chimpanzee's and much bigger than that of the Black Skull (McHenry, 1988) (Figure 4.5). They also had a large powerful jaw, the back teeth (premolar and molars) being extremely large, acting as "grinding stones" for the preparation of hard food objects prior to digestion; yet the front teeth (incisors and canines) were tiny, smaller even than in modern humans (Robinson, 1954, 1972; Tobias, 1967). While their masticatory architecture was well developed, body size was only a little larger than that of most proto-australopithecines, and the body proportions were also similar to that of most proto-australopithecines (McHenry, 1988). This primitive yet specialized hominin, like *Australopithecus*, was a bipedal walker (Susman, 1988, *contra* Robinson, 1972). It appears to have specialized in consuming large amounts of tough plant foods typical of parched environments, unlike contemporary *Homo*, which was more of a generalized opportunistic feeder (see Kay & Grine, 1988; Susman, 1988).

Both *Homo* and *Paranthropus* lived during a period of drier climate and more open vegetation, compared to the wetter conditions existing at the time of *Australopithecus* (Reed, 1997). It is likely that *Paranthropus* overspecialized and, with the ever-changing climatic conditions associated with the northern glaciation cycles, failed to adapt to the more generalized dietary regime that was required given the increasing aridity (see Foley, 2002; Pickering, 2001). This overspecialization ultimately led to its extinction.

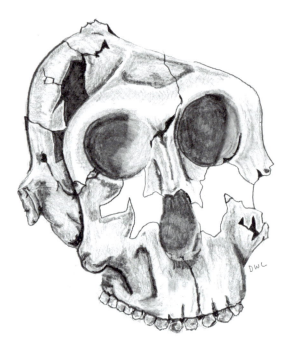

Figure 4.5 ▶   Type specimen of *Paranthropus boisei,* Olduvai Gorge, Tanzania, found by
Mary Leakey in 1959 and dated to around 1.8 million years ago.

The Emergence of the "Rudolfensis Group"

In the 1970s, a remarkable skull, KNM-ER 1470 (Figure 4.6) — known
to the general public simply as "skull 1470" — was discovered at Koobi
Fora, in northern Kenya, in deposits we now know to be from 1.89 Ma
(R.E.F. Leakey, 1973a, 1973b, 1974; Day *et al.*, 1975; B.A. Wood, 1976,
1991). It was originally allocated to *Homo habilis*, the earliest and most
primitive species of our own genus. Doubts began to grow in the 1980s
that 1470 really was *Homo habilis* (it was just too different from the
acknowledged specimens of that species from Olduvai Gorge in Tanzania).
By the late 1980s there was a general acquiescence that it should be placed
in a different species, *Homo rudolfensis* (Alexeev, 1986), and some other
specimens from Koobi Fora were placed in the same species. During the
1990s, some researchers began to question whether it was *Homo* at all
(D.E. Lieberman *et al.*, 1988; partly B.A. Wood, 1991). By the late 1990s,
it was being widely asked what this "*rudolfensis*" actually was and where
it came from.

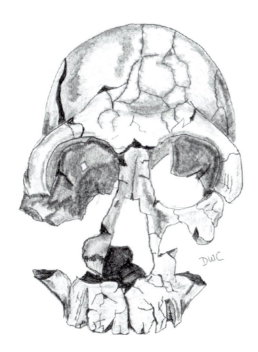

Figure 4.6 ▶   Frontal view of *Kenyanthropus rudolfensis* specimen KNM-ER 1470.

The recent discovery of *Kenyanthropus platyops* has supplied a possible answer: Here, in this contemporary of *Praeanthropus afarensis*, was the by-now eagerly sought ancestor to *rudolfensis*! If this is indeed the answer, then of course *Homo rudolfensis* should be transferred out of *Homo* and placed within its ancestor's genus, *Kenyanthropus*, and called *Kenyanthropus rudolfensis* (see D.E. Lieberman, 2001; partly M.G. Leakey *et al.*, 2001). Its removal from *Homo* would vindicate the arguments put forward originally by Alan Walker, who in the 1970s argued that 1470 was no more than a slightly more derived species of *Australopithecus*, with a larger brain (Walker, 1976; Walker & Shipman, 1996; D.E. Lieberman *et al.*, 1996).

Our phylogenetic analyses (discussed in the next chapter) consistently indicate that the *rudolfensis* group shares a sister-group relationship with *Kenyanthropus platyops*. If this truly reflects the evolutionary history of this hominin, it should be considered a species of this genus. These same analyses also indicate that *Kenyanthropus* shared a sister-group relationship with the species of *Homo*, indicating that they, like *Paranthropus* and *Australopithecus*, are hominins, as opposed to mere hominids.

This scheme confirms that those more than 3-million-year-old species usually allocated to *Australopithecus* do not share a particularly close relationship with the later hominins and probably had little if anything directly to do with their evolution. It also indicates that at least three genera coexisted in Africa around 2.3 Ma: *Paranthropus, Kenyanthropus,* and early *Homo.* They may have last shared a common ancestor as early as 3.5 million years ago.

The Emergence of Earliest *Homo*

The first member of our own lineage occurs in East Africa and is called *Homo habilis* (meaning "handy man"). First specimens of this species were discovered by the Leakeys at Olduvai Gorge in the mid-1960s and date to 1.8 Ma (L.S.B. Leakey *et al.*, 1964). Other probable specimens have turned up in Koobi Fora at 1.89 Ma (see B.A. Wood, 1991) (Figure 4.7). At Olduvai, the species survived to 1.6 Ma (Tobias, 1991). And there is a maxilla indistinguishable from *Homo habilis* from Hadar, dating to 2.3 Ma (Kimbel *et al.*, 1997). From South Africa, specimens from Swartkrans and from the upper levels at Sterkfontein (later than the *Australopithecus* levels) have tentatively been allocated to this species, though they are different in some aspects and may represent a new species of *Homo* (Tattersall & Schwartz, 2000).

This early representative of our own genus still retained a number of australopithecine-like features; clearly it had not yet developed the more complex physical and behavioral repertoire of its descendant *Homo ergaster* (see later). In its daily rate of molar dentine formation, *H. habilis* is closer to the proto-australopithecine and extant ape condition, with almost three times the amount of deposition per day, compared to the reduced amounts observed in later hominins (M.C. Dean, 2000). *Homo habilis* also had body proportions that are reminiscent of *Australopithecus*: Its legs were relatively short; its arms were relatively long. Its arm and hand bones suggest to some that it had the apelike ability of arboreal loco-motion (Johanson *et al.*, 1987). If so, it was adapted to tree climbing while also being able to walk bipedally, just as has been proposed for *Australopithecus*. Indeed, Hartwig-Scherer and Martin (1991) suggested that if the postcranial specimens allocated to *H. habilis* (OH 62) really belong to this species, then *H. habilis* was even more primitive than

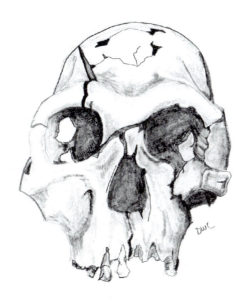

Figure 4.7 ▶   Koobi Fora *Homo habilis* specimen KNM-ER 1813.

Pr. afarensis in its postcranial morphology (see also B.A. Wood & Richmond, 2000). In other features, it is distinct from its australopithecine ancestors, and in the phylogenetic analyses discussed in the next chapter, it clearly belongs to the *Homo* clade. It has a larger cranial capacity, between 510 and 680 cc, and it appears to have been a regular toolmaker and user; fossil specimens are regularly found associated with primitive stone tools, hence its name. Even the earliest specimen, the Hadar maxilla, is associated with stone tools (see Tobias, 1991; B.A. Wood, 1991; Kimbel *et al.*, 1997; Tattersall & Schwartz, 2000; Dunsworth & Walker, 2002).

With the removal of the "*rudolfensis* group" from the genus *Homo,* there is now only one species left that is a likely representative of the next stage of human evolution. This species is *Homo ergaster* (Groves & Mazak, 1975), which probably evolved from *H. habilis*, although the earliest *H. ergaster* remains at Koobi Fora overlap in time with the latest *H. habilis* (see B.A. Wood, 1991). Until recently, many of the specimens allocated to this species had been placed within the species *Homo erectus*; many researchers today, however, believe that the African specimens previously allocated to *H. erectus* are quite distinct from the Asian fossils, which constitute the "true" *H. erectus* (see Andrews, 1984; Stringer, 1984; B.A. Wood, 1984; see partly

Groves, 1989a; B.A. Wood, 1991; B.A. Wood & Richmond, 2000). Recently, three skulls and a jaw from Dmanisi, in the Republic of Georgia, have been ascribed to *H. ergaster* (Gabunia *et al.*, 2000a; Vekua *et al.*, 2002; see also partly Bräuer & Schultz, 1996). If this allocation is correct, then *H. erectus* may not, strictly speaking, have been the first to leave Africa, but it was still almost certainly the first to come out and stay out.

The Earliest Tool Users and Toolmakers and Early Hominin Behavior

Recognizable stone tools first appear in the archaeological record around 2.5 Ma in east Africa; they are associated with two hominin groups, *Paranthropus* and early *Homo* (see Susman, 1988; Schick & Toth, 1993; Kimbel *et al.*, 1997; Deacon & Deacon, 1999; Pickering, 2001) (Figure 4.8). These early stone tools are referred to as the "Oldowan industry" because they were first recognized from localities within Olduvai Gorge. So far the earliest stone tools are those found at the Gona and Bouri sites, Hadar, Ethiopia, dating to around 2.5 Ma; from Lokalelei, West Turkana, Kenya, dating to around 2.4 Ma; from Koobi Fora dated to around 1.9 Ma; and from Olduvai Gorge, dating to around 1.8 Ma (M.D. Leakey, 1971, 1994; Potts, 1988, 1996; Isaac & Isaac, 1997; Kimbel *et al.*, 1997; Semaw *et al.*, 1997; Heinzelin *et al.*, 1999).

The earliest tools from Gona comprise simple cores, whole flakes, and flaking debris. There is evidence of ongoing flake scars on the cores, indicating that the early hominins (*H. habilis?*) had mastered the skills of basic stone knapping (Semaw *et al.*, 1997; Kimbel *et al.*, 1997). Indeed, Semaw *et al.* (1997) argue that the material they recovered from the Gona localities show surprisingly sophisticated control of stone fracture mechanics, which they suggest is equivalent to much younger Oldowan assemblages of around 1.8 Ma.

The nearby stone tool localities around Bouri, which also date to around 2.5 Ma, are said to be associated with the *"Australopithecus" garhi* (Heinzelin *et al.*, 1999), though the fossil hominid remains were not found at the same locality as the tools. It is also from this region that we have early *Homo* dating to 2.3 Ma (Kimbel *et al.*, 1997), and it cannot be dismissed that these tools are from the hand of *Homo*. Mammal remains from this time zone have also been discovered with cut marks and percussion marks said to be made by stone tools. Further, Heinzelin *et al.* (1999)

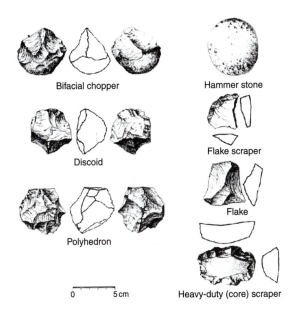

Bifacial chopper   Hammer stone

Discoid

Flake scraper

Flake

Polyhedron

0   5 cm

Heavy-duty (core) scraper

Figure 4.8 ▶ Two types of typical Oldowan tools from Koobi Fora. Top is a classic bifacial chopper; bottom is a more specialized heavy-duty tool (core scaper). From Schick and Toth (1993), p. 113.

suggest that the bone modifications indicate that large mammals were disarticulated and defleshed, and that their long bones were broken open and the marrow extracted. The raw material for tool manufacture was not local and had to be transported into the area, as happened much later at Olduvai (see Potts, 1988). Heinzelin *et al.* (1999) conclude that the major function of the earliest known tools was meat and marrow processing of large carcasses.

It is clear that by around 2.5 Ma a distinct behavioral shift had occurred in food acquisition and processing. Oldowan toolmakers were beginning to unlock the vast amount of nutrition stored up in large mammals' carcasses. It is with the first appearance of the genus *Homo* that we see evidence of vastly increased stone tool distribution throughout the landscape, associated with large mammal remains, which provide evidence of butchering. All evidence suggests that this behavioral shift can be associated with the emergence of *Homo*, as opposed to *Paranthropus*, whose diet was more specialized, focusing on tough, gritty foods that have left many scratch marks and pits on their grinding, stonelike molar teeth (Kay & Grine, 1988). While some tool use may have occurred in *Paranthropus, Homo* was the first to shift dramatically to a dependence on this new technology. So, while *Paranthropus* appears to have used its grindstone-like molars to process its hard, gritty foods, early *Homo* shifted to a stone tool technology, not only to help in food processing, but perhaps also to help in its acquisition.

Is there any additional archaeological evidence to support the Central Place Foraging model (Isaac, 1976, 1978, 1989), the Scavenging hypothesis (Binford, 1981), or the Stone Tool Caching model (Potts, 1988) for later Pliocene early or Pleistocene *Homo*? The site of FLK North-6, located within the main gorge at Olduvai, in the upper part of Bed I, dated to around 1.7 Ma, has been interpreted as an early hominid butchering site (M.D. Leakey, 1971). The excavations in trenches IV and V revealed the remains of an almost complete elephant skeleton, which has been described by M.D. Leakey,

> where an elephant was cut up by early man, who may have come across it accidentally, or deliberately driven it into a swamp to be slaughtered. The tools found nearby would seem to represent those used for cutting the meat off the carcass (*1971:64*).

Based solely on the association of numerous stone tools and elephant remains, Leakey argued that they are contemporary and concluded that they are part of a prehistoric residue of activity. While M.D. Leakey (1971) discusses bone modification on mammalian remains, no mention was made of human-induced modifications to the elephant remains. Thus the "butchery" site is based purely on the association of tools and elephant remains, which must be viewed with some caution.

This type of associative analysis of early hominin behavior can also be seen in Binford's (1981) "residual" analysis of early hominin debris from Olduvai. Binford concentrated on distinguishing between carnivore and hominin behavioral debris patterns. In his approach, any archaeological pattern that cannot be attributed to carnivores (as defined by extant carnivore residual patterns) is assigned to hominin activity. The major problem with this approach is the factoring out of "carnivore debris patterns." It does not allow for any overlap between the multiple use of animal carcasses by both carnivores and hominids (Cameron, 1993b). Also, because Binford could explain such patterns using only extant carnivores, he failed to appreciate possible extinct and other extant carnivore patterns that may have contributed to the archaeological bone accumulation. For example, what about the felids *Dinofelis* and *Megantereon* and the hyaenid *Euryboas* (Groves, 1989a)? In his multivariate analysis of the assemblage composition present at

FLK North-6, Binford argued that the site, like FLKNN-2, represents a carnivore den assemblage. This is based on the similarity of Q-mode factor loadings between these sites and those of modern carnivore den assemblages.

Potts (1988) questioned Binford's carnivore versus hominin bone pattern assemblage analysis. He argued that the large degree of variability between contemporary carnivore debris patterns (not to mention extinct patterns) would make it very difficult to identify the phenomenon responsible for the debris observed in the archaeological record. He stated (p. 135) that

> our current knowledge of carnivore bone accumulations and damage includes significant variation in the species observed (e.g., spotted hyenas versus wolves) and in observation conditions (e.g., bones fed to zoo animals versus bones recovered from natural dens and kill sites). Even within species, bone damage and selection of parts for transport may vary with environmental factors. Although there may be features in common to carnivore bone accumulations, it is still unclear how well these features distinguish carnivore collections from bones collected and modified by hunter-gatherers.

Potts (1986, 1988), Potts and Shipman (1981), Bunn (1981, 2002), and Bunn and Kroll (1986) all conducted microscopic analyses of fossilized bone from Olduvai and all were able to identify stone tool cut marks as well as carnivore tooth marks on bone fragments. These analyses have refuted Binford's claim that FLK North-6 is solely representative of a prehistoric carnivore den. Potts's analysis has clearly shown that hominins *and* other carnivore species were involved in utilizing the remains at Olduvai.

Binford (1981, 1985, 1988) also argued against Isaac's Central Place Foraging hypothesis, which is the model we think is the most likely behavioral repertoire for early *Homo*. (See previous chapter.) Binford believed that early *Homo* was scavenging the leftovers of other carnivores and so had access only to low-food-utility items that carnivores chose not to eat, and had to scavenge these items quickly before moving off to a more protected area away from the kill site. Most Olduvai archaeological sites, he contended, if they can be attributed to hominin activities at all, represent optimal scavenging activities. Potts (1986) refuted Binford's argument and established that the bone formations had accumulated over a 5- to 10-year period or even more. This is an important point, which indicates that some form of central place foraging was indeed being used by the early hominins. It is clear that during this period, hominins contributed to these concentrated scatters, because tool cut marks can be identified on many pieces of bone that accumulated over long periods. Specific locations were being used by early *Homo* over a period of years, which indicates to us that, again, Isaac's Central Place Foraging model is the most likely explanation (Cameron, 1993a; Cameron & Groves, 1993). The processes and likely reasons for this dietary and behavioral shift will be examined in later chapters.

Paleobiogeography

The available evidence indicates that the hominins evolved in northern and eastern Africa before dispersing south into southern Africa (Strait & Wood, 1999). The early proto-australopithecines are all found in East Africa and Chad, ranging from between 4.2 and 2.5 Ma. The southern dispersal of *A. africanus* occurred between 3.5 and 3.0 Ma. It first turns up in Sterkfontein in Member 2 deposits, about 3.5 Ma, where there is an apparently complete skeleton, still in the process of being carefully excavated. The expansion of the australopithecine range into southern Africa appears to have been accompanied by the emergence and dispersal of *Kenyanthropus platyops* in eastern Africa around 3.5 Ma. The emergence of *Homo habilis* first occurred in east Africa by 2.3 Ma, but by 1.8 Ma *H. habilis* was possibly also present in southern Africa. By 2.4 Ma *K. rudolfensis* had moved south into central Africa, just north of the Malawi Rift. Following a similar temporal and spatial dispersal with *Homo habilis*, the species of *Paranthropus* first appear in east Africa around 2.5 Ma with *P. walkeri*, which then gave rise to the later species *P. boisei*, which survived in east Africa for almost another million years, and another species of *Paranthropus* (*P. robustus*) dispersed to southern Africa, becoming extinct around 1.0 Ma.

The dispersal of the hominins from north to south during a number of "migratory" phases strongly correlates with a number of large-scale mammal faunal migrations during the same periods. The evolutionary branching of the hominins — their emergence, proliferation of species, and extinction times — closely resembles the pattern observed in the rise and fall of other large mammal groups as well as of rodents (Vrba, 1985, 1999). In other words, the early evolution and dispersal patterns of the earliest hominins and other large mammals in Africa were the result of a similar response to environmental perturbations (Strait & Wood, 1999; Vrba, 1999). Between the major radiations of early hominin species, between 3.0 and 2.5 Ma, the paleohabitats of the hominins in East Africa were undergoing significant and prolonged periods of climate change. Most proto-australopithecine species appear to have occupied forest and forest edge or gallery forest, though there is also some evidence, especially at Hadar and Koobi Fora, for a more diverse and mosaic paleohabitat over time. However, at the time of the emergence of *Kenyanthropus, Paranthropus,* and *Homo* around 2.5 Ma, the climate had changed to a more arid state and palaeohabitats were now largely dry savanna, open savanna woodland, and/or arid, semiarid steppe scrub (Vrba, 1999). A similar trend was also occurring in southern Africa (McKee, 1999). So the emergence of the new genera correlates with increasing aridity within the African continent. *Paranthropus* appears to have focused more on the hard fruits and nuts (as suggested by its unique robust "nutcracker" masticatory apparatus), thus adapting to a more specific habitat and dietary niche, while later *K. rudolfensis* and early *Homo* were more generalized in their dietary requirements, probably more opportunistic, taking advantage of a broad range of food types, including meat in the case of *Homo* at least. Clearly the early species of *Homo* were able to outcompete the later species of *Kenyanthropus,* for it soon disappears from the hominin fossil record. It is only with the later *Homo* species, *H. ergaster,* however, that finally the hominins break out of Africa into Eurasia proper.

CHAPTER 5

A Systematic Scheme for the Pliocene and Early Pleistocene Hominids

The previous two chapters described the emergence of the late Miocene, Pliocene, and early Pleistocene hominid groups without examining in any detail their likely phylogeny. This chapter will directly address issues of phylogeny and taxonomy by generating a number of phylogenetic analyses of the Miocene and Plio-/ Pleistocene hominids. Overall this chapter will examine: (1) what the systematic significance of Sahelanthropus *is in relation to the Miocene and Pliocene hominids; (2) how the inclusion of* A. anamensis *and* A. garhi *affects the often-commented-upon paraphyletic status of* Australopithecus; *(3) what the phylogenetic relationship of* Kenyanthropus *is to other Pliocene hominids; and (4) how these relationships should be reflected in taxonomy.*

As discussed in the previous chapters, the discovery and description of the new Pliocene genus *Kenyanthropus* (M.G. Leakey *et al.*, 2001) and the late Miocene hominid *Sahelanthropus* (Brunet *et al.*, 2002) have thrown into question the origin of the genera *Australopithecus, Paranthropus*, and *Homo*, not to mention the evolutionary significance of the proto-australopithecines, including *Praeanthropus*, to them. While numerous parsimony analyses on Pliocene and early Pleistocene hominids have been generated (B.A. Wood & Chamberlain, 1986; Chamberlain & Wood, 1987;

Skelton *et al.*, 1986; B.A. Wood, 1991; Skelton & McHenry, 1992, 1998; D.E. Lieberman *et al.*, 1996; Strait *et al.*, 1997), they have tended to focus on the monophyletic status (all share a common ancestor to the exclusion of other taxa) of the species within *Paranthropus* (*P. walkeri*, *P. boisei*, and *P. robustus*) as well as the phylogenetic significance of *Pr. afarensis* and *A. africanus*. While some "consensus" has been reached with regard to the monophyletic status of the *Paranthropus* (see B.A. Wood, 1991; Strait *et al.*, 1997; Strait & Grine, 1998, 2001; Cameron, in press b), the paraphyletic status (do not share a common ancestor and thus are not considered to be particularly closely related because they appear to have their own distinct evolutionary lineage) of australopithecine species is still very much a hot issue.

Whether species currently assigned to *Australopithecus* form a monophyletic or paraphyletic group is extremely important, because if they are shown not to share a common ancestor to the exclusion of all other hominids, then they cannot be considered as belonging to the same genus unless all are included in *Homo*. This obviously has important implications for the phylogeny of early Pliocene hominids and later hominins. M.G. Leakey *et al.* (2001), while recognizing the paraphyletic status of *Australopithecus*, take the conservative, "grade-sensitive" approach and argue that the currently recognized australopithecine taxa represent stem species to this genus, all sharing a suite of key primitive features. Strait *et al.* (1997), however, suggest that the paraphyletic status of specimens previously allocated to *Australopithecus* means that *A. afarensis* should be removed from *Australopithecus* and placed within the genus *Praeanthropus*.

The reallocation of *afarensis* to *Praeanthropus* goes some way to resolving the paraphyletic nature of the australopithecines, as is frequently commented upon by researchers (see B.A. Wood & Chamberlain, 1986; Groves, 1989a; D.E. Lieberman *et al.*, 1996; Strait *et al.*, 1997; B.A. Wood & Collard, 1999; B.A. Wood & Richmond, 2000; M.G. Leakey *et al.*, 2001; D.E. Lieberman, 2001; Strait & Grine, 2001). The paraphyletic nature of *Australopithecus*, however, needs to be further reexamined, given the recent recognition of *A. garhi* (Asfaw *et al.*, 1999), *A. anamensis*, and the species sometimes categorized as *Australopithecus* or *Homo rudolfensis* (which we tentatively place in *Kenyanthropus*).

Phylogenetic Systematics

The phylogenetic history of any species will impact its ability to adapt to changing environmental-ecological conditions. A major problem in identifying phylogenetically significant anatomical features is the impact that function has on the development of morphological form. In order to differentiate between functionally and phylogenetically significant features, it is important to identify homologies and homoplasies within and between assumed closely related genera.

The definition of homology used in most paleontological studies is the phylogenetic concept. Homology implies shared common ancestral *and* shared derived characters; both are, of course, relative to a given point. The brain, for example, is a homologous character because it occurs in the same position in the body, protected by the skull and linked to the major sense organs according to a common pattern (R.D. Martin, 1990). While expansion and development of the frontal and temporal lobes of the brain may be considered synapomorphic characters uniting the fossil and extant hominins, a phylogenetic interpretation of homology is therefore to equate homology with shared ancestry. Homoplasy is the result of either morphological convergence or parallelism and not the result of immediate shared common ancestry.

How does one identify possible homologies and homoplasies? The best approach to tackling this problem is to use phylogenetic systematics (cladistics). Phylogenetic systematics is concerned with identifying the evolutionary branching sequence and establishing sister-group relationships between taxa. The allocation of fossil taxa to specific groups cannot be based on overall morphological similarity, because of the primitive retention of features as well as parallel evolution. In order to work out the phylogeny of a specific group, it is important that we reconstruct phylogenies independent of previously held assumptions.

Phylogenetic systematics enables us to identify likely homologies and homoplasies without reference to a previously constructed phylogeny, for it works in the following way: (1) Always assume that characters at the first instance are homologies; (2) use an outgroup comparison to distinguish more general (primitive) from more specialized (derived) features; (3) group species on the basis of shared derived features (synapomorphies); (4) in the event of conflicting information, tend to choose the phylogenetic relationships that are supported by the largest number of features; and (5) interpret inconsistent results, post hoc, as homoplasies. Thus homologies, which determine phylogenetic relationships, are identified a priori, without reference to a phylogeny, and later confirmed as homologies or reinterpreted as homoplasies (see Wiley, 1981; Brooks & McLennan, 1991).

Phylogenetics is based on the allocation of taxa to groups based on the recognition that they share unique characters (synapomorphies) that distinguish them from their ancestors as well as more distantly related groups. The identification of these derived characters is based on the study of an outgroup. An *outgroup* is a group of closely related taxa that are not directly part of the study group. If we were studying hominids, for example, the obvious outgroup would be the early Miocene hominoids and the Old World monkeys.

Synapomorphies establish sister-group relationships between taxa, while primitive features that are shared by at least two outgroups are assumed to be symplesiomorphies and cannot be used to allocate species to a clade. For example, when examining the New World and Old World monkeys, an obvious symplesiomorphic feature is the *presence* of a tail; but when the great apes are included in such a study, the *absence* of a tail is a synapomorphy linking all primates without tails as hominoids. When looking at a finer resolution of detail, let's say groups *within* the hominoids, the absence of a tail is now of no phylogenetic significance because all hominoids are defined by not having a tail. The absence of a tail in hominoids is meaningless in trying to determine which species are closely related. In this case the absence of a tail becomes a primitive, or plesiomorphic, feature of the hominoids. A synapomorphy of the hominins, however, might be an increase in brain size (relative to body weight).

Recently Collard and Wood (2000, 2001a, 2001b) have published a number of influential papers using parsimony analysis to reconstruct the phylogeny of extant great apes and papionins based on craniofacial morphology. These studies are used to help determine the consistency of phylogenetic trees generated from morphological information against trees generated from molecular data. In all cases the morphological analyses fail to reproduce the molecular tree. They conclude that the four regions of the craniofacial complex that they studied were equally affected by homoplasy and were, therefore, "equally unreliable for phylogenetic reconstruction" (2001a:167). We would suggest, however, that the inclusion of closely related fossil species is essential in helping to determine the correct polarity of characters. This is clearly the case in the significant impact that cranial expansion has had on hominin craniofacial morphology. Thus the recognition of "intermediate" stages in evolutionary development, which may be missing from extant studies alone, are likely to be crucial in helping to define the polarity of characters and their phylogenetic significance.

Fossils are the only direct morphological evidence that we have pertaining to the evolution of form, which enables us to examine likely selective process associated with adaptations. Hennig (1966) originally argued that the inclusion of fossil material within any phylogenetic analysis is crucial because it allows us to determine the likely "true" polarity of characters, which enables us to identify homoplasies and homologies (also see Wiley, 1981). The use of extant taxa alone will provide only a small sample of all the character combinations that have existed, and the most stringent test of character homology comes from including fossil and extant taxa in an analysis (Harvey & Pagel, 1998). Indeed, fossils that lie near the base of a cladogram are likely to be crucial, because these will tend to have retained more ancestral characters that have been modified in the later fossils and extant descendants (see Donaghue et al., 1989; M.V.H. Wilson, 1992; Kemp, 1999).

Gauthier et al. (1988), in their study of tetrapods, demonstrated that the inclusion of fossil taxa made an appreciable difference to the "correct interpretation" of the phylogeny, as suggested by extant molecular studies. For example, the use of 109 morphological characters from extant taxa alone suggested that mammals were the sister group to crocodiles plus birds and that turtles were the sister group of that clade. But when fossil taxa were included, mammals were placed as a sister group to all other tetrapods, and it was lizards that were placed as a sister group to crocodiles and birds (see also Donaghue et al., 1989; Novacek, 1992; M.V.H. Wilson, 1992; A.B. Smith, 1994; Kemp, 1999).

It is crucial that, prior to any study, the characters considered for analysis are themselves examined to determine their likely phylogenetic value. This includes determining whether characters are redundant. Character redundancy is the presence of characters whose overall appearance are intricately related and thus should be considered as representing just one character, such as robust canines and developed canine juga. In order to help identify redundancy, we must understand the functional and developmental significance of the phenotypic features that are being considered for analysis. Strait (2001) demonstrates such an approach, proposing a method for the integration of phenotypic characters into functional and developmental complexes in an examination of the hominin cranial base. His method is based on the construction of a data matrix, a functional/structural analysis of metric values, a character analysis to help identify phylogenetic independence of characters within the matrix (with functional subsets identified), followed by a parsimony analysis of the proposed integrated complexes. In using the cranial base as an example, Strait was able to show, using factor analysis and the resulting generated correlation matrices, how some characters are correlated. From this he was able to identify three hierarchical inferences of integration: uniform, compatible, and function-ally weighted complexes. From the character analysis, a number of proce-dures followed, including character deletion, replacing character states with a binary system generated from parsimony analysis of functional "subsets," and the differential weighting of some characters. Strait was able to apply this method successfully to the cranial base, although he found very few characters that could be considered truly integrated.

While it was tempting to try to adopt aspects of Strait's method here, it was finally deemed unwise because it would require the construction of a number of hypothesized and untestable integrative complexes that could not be substantiated by factor analysis (due to missing variables). Indeed, as Strait warns, "uncritical hypotheses of integration could also have a deleterious effect on phylogenetic analyses. . . . [T]he concepts of integra-tion should be used in phylogenetic analysis only if hypotheses of integra-tion can be adequately tested" (2001:294). We explain in the appendix our attempt to identify likely redundant/duplication of features.

Skelton and McHenry (1992, 1998), believing that the masticatory apparatus is dominated by homoplasies, attempted to remove the "bias" of this complex by reducing masticatory features in their parsimony analyses. In doing so, they concluded that *P. walkeri* should be removed from the robust australopithecine clade, because the synapomorphies uniting these taxa were demonstrated to be largely masticatory characters. They suggest that, when this "bias" is removed, KNM-WT 17000 becomes a more primitive hominin rather than part of a *Paranthropus* clade. Characters helping to define the masticatory apparatus, especially at the species or genus level, we would argue, however, must be considered as part of the phylogenetic history of the taxa concerned, for there is no evidence to sug-gest that the masticatory apparatus is more prone to homoplasy than any other anatomical region (see Strait *et al.*, 1997; Collard & Wood, 2000, 2001a, 2001b). The studies of Collard and Wood have concluded that the phylogenetic information supplied by the masticatory complex should not be underestimated. They show that characters *not* directly associated with the masticatory apparatus are no more reliable for phylogenetic recon-struction than are characters often used to help define this complex (see also Strait *et al.*, 1997). This can be tied to the concept of phylogenetic niche conservatism.

The principle of phylogenetic niche conservatism is based on the proposition that past and present members of a lineage are likely to have occupied similar environments, because only those species that are best suited to particular environments are likely to survive (Harvey & Pagel, 1998). For example, the masticatory correlates between *Paranthropus* species of eastern and southern Africa suggest that not only did these species at some time in the past, share a similar environmental niche, but they also shared an immediate common ancestor. Their features must have at least partially evolved as a common adaptive response to similar adaptive pressures that were also operating on the last common ancestor. In other words, it is likely that the speciation event resulting in *P. robustus* and *P. boisei* was not related to a dramatic shift from a forest to a desert environment (thus resulting in a functionally distinct morphology), because it is unlikely that any ancestral population could survive such a dramatic environmental shift. It is more likely that an ancestral population moved into a slightly more or less wooded environment, because this is the adaptive strategy of its immediate common ancestor. This pattern of speciation is also closely aligned with vicariance, in which the ancestral species extends its range under favorable and specific environmental circumstances and then becomes geographically restricted to paleohabitats by geographic and/or environmental events (see Andrews & Bernor, 1999; Strait & Wood, 1999). There is a real danger in trying to "disentangle" processes and patterns that are intimately related. This is because numerous functional "homoplasies" may in fact be homologies. As such, studies that attempt to disentangle function from phylogeny may be "throwing out the baby with the bathwater."

Materials

The taxa and the specimens used to define the species examined are listed in Table 5.1. We have been intentionally conservative in the allocation of specimens to taxa. Given the current taxonomic confusion concerning both OH 9 and OH 62, these are not considered. The same applies to the early *Homo* specimens from Georgia (*H. georgicus*) and the recently reported material from Danakil (Afar) and Bouri (Middle Awash), both from Ethiopia (see Abbate *et al.*, 1998; Gabunia *et al.*, 2000a; Asfaw *et al.*, 2002). Currently only one specimen of *P. walkeri* is recognized here (KNNM-WT 17000), and none of the lower Omo "robust" mandibles are considered. Given the poor preservation of the specimens from Chad allocated to *A. bahrelghazali*, they have not been included in this study.

Because *Sahelanthropus* is currently dated to the Late Miocene of Africa, the outgroup consists of Middle and Late Miocene African genera

TABLE 5.1 ▶ Specimens Included in the Hypodigm of the Hominidae

Kenyapithecus:	Middle Miocene specimens from Fort Ternan, Kenya.
Dryopithecus:	Miocene specimens from Hungary and Spain.
Graecopithecus:	Miocene specimens from northern Greece.
Shelanthropus:	Late Miocene specimens from the Djurab Desert, northern Chad.
Ardipithecus ramidus:	Specimens from Pliocene strata at Aramis, Middle Awash, Ethiopia.
Australopithecus anamensis:	Specimens from Allia Bay, Kenya.
Praeanthropus afarensis:	**AL** 128–23, 162–28, 200–1, 33–125, 58–22, 145–35, 188–1, 198–1, 199–1, 200–1, 207–13, 266–1, 277–1, 288–1, 311–1, 333w–1, 33w–12, 333w–0, 333–1, 333–2, 333–45, 333–105, 400–1a, 417–1, 444–2, **LH** 4
Australopithecus africanus:	**Sts** 5, 17, 19, 20, 26, 67, 71, 52a, **Stw** 13, 73, 252, 505, **TM** 1511, 1512, **MLD** 1, 6, 9, 37/38
Australopithecus garhi:	**BOUV-VP**-12/130
Kenyanthropus platyops:	**KNM-WT** 40000, 40001, 38350
Paranthropus walkeri:	**KNM-WT** 17000
Paranthropus robustus:	**SK** 6, 12, 13/14, 23, 34, 46, 47, 48, 49, 52, 65, 79, 83, 848, 1586, **SKW** 5, 8, 11, 29, 2581, **TM** 1517, **DNH** 7
Paranthropus boisei:	**OH** 5, **KNM-ER** 403, 404, 405, 406, 407, 725, 727, 728, 729, 732, 805, 810, 818, 1803, 1806, **KGA** 10–525
Kenyanthropus rudolfensis:	**KNM-ER** 1470, 1590, 3732, 3891, 1482, 1483, 1590, 1801, 1802, 3950
Homo habilis:	**OH** 7, 13, 16, 24, **KNM-ER** 1805(?), 1813, 1478, 1501, 1502, 3735
Homo ergaster:	**KNM-ER** 730, 820, 992, 1812, 1507, 3733, 3883, **KNM-WT** 15000, **OH** 22
Homo sapiens:	Modern human crania from the ANU Bioanthropology Laboratory

Note: (?) indicates considerable debate surrounding specimen but allocation to species accepted here.

as well as *Pongo* and *Gorilla*. The selection of the Miocene genera is based on their hominid status, as well as their proposed "African-like" hominid morphology (Cameron, in press [a]). Extant *Pongo* and *Gorilla* are included in the outgroup because their evolutionary emergence predates the appearance of the late Miocene and Pliocene fossil hominids/ hominins. The fossils that are believed to be part of the Ponginae — i.e., *Sivapithecus*, *Ankarapithecus*, and *Lufengpithecus* — have been excluded, given their derived morphological condition relative to the more "conservative" African ape condition, though extant *Pongo* is included. Because the emergence of the *Pan* lineage occurs within the time frame of the ingroup — i.e., perhaps after the emergence of *Sahelanthropus* — it is included as part of the ingroup.

Wood (1991), Skelton and McHenry (1992), Strait *et al.* (1997), and Strait and Grine (2001) have published the most recent interpretations of hominin phylogeny, which also provide detailed definitions and discussions of the characters used. This study tends to focus on the characters used by Strait *et al.* (1997) with modifications provided in Strait and Grine (2001). Data provided by Wood (1991) have been used to construct and define a number of characters. While Strait *et al.* (1997) is used as the starting point for this study, this is not to say that we agree with all of the character assignments provided by them. In some cases a different interpretation is provided, not only of the character state, but also of its condition within the taxa concerned; for example, it is considered here that anterior pillars (character 62) are variable within *P. robustus*, given the description of the recently published specimen from Drimolen (Keyser, 2000). Also, unlike previous studies, middle and late Miocene hominids are used to define the outgroup, and the new taxa, *Sahelanthropus* and *Kenyanthropus*, are considered for the first time. In some cases, the inclusion of these additional taxa has necessitated a review of some character states. A number of original characters, such as character 7 (temporal fossa size [as an index]), is also included. A total of 92 cranial, facial, dental, and mandibular characters was finally chosen for parsimony analysis distributed through 20 taxa (see Appendices). Characters that are known to be highly sexually dimorphic, such as canines, are defined by the male condition only. A number of metric characters were also defined. While in some cases, absolute linear measurements are used by Strait *et al.* (1997) to define character states, wherever possible, these have been converted into indices (see Appendices).

Method

The parsimony analyses were generated using the computer program PAUP* version 4.0 beta (Swofford, 1998). The phenotypic data provided in Appendix Table 1 were analyzed in two separate ways in order to determine the impact that different data sets may have on resulting topologies. First, all 92 characters were analyzed, and second, only those characters preserved in *S. tchadensis* or *K. platyops* were analyzed. Each data set was analyzed using the following procedures: (1) a strict consensus tree was generated; (2) a number of 50% majority-rule consensus trees was used, with the tree length continually increased up to three additional steps; and (3) a bootstrap analysis with 1,000 replications was generated.

All characters analyzed, except for characters 1, 3, 29, 74, and 77, were treated as ordered. Ordered characters are weighted in the sense that all intermediate stages are considered to have occurred; e.g., a change from

0 to 3 is weighted to reflect the three steps. Following Strait *et al.* (1997), it is also maintained that a change between adjacent states (e.g., between 2 and 3) is treated as a single step in a tree; i.e., equally weighted state changes are used throughout this study.

The strict consensus tree was obtained using the "heuristic" search option. The consensus tree is shown, along with its length, consistency, retention, and rescaled consistency indices. The consistency index (CI) is calculated from the number of homoplasies that must be assumed in the most parsimonious solution. The consistency index for a cladogram, as a whole, is equal to the total number of derived character states scored in the matrix, divided by the number of steps required to produce a tree. As such, the CI decreases as the level of homoplasy increases. Also, the amount of homoplasy will generally increase as the number of genera included also increases (Smith, 1994). The retention index (RI) measures the proportion of terminal genera that retain the character identified as a synapomorphy for that group. For example, if a character identified as a synapomorphy for a clade is present in all the terminal genera, it is given an RI of 1.0. If this same character, however, through later transformation or reversal, is present in only 50% of the terminal genera, its RI will now be just 0.5 (Smith, 1994). The rescaled consistency index (RCI) is calculated by multiplying the CI by the RI (Farris, 1989). It has been argued that the RI and the RCI are more robust in terms of being less sensitive to variations in maximum and minimum tree-length (Farris, 1989; Strait *et al.*, 1997).

In each analysis a number of 50% majority-rule consensus trees was generated. These trees were continually generated by increasing the tree length by one step each time, until they had reached three steps. A 50% majority-rule consensus tree chooses the topologies that appear most often among the alternative cladograms. Only groups that appear in more than a specified percentage of all rival cladograms are used to construct a majority-rule consensus tree. Thus, adapting a 50% cut-off means that any group that appears in the majority-rule consensus tree is found in more than half of the competing cladograms (Schoch, 1986; Smith, 1994).

Finally, a bootstrap analysis was generated. This method of analysis selects characters at random from the data matrix, with replacement, thus constructing a new data matrix of the same dimensions as the original matrix table. The new data matrix may include the same character more than once, while others may be deleted altogether, because characters are chosen at random with a replacement option. This new matrix table is used to calculate a topology, and the process is repeated many times. In the analyses to follow, 1,000 replications were requested. Thus, a branch appearing in only 250 replicates represents just 25% (Noreen, 1989; Smith, 1994; Swofford, 1998).

Results

1. Analyses with all 92 Characters Included

The strict consensus tree of all 92 characters is shown in Figure 5.1, generated from 8 trees (tree length = 387; CI = 0.485; RI = 0.629; and RC = 0.302). From this analysis, after the divergence of the outgroup, *Pan* emerges, followed by *Ardipithecus*, then the *anamensis* group, followed by *Praeanthropus*. At this point we see a polytomy containing four clades. *Sahelanthropus* shares a common ancestor with the *garhi* group, with an expanded hominin clade containing *Australopithecus*, *Kenyanthropus*, and

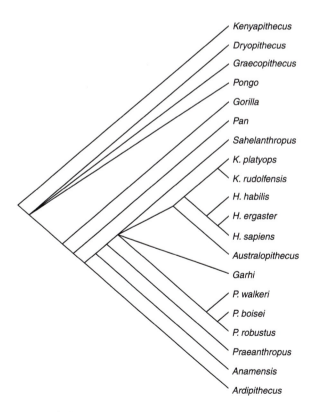

Figure 5.1 ▶ Strict Consensus Tree of 92 characters (see text for details).

Homo, and another consisting of *Paranthropus* species. Within the expanded hominin clade *Australopithecus* splits off, followed by two sub-clades, one containing species of *Kenyanthropus* and the other the three species of *Homo*.

Following this analysis, a 50% majority rule consensus tree was requested, increasing the tree length by an additional step; i.e., to 388. This resulted in the generation of 86 trees, the consensus of which is shown in Figure 5.2. This tree is the same as the consensus tree just discussed. The replication values for this scheme are robust, even the lowest value is still relatively well supported; i.e., the placement of the *Paranthropus* clade with the *Homo* group in 62% of cases. The next analysis included the tree length increased by 2 extra steps (i.e., 389) resulting in the topology shown in Figure 5.3, from 502 likely trees. This tree again reproduces the consensus tree, and replication values are relatively high. The final analyses increased the tree length by

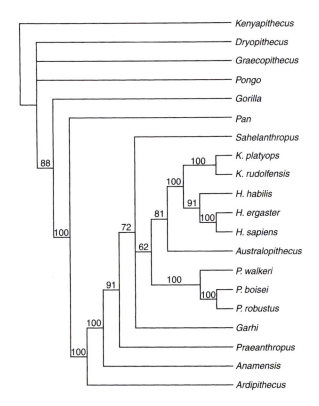

Figure 5.2 ▶   Majority-rule topology, one step beyond consensus (see text for details).

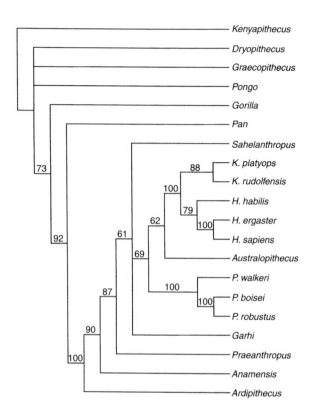

Figure 5.3 ▶ Majority-rule typology, two steps beyond consensus (see text for details).

3 steps (i.e., 390), resulting in the tree shown in Figure 5.4, from 2,096 possible trees. This tree confirms the previous topologies, the only exception being that *Praeanthropus* is now shown to share a last common ancestor with *Sahelanthropus*, the *garhi* group, and more derived hominins (or to interpret it another way, its relationship to these hominins is now unresolved).

To further test these relationships, a bootstrap analysis was generated requesting 1,000 runs. The tree generated is shown in Figure 5.5. This topology is similar to those previously generated. The only difference in the position of the later hominids is that *Ardipithecus* and the *anamensis* group now share a sister-group relationship to the exclusion of all other hominids. Following this, we observe the emergence of *Praeanthropus*, followed by the *garhi* group. *Sahelanthropus* then emerges, followed by *Australopithecus*. *Kenyanthropus*, *Homo*, and *Paranthropus* share a common ancestor to the exclusion of all other hominids, while *Kenyanthropus* and *Homo* share a last

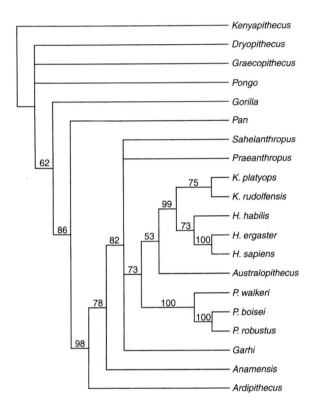

Figure 5.4 ▶   Majority-rule typology, three steps beyond consensus (see text for details).

common ancestor to the exclusion of *Paranthropus*. The replication values, however, are not particularly strong for most suggested relationships (to perhaps be expected given that it is a bootstrap analysis), the only exception being the monophyletic status of species within *Paranthropus* and the sister-group relationship of *H. ergaster* and *H. sapiens*.

In all of these analyses, the paraphyletic status of *Australopithecus* and the monophyletic status of *Paranthropus* are confirmed (with *P. boisei* being sister to *P. robustus*). A *Homo* clade containing all three species of *Homo* as well as another containing both species of *Kenyanthropus* is retained in all analyses. The position of *A. africanus,* representing a basal hominin to both of these clades to the exclusion of all other taxa, is also confirmed. The position of *Sahelanthropus* and members of the *garhi* group remains problematic, though both appear to be more derived in the hominin direction as opposed to being more "primitive" Pliocene hominids.

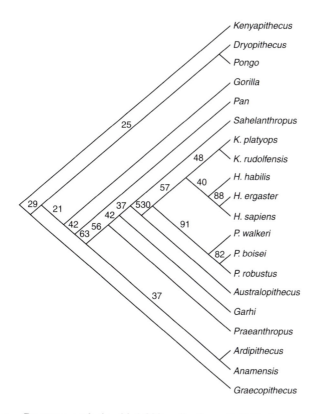

Figure 5.5 ▶ Bootstrap analysis with 1,000 replications of all 92 characters (see text for details).

2. Analyses of 52 Characters Preserved in Either *S. tchadensis* and/or *K. platyops*

The analyses generated here have removed all characters from the analysis that are not preserved in either *S. tchadensis* or *K. platyops*. The forty characters deleted from this analysis are 3, 7, 9, 10, 11, 13, 16, 20, 23, 25, 27, 28, 30, 32, 35, 37, 40, 41, 45, 47, 53, 54, 57, 60, 61, 64, 69, 72, 73, 74, 75, 77, 78, 79, 80, 83, 85, 86, 87, and 92.

The strict consensus tree is shown in Figure 5.6, generated from 20 trees, where tree length = 234; CI = 0.453; RI = 0.601; and RC = 0.272. The emergence of *Pan* and *Ardipithecus* is followed by *Sahelanthropus*, the *anamensis* group, *Praeanthropus*, the *garhi* group, and the basal members of the *Paranthropus* clade and the expanded hominin clade all sharing a common ancestor. *Paranthropus* forms a monophyletic group, while

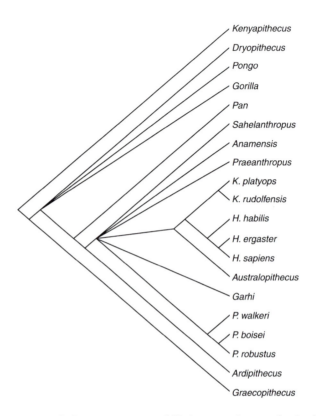

Figure 5.6 ►   Strict consensus tree of 52 characters (see text for details).

the "australopithecines" are paraphyletic. *Australopithecus* is basal to the *Kenyanthropus* and *Homo* clades.

The 50% majority-rule consensus was generated by increasing the tree length by one extra step, resulting in 52 possible trees (Figure 5.7). The emergence of *Pan* and then *Ardipithecus* is followed by the *garhi* and *anamensis* groups as well as *Sahelanthropus* and *Praeanthropus*, all sharing a common ancestor (or unresolved), with a clade containing the *Paranthropus* lineage, and the more derived hominins, including *Australopithecus*. All of the percentage replication values can be considered robust. As in previous analyses, *Australopithecus* is sister to the two sub clades containing *Kenyanthropus* and *Homo*, and the replication values are strong for this clade (100%). The same strong support is also provided in the values of the *Paranthropus* clade (100%). The next analysis increased the tree length by two additional steps, resulting in the generation of 1,619 possible trees (Figure 5.8). This scheme

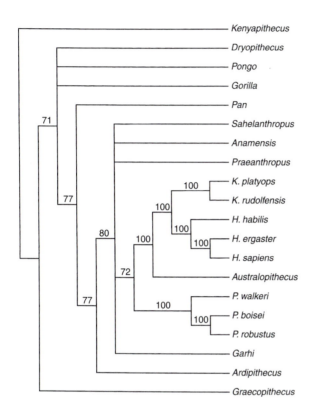

Figure 5.7 ► Majority-rule typology, one step beyond consensus (see text for details).

is the same as that just discussed, the only difference being slightly reduced replication values (to be expected). The last analysis increased the tree length by three additional steps, resulting in the generation of 8,000 likely trees. The consensus tree for this analysis is shown in Figure 5.9. The only difference in this tree from those discussed previously is the repositioning of *Ardipithecus* to the unresolved status of *Sahelanthropus*, *Praeanthropus*, and the *garhi* and *anamensis* groups. Again the replication values are relatively robust.

Figure 5.10 represents the bootstrap analysis (1,000 replications) generated on the 52 characters preserved in either *Sahelanthropus* and/or *K. platyops*. This analysis seems to confirm the more derived status of *Sahelanthropus* and *Australopithecus*, but with very low footstrap values.

All of these analyses confirm the paraphyletic status of the australopithecines, while supporting the monophyletic status of *Paranthropus*

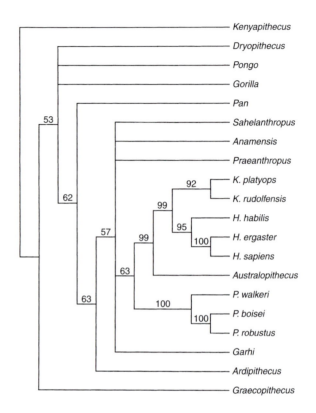

Figure 5.8 ▶ Majority-rule typology, two extra steps beyond consensus (see text for details).

species. The expanded *Homo* clade is retained, with *H. habilis* being placed as a basal member of *Homo*. *Australopithecus* is maintained as representing the basal group from which *Kenyanthropus* and *Homo* emerge. The phylogenetic position of the other Plio/Pleistocene hominids remains largely unresolved, and their phylogenetic relationship to the hominins is obscure.

Inferred Phylogenetic Relationships

The analyses conducted here consistently support certain phylogenetic relationships. The most consistent result is that the "australopithecines" are paraphyletic. Next, *Australopithecus africanus* is more derived in the

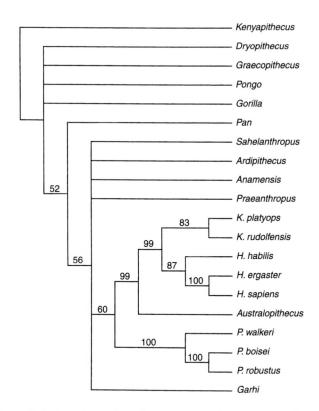

Figure 5.9 ▶ Majority-rule typology, three extra steps from consensus (see text for details).

hominin direction (relative to the other species generally allocated to *Australopithecus*) and in all cases is basal to the clade containing *Kenyanthropus* and *Homo*. *Homo habilis* is usually considered the basal species to the *Homo* clade, while *P. walkeri* is basal to *Paranthropus*. The "1470 group" is confirmed as being a species of *Kenyanthropus* (*K. rudolfensis*) as they consistently form a sister-group relationship to the exclusion of all other taxa. The monophyletic status of *Paranthropus* is in all cases confirmed. *Ardipithecus* and the *anamensis* group are shown in almost all cases to reflect the basal Plio/Pleistocene hominid condition, with the Miocene *Sahelanthropus* being derived from it. The phylogenetic status of *Sahelanthropus, Praeanthropus*, and the *garhi* group, however, remains far more problematic, though overall they seem to represent a

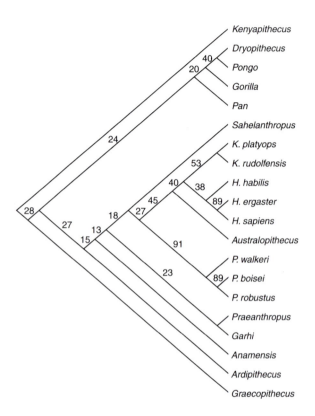

Figure 5.10 ▶   Bootstrap analysis, 1,000 replications of 52 characters (see text for details).

bridge between the more primitive Pliocene hominid represented by *Ardipithecus* and the *anamensis* group and the more derived Plio/Pleistocene hominins represented by *Paranthropus, Australopithecus, Kenyanthropus,* and *Homo.*

There are three courses that could be adopted in interpreting these results in terms of systematics. One is to simply accept generic paraphyly as an "occupational hazard" in paleoanthropology; this would have the benefit of stabilizing the taxonomy (and so the all-important nomenclature), at least for a while, but at a cost of flouting the principles of phylogenetic systematics (Wiley, 1981; also see Harrison, 1993, for a detailed discussion of this issue).

The second course is much more radical: Refer all to a single genus, for which the prior available name would, of course, be *Homo.* This would have the superficial drawback that finer interrelationships among species would no longer be defined by generic names. But, as we have seen, this is not possible anyway, given the apparently irreducibly paraphyletic nature of some widely recognized genera. However, it would stabilize generic nomenclature, and it would in addition mean that a reshuffle of species between genera would not occur each time anyone discovers apparently new evidence for their affinities. Groves (2001) has supported giving higher categories some objectivity by allocating a time depth to each: In the case of a genus, this would mean some 4 to 6 million years. Under such a scheme, it is clear that all hominin genera, even *Ardipithecus* (and perhaps *Pan*), would be synonymized with *Homo*, and the tribe Hominini would become unnecessary. CPG favors this proposal (see Table 5.2).

The third course would be to erect further genera, one for each major clade. This approach has been argued and adopted by Cameron (1997a, 1998, 2001, in press) in his analysis of the Miocene hominids. Most of the hominin genera would thereby be monotypic. This is also in agreement with Collard and Wood (1999), Wood and Collard (1999), and Strait and Grine (2001), who argue that genera should ideally correspond to both the grade and clade concepts. This means that a species of the same genus must be both monophyletic and adaptively distinct from species that belong to other genera. This approach has the advantage that a certain stability is introduced to hominid classification, and that the principles of phylogenetic systematics would not be flouted. This is the scheme adopted here. DWC favors this proposal (see Table 5.3).

Cameron's taxonomic scheme supports the generic distinction of the Hadar and Laetoli Pliocene hominins from *Australopithecus*. This partly follows Day *et al.* (1980), Harrison (1993), and Strait *et al.* (1997), who referred them to the genus and species *Praeanthropus africanus*, though for reasons outlined in Groves (1999), Strait (2001), and Strait and Grine (2001), the species should now be called *Pr. afarensis*. The paraphyletic nature of the australopithecines has been recognized by others, including Chamberlain and Wood (1987), Groves (1989), Wood (1991), Skelton and McHenry (1992), Lieberman *et al.* (1996), Strait *et al.* (1997), Strait (2001), and Strait and Grine (2001). With Strait and Grine (2001), a generic distinction would be warranted for the hominins currently allocated to *A. anamensis* and *A. garhi*, which cannot be considered species of *Australopithecus*. As such, *Australopithecus africanus* would represent the only species within this genus. Nor does this study support the scheme suggested by Asfaw *et al.* (1999), that members of the *garhi* group represent a likely immediate ancestor to early *Homo*.

TABLE 5.2 ▶ Colin's Mio-/Pliocene Hominid Taxonomy

Hominidae
   Kenyapithecinae
   Ponginae
   Dryopithecinae
   Homininae
        Gorillini
             Graecopithecus
             Gorilla
                  Gorilla gorilla
                  Gorilla beringei
      Hominini
        Pan
             Pan troglodytes
             Pan paniscus
       Orrorin
             Orrorin tugenensis
       Homo
          (stem group)
             Homo kadabba
             Homo ramidus
          (australopithecine group)
             Homo anamensis
             Homo bahrelghazali
             Homo afarensis
             Homo garhi
             Homo africanus
          (paranthropine group)
             Homo walkeri
             Homo boisei
             Homo robustus
          (kenyapithecine group)
             Homo platyops
             Homo rudolfensis
          (habiline group)
             Homo habilis
          (erectine group)
             Homo ergaster
Hominidae indet.:
   Sahelanthropus tschadensis
   Lothagam hominid

DWC further considers that only species within *Paranthropus,* *Australopithecus,* *Kenyanthropus,* and *Homo* can be considered true hominins. *Ardipithecus* and members of the *anamensis* group are simply plesiomorphic Pliocene hominids, whose evolutionary history likely

TABLE 5.3 ▶ Dave's Mio/Pliocene Hominid Taxonomy

Hominidae
    Kenyapithecinae
    Ponginae
    Dryopithecinae
    Oreopithecinae
    Gorillinae
        Graecopithecini
            Graecopithecus
        Gorillini
            Gorilla
                G. gorilla
    Paninae
        Panini
            Pan
                P. troglodytes
                P. paniscus
    Homininae
        Hominini
            Orrorin
                O. tugenensis
            Sahelanthropus
                S. tschadensis
            Garhi deme
            Australopithecus
                A. africanus
            Paranthropus
                P. walkeri
                P. boisei
                P. robustus
            Kenyanthropus
                K. platyops
                K. rudolfensis
            Homo
                H. habilis
                H. ergaster
Hominidae indet.
    Lothagam hominid
    Ardipithecus ramidus
    Praeanthropus afarensis
    Anamensis hominids
    Bahrelghazali hominids

ended in extinction, with no direct contribution to the evolution of the later hominins. Whether *Sahelanthropus*, *Praeanthropus*, or members of the *garhi* group can be considered hominins in this scheme remains unresolved, though *Sahelanthropus* does appear from the available evidence to be closer phylogenetically to the later hominins than to *Pan* or *Ardipithecus*. Given the early dates for *Sahelanthropus*, this result must be considered of some significance to later human evolution.

The Evolution of Hominin Craniofacial Morphology

Given the consistent relationships observed between hominid groups, discussed previously, we present Figure 5.11 as reflecting the most likely phylogenetic scheme for the hominids. The evolution of the hominin craniofacial complex will be discussed in relation to the nodes shown. The definition of these nodes, as well as the "assumed" character polarity at certain nodes (i.e., a result of missing characters), has been generated from

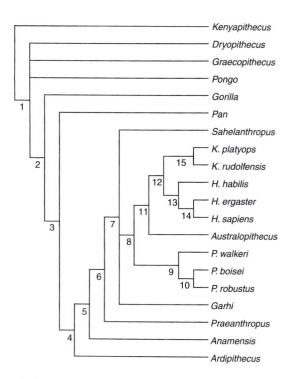

Figure 5.11 ▶   Phylogenetic scheme of the hominids adopted here (see text for details).

the describe tree option in PAUP (Swofford, 1998). When a character is stated as being "relatively narrower," "relatively deeper," etc., this is compared to the condition at the previous node, unless stated otherwise (see Table 5.4). From this discussion a number of adaptive features will be identified for the craniofacial complex. These adaptive features will be associated with the evolution of derived features from the primitive hominid condition.

TABLE 5.4 ▶ List of Characters Used to Define the Nodes in Figure 5.11

Implied Primitive Condition of the Hominids

(1)	Torus intermediate development
(2)	Supraorbital intermediate thickness
(3)	Glabella intermediate swelling
(4)	Supraorbital sulcus absent
(5)	Temporal lines with strong anteromedial incursion
(6)	Sagittal crest in males developed
(7)	Temporal fossa size is large
(8)	Postorbital constriction is intermediate
(9)	Parietal usually no overlap at asterion
(10)	Parietal overlap when present is not extensive
(11)	Asterionic notch present
(12)	Compound temporal crest is present
(13)	Supraglenoid gutter of intermediate width
(14)	Mastoid reduced lateral inflation
(15)	Temporal squama pneumatization extensive
(16)	External cranial base is extended
(17)	Nuchal plane inclination steep
(18)	Anterior tympanic edge medial to porion
(19)	External auditory meatus is small
(20)	Articular tubercle is large
(21)	Petrous is sagittally orientated
(22)	Cranial base is intermediate in breadth
(23)	Basioccipital is long
(24)	Glenoid fossa is intermediate in depth
(25)	Glenoid fossa is large in size
(26)	Postglenoid process is large and unfused with tympanic
(27)	TMJ distanced from dental complex
(28)	Eustachion process prominent
(29)	Tympanic tubular in shape
(30)	Vaginal process of tympanic small
(31)	Digastric muscle insertion broad and shallow
(32)	Longus capitis insertion long and oval
(33)	Foramen magnum oval shaped
(34)	Basion posterior to bi-tympanic
(35)	Foramen magnum inclination inclined posterior

TABLE 5.4 ► Continued

(36)	Cranial capacity <500 cm^3
(37)	Cerebellar morphology with lateral flare and posterior protrusion
(38)	Occipitomarginal sinus of low frequency
(39)	Facial hafting intermediate
(40)	Interorbital breadth intermediate
(41)	Frontal sinus is developed
(42)	Lacrimal fossa within orbit
(43)	Mid face prognathic
(44)	Upper facial breadth intermediate
(45)	Mid-facial breadth intermediate
(46)	Anterior zygomatic insertion at M^1
(47)	Masseter origin is low
(48)	Palate prognathism is strong
(49)	Subnasal angle is low
(50)	Incisor prognathism beyond bi-canine
(51)	Anterior palate intermediate depth
(52)	Palate is thin
(53)	Palate breadth is intermediate
(54)	Nasal bones projected and tapered
(55)	Nasal keel absent
(56)	Inferolateral orbital margin rounded
(57)	Orbital fissure configuration is round
(58)	Maxillary trigon absent
(59)	Malar near vertical in orientation
(60)	Malar diagonal length intermediate
(61)	Infraorbital foramen within upper 50% of malar height
(62)	Anterior nasal pillars absent
(63)	Nasal entrance stepped
(64)	Incisive canal undeveloped
(65)	Nasoalveolar clivus convex
(66)	Orbital is oval-rhomboid
(67)	Inferior orbital margin aligned to superior nasal margin
(68)	Nasal clivus intermediate length
(69)	Maxillary sinus is large
(70)	Zygomatic insertion is high
(71)	Canine fossa is deep
(72)	Mandibular symphysis recedes
(73)	Symphyseal robusticity intermediate
(74)	Mandibular tori similar development
(75)	Mandibular corpus robusticity intermediate
(76)	Mandible premolar orientation U-shaped
(77)	Mental foramen opening is variable
(78)	Mental foramen hollow present
(79)	Mandibular extramolar sulcus broad
(80)	Upper incisor intermediate size
(81)	Upper incisor heteromorphy developed
(82)	Male canines robust and daggerlike

TABLE 5.4 ▶ Continued

	(83)	Upper premolar complex, relative to molar complex intermediate in size
	(84)	Molars small
	(85)	Cusps close to edge
	(86)	P_3 metaconid absent
	(87)	P_3 mesiobuccal expansion strongly developed
	(88)	Molar enamel intermediate thickness
	(89)	Molar lingual cingulum weak/absent
	(90)	M^2 broader than long
	(91)	Upper molars cusps and inflated, limited cristae
	(92)	dM_1 MMR absent, protoconid anterior, fovea opening
Node 2:	(1)	Torus strong and barlike
	(4)	Supraorbital sulcus moderately developed
	(51)	Anterior palate is shallow (homoplasy)
	(66)	Orbital shape is circular-rhomboid
	(74)	Mandibular inferior torus weaker
	(81)	Upper incisor heteromorphy intermediate
	(88)	Molar enamel thin (homoplasy)
	(91)	Molar cusps and cristae well developed (homoplasy)
Node 3:	(5)	Temporal lines with moderate anteromedial incursion
	(6)	Sagittal crests in males weakly expressed
	(7)	Temporal fossa size is intermediate
	(21)	Petrous orientation is intermediate
	(24)	Glenoid fossa is shallow
	(25)	Glenoid fossa is intermediate in size
	(59)	Malar anterior slope
	(77)	Mental foramen opens anterosuperiorly (homoplasy)
	(79)	Mandibular extramolar sulcus is narrow
Node 4:	(3)	Glabella broad in area, but not inflated
	(4)	Supraorbital sulcus of intermediate development
	(16)	External cranial with increased flexure
	(17)	Nuchal plane inclination intermediate
	(20)	Articular tubercle is small
	(23)	Bassioccipital is intermediate in length
	(28)	Eustachion processes weak/absent
	(32)	Longus capitus small and circular
	(34)	Basion close to bi-tympanic
	(35)	Foramen magum inclination horizontal
	(38)	Occipitomarginal sinus intermediate frequency
	(40)	Interorbital breadth broad (homoplasy)
	(49)	Subnasal angle is intermediate (homoplasy)
	(53)	Palate is broad (homoplasy)
	(54)	Nasal bones projected and expanded
	(64)	Incisive canal intermediate development
	(68)	Nasal clivus long
	(73)	Symphysis is robust

TABLE 5.4 ▶   Continued

	(82)	Male canines intermediate
	(84)	Molars large (homoplasy)
	(85)	Lingual cusps at margin — buccal cusps internal
	(87)	P$_3$ mesiobuccal expansion moderately developed
Node 5:	(77)	Mental foramen opens laterally
	(88)	Molar enamel thick (homoplasy)
	(91)	Molar cusps and inflated, limited cristae (homoplasy)
	(92)	dM$_1$ MMR slight, protoconid anterior, fovea opening
Node 6:	(20)	Articular tubercle is large (homoplasy)
	(46)	Anterior zygomatic insertion at M^1/P^4
	(72)	Mandibular symphysis recedes, intermediate
	(75)	Mandibular corpus is robust (homoplasy)
	(83)	Upper premolar/molar size marked difference (homoplasy)
	(86)	P$_3$ metaconid infrequent (homoplasy)
	(87)	P$_3$ mesiobuccal expansion variable
Node 7:	(11)	Asterionic notch absent
	(14)	Mastoids with lateral inflation
	(17)	Nuchal plane inclination weak
	(47)	Masseter origin is high (homoplasy)
	(61)	Infraorbital foramen location variable
	(65)	Nasoalveolar clivus flat
	(72)	Mandibular symphysis near vertical
	(73)	Symphysis extremely robust
	(75)	Mandibular corpus is extremely robust
	(78)	Mental foramen hollow variable
	(79)	Mandibular extramolar sulcus intermediate in development
	(85)	Lingual and buccal cusps internal
	(86)	P$_3$ metaconid present
	(87)	P$_3$ mesiobuccal expansion weak to absent
Node 8:	(12)	Compound crest is variable
	(21)	Petrous orientation is coronal
	(22)	Cranial base is broad
	(24)	Glenoid fossa is intermediate in depth (homoplasy)
	(26)	Postglenoid process moderate and maybe fused/unfused to tympanic
	(29)	Tympanic crest with vertical plate (homoplasy)
	(36)	Cranial capacity intermediate, 428–550 cm^3
	(48)	Palate prognathism is intermediate
	(63)	Nasal entrance intermediate slope
	(76)	Mandibular premolar orientation is parabolic
Node 9:	(3)	Glabella inflated
	(5)	Temporal lines with strong anteromedial incursion (homoplasy)
	(6)	Sagittal crests in males developed (homoplasy)
	(7)	Temporal fossa size is large (homoplasy)
	(8)	Postorbital constriction is developed (homoplasy)
	(9)	Parietal with overlap at asterion variable
	(25)	Glenoid fossa is large in size (homoplasy)

TABLE 5.4 ▶ Continued

	(33)	Foramen magnum heart shaped
	(39)	Facial hafting developed (homoplasy)
	(43)	Facial dishing present
	(46)	Anterior zygomatic insertion at P^4
	(50)	Incisor prognathism within bi-canine line (homoplasy)
	(52)	Palate is thick
	(56)	Inferolateral orbital margin not rounded (homoplasy)
	(57)	Orbital fissure configuration is comma shaped (homoplasy)
	(58)	Maxillary trigon present
	(60)	Malar diagonal length is long (homoplasy)
	(61)	Infraorbital foramen within lower 50% of malar height
	(63)	Nasal entrance smooth with overlap (homoplasy)
	(64)	Incisive canal developed (homoplasy)
	(65)	Nasoalveolar clivus is concave
	(71)	Canine fossa is shallow (homoplasy)
	(74)	Mandibular inferior torus more developed (homoplasy)
	(78)	Mental foramen hollow absent (homoplasy)
	(79)	Mandibular extramolar sulcus is broad (homoplasy)
	(85)	Lingual cusps internal — buccal cusps strongly internal
	(88)	Molar enamel hyper thick (homoplasy)
	(91)	Molar cusps and cristae bunodont
	(92)	dM1 thick, protoconid even with metaconid, fovea closed
Node 10:	(10)	Parietal overlap extensive
	(15)	Temporal squama pneumatization variable (homoplasy)
	(16)	External cranial base flexed (homoplasy)
	(18)	Anterior tympanic edge aligned or lateral to porion
	(19)	External auditory meatus is large (homoplasy)
	(23)	Bassioccipital is short (homoplasy)
	(26)	Postglenoid process small and maybe fused/unfused to tympanic
	(30)	Vaginal process of tympanic is large (homoplasy)
	(34)	Basion anterior to bi-tympanic
	(37)	Cerebellar tucked (homoplasy)
	(38)	Occipitomarginal sinus high frequency (homoplasy)
	(40)	Interorbital breadth extremely broad
	(44)	Upper facial breadth broad (homoplasy)
	(82)	Male canines reduced (homoplasy)
Node 11:	(1)	Torus intermediate development (homoplasy)
	(6)	Sagittal crests in males usually absent
	(12)	Compound crest is absent (homoplasy)
	(13)	Supraglenoid width narrow
	(14)	Mastoids have reduced lateral inflation (homoplasy)
	(51)	Anterior palate is deep (homoplasy)
	(60)	Malar diagonal length is short
	(67)	Inferior orbital margin well below superior nasal margin (homoplasy)
Node 12:	(5)	Temporal lines with weak anteromedial incursion
	(15)	Temporal squama pneumatization variable (homoplasy)
	(16)	External cranial base flexed (homoplasy)

TABLE 5.4 ▶ Continued

	(20)	Articular tubercle is small (homoplasy)
	(23)	Bassioccipital is short (homoplasy)
	(31)	Digastric muscle insertion deep and narrow
	(36)	Cranial capacity increased, 509–675 cm^3
	(37)	Cerebellar tucked (homoplasy)
	(38)	Occipitomarginal sinus low frequency
	(47)	Masseter origin is low (homoplasy)
	(49)	Subnasal angle is high (homoplasy)
	(54)	Nasal bones not projected
	(59)	Malar near vertical orientation (homoplasy)
	(61)	Infraorbital foramen within upper 50% of malar height (homoplasy)
	(63)	Nasal entrance stepped (homoplasy)
	(74)	Mandibular tori undeveloped
	(75)	Mandibular corpus is robust (homoplasy)
	(79)	Mandibular extramolar sulcus is narrow (homoplasy)
	(84)	Molars size intermediate
	(85)	Lingual cusps at margin — buccal cusps approaching internal (homoplasy)
Node 13:	(4)	Supraorbital sulcus moderately developed (homoplasy)
	(15)	Temporal squama pneumatization reduced (homoplasy)
	(26)	Postglenoid process small and fused to tympanic
	(27)	TMJ reduced from dental complex
	(46)	Anterior zygomatic insertion at M^1 (homoplasy)
	(68)	Nasal clivus is intermediate in length (homoplasy)
	(78)	Mental foramen hollow absent (homoplasy)
Node 14:	(24)	Glenoid fossa is variable in depth
	(30)	Vaginal process of tympanic is large (homoplasy)
	(36)	Cranial capacity large, 750–1,100 cm^3 (homoplasy)
	(38)	Occipitomarginal sinus intermediate frequency (homoplasy)
	(48)	Palate prognathism is weak
	(57)	Orbital fissure configuration is comma shaped (homoplasy)
	(63)	Nasal entrance smooth with overlap (homoplasy)
	(64)	Incisive canal developed (homoplasy)
	(82)	Male canines reduced (homoplasy)
	(83)	Upper premolar/molar size significantly different (homoplasy)
	(84)	Molars small (homoplasy)
Node 15:	(2)	Supraorbital thick
	(3)	Glabella inflated (homoplasy)
	(29)	Tympanic tubular in shape (homoplasy)
	(44)	Upper facial breadth broad (homoplasy)
	(50)	Incisor prognathism within bi-canine line (homoplasy)
	(60)	Malar diagonal length is long (homoplasy)
	(70)	Zygomatic insertion is low (homoplasy)
	(81)	Upper incisor heteromorphy reduced

Apomorphic Condition of the Hominids

Kenyapithecus:	(51)	Anterior palate is shallow
	(69)	Maxillary sinus of intermediate development (homoplasy)

TABLE 5.4 ► Continued

	(70)	Zygomatic insertion is low (homoplasy)
	(74)	Mandibular inferior torus stronger (homoplasy)
	(77)	Mental foramen opens anterosuperiorly
	(88)	Molar enamel thick
	(90)	M² square
Dryopithecus:	(12)	Compound temporal crest is absent
	(24)	Glenoid fossa is deep
	(26)	Postglenoid process moderate and maybe fused/unfused to tympanic
	(29)	Tympanic crest with vertical plate
	(40)	Interorbital breadth broad (homoplasy)
	(49)	Subnasal angle is intermediate
	(59)	Malar convex (homoplasy)
	(69)	Maxillary sinus of intermediate development (homoplasy)
	(71)	Canine fossa is shallow (homoplasy)
	(83)	Upper premolar/molar size has marked difference
	(88)	Molar enamel thin
	(89)	Molar lingual cingulum developed
	(91)	Molar cusps and cristae well developed
Graecopithecus:	(1)	Tori moderately developed
	(3)	Glabella depressed
	(4)	Supraorbital with mid-sulcus
	(40)	Interorbital breadth broad (homoplasy)
	(42)	Lacrimal fossa within infraorbital region
	(43)	Mid-face strongly concave
	(44)	Upper facial breadth extremely broad
	(51)	Anterior palate is deep (homoplasy)
	(53)	Palate is broad
	(55)	Nasal keel present
	(66)	Orbital shape rectangular, broader than high (homoplasy)
	(67)	Inferior orbital margin well above superior nasal margin (homoplasy)
	(70)	Zygomatic insertion is low (homoplasy)
	(71)	Canine fossa is shallow (homoplasy)
	(75)	Mandibular corpus is robust
	(76)	Mandibular premolar orientation is V-shaped
	(77)	Mental foramen opens posteriorly
	(83)	Upper premolar/molar size significantly different
	(84)	Molars large
	(86)	P₃ metaconid infrequent
	(88)	Molar enamel hyperthick
Pongo:	(1)	Rimlike orbital tori
	(3)	Glabella cannot be defined
	(40)	Interorbital breadth narrow
	(41)	Frontal sinus absent
	(43)	Mid-face slightly concave
	(44)	Upper facial breadth narrow
	(47)	Masseter origin is high
	(51)	Anterior palate is deep (homoplasy)

TABLE 5.4 ► Continued

	(63)	Nasal entrance smooth with overlap
	(64)	Incisive canal developed
	(73)	Symphysis is gracile
	(74)	Mandibular inferior torus stronger (homoplasy)
	(80)	Upper incisor large
	(91)	Molar cusps with enamel wrinkling
Gorilla:	(2)	Supraorbital thick (homoplasy)
	(3)	Glabella inflated (homoplasy)
	(4)	Supraorbital sulcus well developed
	(8)	Postorbital constriction is developed
	(13)	Supraglenoid gutter broad
	(18)	Anterior tympanic edge aligned or lateral to porion
	(19)	External auditory meatus is large
	(39)	Facial hafting developed
	(46)	Anterior zygomatic insertion at M^1/M^2
	(55)	Nasal keel present (homoplasy)
	(56)	Inferolateral orbital margin not rounded
	(59)	Malar convex (homoplasy)
	(60)	Malar diagonal length is long
	(66)	Orbital shape rectangular, broader than high (homoplasy)
	(67)	Inferior orbital margin well above superior nasal margin (homoplasy)
	(90)	M^2 square (homoplasy)
Pan:	(6)	Sagittal crests in males usually absent (homoplasy)
Ardipithecus:	(18)	Anterior tympanic edge aligned or lateral to porion (homoplasy)
	(31)	Digastric muscle insertion deep and narrow
Anamensis group:	(74)	Mandibular inferior torus stronger (homoplasy)
Praeanthropus:	(8)	Postorbital constriction reduced
	(38)	Occipitomarginal sinus present (homoplasy)
	(77)	Mental foramen opening is variable (homoplasy)
	(85)	Lingual cusps at margin — buccal cusps approaching internal (homoplasy)
	(91)	Molar cusps and cristae well developed
Garhi group:	(1)	Torus intermediate development (homoplasy)
	(41)	Frontal sinus intermediate development (homoplasy)
	(65)	Nasoalveolar clivus is convex (homoplasy)
Sahelanthropus:	(2)	Supraorbital thick (homoplasy)
	(4)	Supraorbital sulcus not present (homoplasy)
	(6)	Sagittal crests in males absent (homoplasy)
	(22)	Cranial base is narrow
	(39)	Facial hafting developed (homoplasy)
	(46)	Anterior zygomatic insertion at P^4 (homoplasy)
	(49)	Subnasal angle is high
	(68)	Nasal clivus short (homoplasy)
	(88)	Molar enamel intermediate thickness (homoplasy)

TABLE 5.4 ▶ Continued

	(90)	M^2 square (homoplasy)
P. walkeri:	(9)	Parietal overlap at asterion
	(11)	Asterionic notch present (homoplasy)
	(12)	Compound temporal crest is present (homoplasy)
	(13)	Supraglenoid width is broad (homoplasy)
	(16)	External cranial base is extended (homoplasy)
	(24)	Glenoid fossa is shallow (homoplasy)
	(36)	Cranial capacity is small, <500 cm^3 (homoplasy)
	(38)	Occipitomarginal sinus absent (homoplasy)
	(45)	Mid-facial breadth is extremely broad
	(48)	Palate prognathism is strong (homoplasy)
	(70)	Zygomatic insertion is low (homoplasy)
P. robustus:	(7)	Temporal fossa is intermediate in size (homoplasy)
	(8)	Postorbital constriction is intermediately developed (homoplasy)
	(9)	Parietal with no overlap at asterion (homoplasy)
	(15)	Temporal squama pneumatization reduced (homoplasy)
	(26)	Postglenoid process small and fused to tympanic
	(28)	Eustachian process prominent (homoplasy)
	(31)	Digastric muscle insertion deep and narrow (homoplasy)
	(33)	Foramen magnum oval shaped (homoplasy)
	(60)	Malar diagonal length is intermediate (homoplasy)
	(62)	Anterior nasal pillars variable (homoplasy)
	(71)	Canine fossa is deep (homoplasy)
	(83)	Upper premolar/molar size intermediate (homoplasy)
P. boisei:	(2)	Supraorbital thick (homoplasy)
	(24)	Glenoid fossa is deep (homoplasy)
	(29)	Tympanic crest with inclined plate
	(56)	Inferolateral orbital margin rounded (homoplasy)
	(58)	Maxillary trigon variable
	(61)	Infraorbital foramen location is variable
	(80)	Upper incisor reduced
A. africanus:	(21)	Petrous orientation is intermediate (homoplasy)
	(22)	Cranial base is intermediate in breadth (homoplasy)
	(28)	Eustachian process prominent (homoplasy)
	(32)	Longus capitus long and oval (homoplasy)
	(35)	Foramen magnum posteriorly inclined (homoplasy)
	(41)	Frontal sinus intermediate development (homoplasy)
	(45)	Mid-facial breadth is broad
	(54)	Nasal bones projected and tapered (homoplasy)
	(55)	Nasal keel present (homoplasy)
	(62)	Anterior nasal pillars variable (homoplasy)
	(72)	Mandibular symphysis recedes, intermediate (homoplasy)
	(73)	Symphysis is robust (homoplasy)
	(77)	Mental foramen opening is variable (homoplasy)
	(89)	Molar lingual cingulum developed (homoplasy)
K. platyops:	(36)	Cranial capacity <500 cm^3
	(46)	Anterior zygomatic insertion at P^4 (homoplasy)
	(84)	Molars small (homoplasy)

TABLE 5.4 ►   Continued

K. rudolfensis:	(36)	Cranial capacity large, 750–1,100 cm^3 (homoplasy)
	(84)	Molars large (homoplasy)
H. habilis:	(11)	Asterionic notch is variable
	(12)	Compound temporal crest is present laterally
	(14)	Mastoids with lateral inflation (homoplasy)
	(18)	Anterior tympanic edge aligned or lateral to porion (homoplasy)
	(19)	External auditory meatus is large (homoplasy)
	(20)	Articular tubercle is variable
	(22)	Cranial base is intermediate in breadth (homoplasy)
	(34)	Basion variable — at/anterior to bi-tympanic
	(41)	Frontal sinus intermediate development (homoplasy)
	(49)	Subnasal angle is intermediate (homoplasy)
	(51)	Anterior palate is shallow (homoplasy)
	(55)	Nasal keel present (homoplasy)
	(62)	Anterior nasal pillars variable (homoplasy)
	(64)	Incisive canal undeveloped (homoplasy)
	(71)	Canine fossa is shallow (homoplasy
H. ergaster:	(35)	Foramen magnum inclination inclined anteriorly
	(44)	Upper facial breadth broad (homoplasy)
	(60)	Malar length is intermediate (homoplasy)
	(68)	Nasal clivus is long (homoplasy)
H. sapiens:	(1)	Torus weak to absent
	(2)	Supraorbital reduced thickness
	(4)	Supraorbital sulcus not present (homoplasy)
	(7)	Temporal fossa size is small
	(8)	Postorbital constriction is absolutely reduced
	(24)	Glenoid fossa is deep (homoplasy)
	(25)	Glenoid fossa is small in size
	(36)	Cranial capacity very large, >1,400 cm^3
	(39)	Facial hafting reduced
	(63)	Nasal entrance smooth with no overlap
	(65)	Nasoalveolar clivus convex (homoplasy)
	(68)	Nasal clivus is short (homoplasy)
	(72)	Mandibular symphysis with chin
	(73)	Symphyseal robusticity is intermediate (homoplasy)
	(75)	Mandibular corpus robusticity intermediate (homoplasy)
	(77)	Mental foramen opens posteriorly (homoplasy)

Primitive Condition of the Middle and Late Miocene Hominids

The overall primitive condition of the hominids is defined as follows. The supraorbital torus is a strong barlike feature, moderately thick, with intermediate glabella swelling. The frontal bone is enclosed by temporal lines with a strong anteromedial incursion, which in males, at least, usually results in a strong sagittal crest. The parietal has no overlap at asterion, with an asterionic notch. Postorbital constriction is developed and the temporal fossa is large. The cranium has developed compound temporal crests. The supraglenoid gutter is intermediately developed, while the mastoids are not overtly inflated. The temporal squama is extensively pneumatized. The external cranial base is extended. The rear of the cranium at the nuchal plane is defined by a steep inclination. The small auditory meatus tends to be medial to porion, and the petrous bone is sagittally aligned; associated with this alignment is the intermediate breadth of the cranial base. The articular tubercle is large. The basioccipital is long. The glenoid fossa is large in size, though with reduced depth. The postglenoid process is large and unfused to the tubular tympanic, which has a small vaginal process. A eustachian process is present and well developed. The distance between the TMJ and dental complex is increased. The insertion region of the digastric muscle along the base of the cranium is broad and shallow, and the longus capitis muscle insertion is long and oval in shape. The foramen magnum is oval in shape and located posterior to the bi-tympanic line as well as being inclined posteriorly. Overall cranial capacity is small, cerebellar morphology is laterally flaring with posterior protrusion, and an occipitomarginal sinus is infrequent.

The upper face is set high, relative to the frontal, with a developed frontal sinus. Interorbital is intermediate in breadth, and the lacrimal fossa is located within the orbital region as part of the interorbital. The mid face is defined by a well developed premaxilla with a "snout-like" appearance, emphasized by its strong palate prognathism. The upper- and mid-face are intermediate in overall breadth, as well as depth (the inferior orbital margin is aligned to the superior nasal aperture margin). The orbits are oval/rhomboid in shape, and the orbital fissure configuration is round. The inferolateral orbital margin is rounded, and a maxillary trigon is absent. Nasal bones are projected and tapered with no sagittal keel. The zygomatic/malar (viewed laterally) is near vertical in orientation, and its anterior insertion to the alveolar border is relatively high and tends to be at the M^1. The maxillary sinus is large. The infraorbital foramen(-ina) is/are located within the upper 50% of malar height. Diagonal malar length is intermediate. The angle of the convex subnasal region is relatively low, though the incisor alveolar border is prognathic, well beyond the bi-canine line. In terms of length, the subnasal region is intermediate. Anterior pillars do not define the inferolateral nasal aperture borders, though a well-developed canine fossa is present. The nasal entrance is stepped, with an undeveloped incisive canal. The anterior palate is moderately deep, while in breadth the palate is intermediate. The palate is thin. The upper incisor complex is of intermediate size, with developed incisor heteromorphy. Male canines are robust and daggerlike. Upper premolar complex, relative to the molar complex, is intermediate in size. Molars are relatively small, broader than long, with intermediate enamel thickness, and weak lingual cingulum. Cusps do not crowd the occlusal surface and are inflated with limited cristae development.

The mandibular symphysis recedes and is moderately robust. Superior and inferior mandibular tori are of similar development. The corpus is moderately robust. The mandible is U-shaped. The mental foramen opening is variable, located within a shallow hollow. The mandibular extramolar sulcus is broad. The P_3 does not have a metaconid, and it has a developed mesiobuccal expansion.

In summary, the primitive hominid condition is characterized by strong neuro-orbital disjunction, which is emphasized by the features associated with the development of the supraorbital, postorbital constriction, and the low cranium, relative to facial height (the brain is pushed back from the face). Partly associated with these characters is the klinorynchous condition, with anterior cranial base extension, which can explain to varying degrees the development of the prognathic "snoutlike" premaxilla and the subnasal morphology (Weidenreich, 1941, 1943, 1951; Shea, 1985, 1988, 1993; Lieberman, 2000). The face and cranial base are of intermediate breadth. The posterior position of the foramen magnum, orientation of the petrous, and the steep nuchal plane can be associated with the pattern of suspension and "knuckle walking," as opposed to bipedal locomotion. The heteromorphic status of the incisors as well as molar occlusal morphology and relatively thin molar enamel can be related to dietary requirements, though as discussed above, such "functional" features cannot be dismissed outright in terms of phylogenetic significance, given the concept of "phylogenetic niche conservatism."

Node 2

Following on from the primitive hominid condition is the node representing the last common ancestor with *Gorilla*. From the primitive hominid condition, we now see the emergence of a strong barlike torus, with a developed frontal sulcus. The palate has decreased in depth and the orbits are now circular-rhomboid in shape. Upper incisor heteromorphy has decreased and molar enamel has decreased in thickness. Molar cusps and cristae have increased in overall development and, finally, the mandibular torus is now weaker than the superior torus.

The emergence of *Gorilla* is defined by a number of apomorphies. The supraorbital is thick, the glabella is inflated, and the sulcus has increased in depth. There is an absolute increase in postorbital constriction. The supraglenoid gutter has increased in breadth. Anterior tympanic is now aligned with porion and the external auditory meatus has increased in size. Facial hafting is exaggerated; i.e., the cranium is set low, relative to the face. Nasal bones are now defined by a developed nasal keel. Orbital shape is now rectangular, broader than long, and the inferolateral orbital margin is rounded. Facial depth has increased (inferior orbital margin is well above superior nasal aperture margin). The zygomatic bone is convex and is now inserted more posteriorly. The diagonal malar length has increased in size, and finally the molars tend to be square.

Overall, much of the condition observed in *Gorilla* is simply an exaggeration of the primitive hominid condition defined above — an increase in neuro-orbital disjunction, which is associated with a likely increase in extension of the cranial base. This will impact increased development of the temporal fossa, facial hafting and development of the supraorbital region, and the facial frame in general (see Shea, 1985, D. Lieberman, 2000). While we can recognize an overall pattern, describing the reason underlining its development is a much more difficult task. It may be associated with a requirement to increase masticatory apparatus by increasing the temporalis muscle (increased postorbital constriction and temporal fossa), which contributed to the overall morphological form observed. Masticatory considerations probably also lie behind the reduction in molar enamel thickness, which will assist in defining more developed cusps and cristae, for an increased "shearing action" as well as increasing foveae depth to assist in the collection of juices from plant material (see Teaford, 2000). It remains possible that *Graecoptheus* belongs to the *Gorilla* lineage.

Node 3

Following on from the previous node is the emergence of the chimpanzee lineage from the later Plio/Pleistocene hominids. Nine derived features define this hominid. The temporal lines are defined by a more moderate anteromedial incursion of the frontal bone, which is clearly influencing the reduced development of sagittal crests in males. Temporal fossa size has decreased. Petrous orientation is less sagittally aligned. The glenoid fossa is more shallow and reduced in overall size. The zygomatic/malar tends to have an anterior slope. The mental foramen of the mandible opens anterosuperiorly and the extramolar sulcus has decreased in size. The immediate common ancestor to the chimpanzee and later hominids has a reduced temporalis and TMJ development, obviously associated with dietary and masticatory considerations.

Node 4

This node represents that last common ancestor to *Ardipithecus* and the other more derived hominids. This hypothetical ancestor has an increase in cranial base flexure, and reduction in nuchal plane inclination, articular tubercle, eustachian process, and longus capitis insertion. The basioccipital is reduced in length. Foramen magnum is located more anteriorly and is inclined in a more horizontal position. There is an increase in the frequency of the occipitomarginal sinus. In terms of upper facial features, glabella is broad but not inflated, and the supraorbital sulcus is reduced. Interorbital breadth has increased and nasal bones are projected and expanded. The palate has also increased in breadth. The nasal clivus has increased in length, and its height and the incisive canal are more developed. Upper male canines are reduced in size and there has been an increase in upper molar size. The buccal cusps tend to crowd the occlusal surface. The mandibular symphysis is more robust and the P_3 mesiobuccal expansion has decreased.

At this point in human evolution, we see increased flexure of the cranial base. This may be associated with the foramen magnum now moving in a more anterior position and a reduction in lower facial prognathism (and increased palate breadth). At the same time, we see a tendency for male canines to be smaller in size and less daggerlike, while the molars have increased in size. *Ardipithecus* has two apomorphies — the auditory meatus is now aligned to porion (suggestive of increased cranial base breadth?), and the digastric muscle insertion is now deep and narrow. So far all we can say is that in its preserved morphology, *Ardipithecus* appears to reflect a distinct pattern of digastric muscle development to that observed in the outgroup, which may also be associated with the differential development of its broader(?) cranial base.

Node 5

This node defines the last common ancestor between the *anamensis* group and other hominins, defined by three derived features: the mandibular mental foramen opens laterally; molar enamel has increased in thickness (which is related to the next feature in that cusps and cristae are inflated); and the deciduous molar mesial marginal ridge is slight, with protoconid set anterior. The *anamensis* hominid is defined by only one apomorphy, the mandibular inferior torus is strongly developed; the lack of additional apomorphies is clearly related to low specimen numbers and poor preservation.

Node 6

At this node, the last common ancestor between *Praeanthropus* and later hominins, the articular tubercle again increases in size (homoplasy), and the zygomatic has moved anteriorly. The difference between the premolar and molar complexes has increased. The mandibular symphysis is less receding and the corpus has increased in overall robusticity. Finally, P_3 metaconids start to appear and the expansion of the mesiobuccal corner is variable.

Praeanthropus afarensis is marked by a continued reduction in postorbital constriction and an occipitomarginal sinus is present. Also, molar cusps and cristae are more developed, and the buccal cusps continue to move medially away from the buccal edge. There is a continuation of the hominin condition of reduced neuro-orbital disjunction, emphasized by a further reduction in postorbital constriction. The differential pattern of occlusal morphology can be associated with an increase in shearing action, possibly associated with increased consumption of vegetative material from that of its contemporaries and hypothetical ancestor.

The neurological configuration in *Praeanthropus* has been used to support a close phylogenetic relationship between it and *Paranthropus* (see Skelton *et al.*, 1986). As we will see, however, this morphology is not present in the presumed ancestor of the *Paranthropus* clade (*P. walkeri*) and thus must be considered a homoplasy. Indeed, the evolution of this feature can be explained as the result of a number of likely developmental demands not closely associated with phylogenetic considerations. For example, the development of this vascular pattern has been equated with the changing gravitational pressures associated with bipedalism (Falk & Conroy, 1984; Falk, 1986, 1988). It is suggested that this early pattern would enable increased flow of blood to the vital organs, given the changed posture. Tobias (1967), however, suggests that it may be associated with the early growth during ontogeny of the cerebellum, which may have forced blood into the marginal sinus system, then becoming the established path of blood supply during adulthood.

Node 7

This node represents the common hypothetical ancestor of the hominins (*sensu* DWC), including *Sahelanthropus* and the *garhi* group, suggesting that the origin of the hominins is significantly deeper in time then previously considered (if we truly accept *Sahelanthropus* as being a basal hominin, which DWC does but CPG does not), and that the proto-australopithecines had little if anything to do with the origins of the hominins and represent an additional Plio/Pleistocene hominid radiation event, resulting in the extinction of numerous hominid species.

This basal hominin is defined by the emergence of an asterionic notch. Also, the mastoids are inflated, and the nuchal plane is less inclined. The origin of the masseter is high and the nasal clivus is flat. Following on from the previous node, the lingual cusps and buccal cusps crowd the occlusal surface. The mandibular symphysis is near vertical and the mandible as a whole is robust. The extramolar sulcus has increased in breadth. Finally, P^3 metaconids are present and the mesiobuccal expansion is weak to absent. *Sahelanthropus* has a thick supraorbital, no frontal sulcus, and no sagittal crest. The cranial base is narrow and facial hafting is similar to that observed in *Gorilla* (suggested by DWC to be a homoplasy). The zygomatic has moved even farther anteriorly and the nasal clivus is short and high. Molar enamel has increased, and the upper molars tend to be square in shape (another homoplasy shared with *Gorilla*). The *garhi* hominins have a reduction in torus robusticity and development, with increased development of the frontal sinus, and the nasal clivus is convex.

Node 8

This node represents the last common ancestor to the *Paranthropus* clade and the more derived hominins (*Australopithecus, Kenyanthropus*, and *Homo*). Compound crests are variable in development and the petrous is more coronally orientated. This is clearly associated with increased breadth of the cranial base. The glenoid fossa has increased in depth, though the postglenoid process has decreased in size. The tympanic crest is defined with a vertical plate. Cranial capacity has increased and palate prognathism is reduced. Finally, the nasal entrance is defined by a slope (as opposed to step) and the mandible is more parabolic in shape.

From the morphological condition emerging from nodes 7 and 8, we can observe an increase in cranial capacity, while also documenting a reduction in postorbital constriction and temporal fossa size, a less inclined nuchal plane, and a realignment of the petrous. These later features may be associated with a reduction in temporalis and differential development of neck musculature. Indeed, increasing coronal orientation of the petrous bone, reduced basioccipital length, and repositioning of basion relative to the bi-tympanic have all been equated with the necessary reconfiguration of the cranial base in regards to increased basial flexure, accompanied by a reduction in cranial base length (see Weidenreich, 1943; DuBrul, 1977; Olson, 1985; Aiello & Dean, 1990). As noted by Walker *et al.* (1986) and Dean (1988), however, *P. walkeri*, while having an extended cranial base, is *also* characterized by coronally orientated petrous bones and an anteriorly positioned foramen magnum (see Strait *et al.*, 1997); i.e., the exact opposite of what should be expected. In addition, Cramer (1977) argues that such a "correlation" does not always hold true for extant hominids. Finally, Dean (1988) has also suggested that the orientation of the petrous bone at least may not be correlated with increased cranial base flexion, but rather the result from prenatal flattening of the skull base as the cerebellum expands faster than the posterior part of the cranium is able to elongate.

Node 9

This node represents the basal ancestor to the *Paranthropus* clade. With the emergence of the *Paranthropus* lineage we can see the reemergence of a number of "primitive" features: The temporal lines are again marked by a strong anteromedial incursion, with males having a well-developed sagittal crest. Postorbital constriction has again increased. The glenoid fossa has increased in size and the foramen magnum is now heart shaped. The face is positioned high, relative to the frontal, and glabella is inflated. Facial dishing as well as maxillary trigon are present. The inferolateral orbital corner is not rounded and the orbital fissure is comma shaped. The zygomatic insertion at the alveolar border is set anteriorly at the P^4 or anterior to it. Malar diagonal length is long. The incisor alveolar border is set close to the bi-canine line. The palate is thick, there is now overlap within the clivus and nasal floor, and incisive canal is developed. The nasal clivus itself is concave and the canine fossa is very much reduced. The molar enamel is hyperthick, and associated with this is the flat bunodont occlusal morphology. In the mandible, the inferior torus is being developed, no metal foramen hollow, with a broad extramolar sulcus.

Emerging from this hypothetical ancestor is *P. walkeri*, in which parietal overlap occurs at asterion, and an asterionic notch is present. The compound temporal crests have reappeared and the glenoid fossa is shallow. The supraglenoid width is extremely broad, and the external cranial base is extended. There is a reduction in cranial capacity and there is no occipitomarginal sinus. Mid-facial breadth is also extremely broad. Zygomatic insertion is low and the palate has increased in prognathism. The reemergence of many of these "primitive" features in *P. walkeri* is of functional interest, and suggests either that it occupied a distinct dietary and/or habitat niche from that of its later daughter species from Kenya and South Africa (see below), or that the placement of *Paranthropus* is incorrect, and that homoplasy must be sought elsewhere in its anatomy. This example further emphasizes (if any is needed) the difficulty in trying to explain the developmental reasons behind observed morphological patterns. All such correlations must be considered tentative.

Node 10

This node represents the last common ancestor of *P. boisei* and *P. robustus*. With its emergence, pneumatization of the temporal squama is reduced and the parietals are defined by extensive overlap. External auditory meatus is large, and the anterior tympanic edge is either aligned or lateral to porion. The external cranial base is more flexed and the basioccipital is shorter. There is a continued reduction in the postglenoid process, which is fused to the tympanic. The vaginal process of the tympanic is large. There is increased anterior migration of the foramen magnum beyond the bi-tympanic line. There is an increase in the frequency of the occipitomarginal sinus, and the cerebellar is more tucked. The upper face has increased in overall breadth, including the interorbital. Finally, male upper canines are reduced in size and appearance. From this hypothetical ancestor, both later species of *Paranthropus* emerge.

Node 11

While the *Paranthropus* lineage was evolving its unique set of morphological features, the basal hominin to *Australopithecus, Kenyanthropus,* and *Homo* was also evolving its own set of unique anatomical features.

distinct not only from the *Paranthropus* lineage, but also the earlier Miocene and Plio/Pleistocene hominids. At this point, we can see the emergence of a weaker supraorbital torus with no sagittal crests or compound temporal crests. The supraglenoid gutter is reduced in width. Mastoids have a reduction in lateral inflation. The anterior palate is now deeper and the diagonal malar length has continued to reduce in length. Finally, the face is reduced in depth.

Emerging from this ancestral population is *Australopithecus*, which has a number of apomorphies — mostly homoplasies with related taxa, though there remains a possibility that the branch positions of *Australopithecus* and *Paranthropus* have been switched. The petrous is intermediate in orientation, corresponding with intermediate cranial base breadth. Eustachian process is developed and the longus capitis is long and oval. Foramen magnum is positioned posteriorly. Frontal sinus is reduced. Mid-facial breadth has increased. Nasal bones are projected and tapered, with a developed nasal keel, and anterior nasal pillars are developed. Molar lingual cingulum is developed. Symphyseal recession has increased and is robust in construction.

Node 12

This is the hypothetical common ancestor to the *Kenyanthropus* and *Homo* lineages. There is a continued reduction in the anteromedial incursion of the temporal lines, and the temporal squama is variable in its pneumatization. The external cranial base is more flexed and the articular tubercle is small. The digastric muscle insertion region is deep and narrow, and the basioccipital is short. There is a continued increase in cranial capacity. Occipitomarginal sinus frequency is low and the cerebellar is tucked under. The zygomatic is now in a near vertical orientation, and the masseter origin is low. Nasal bones are not projected and the nasal cavity entrance is stepped. The subnasal region is increased in height. The upper molars are reduced in size. Molar lingual cusps do not crowd occlusal surface, buccal cusps do. Mandibular tori are undeveloped, while the corpus is robust. Mandibular extramolar sulcus is narrow. The evolutionary trend at this point is for the brain to increase in size, while conversely the face and dental complex are reducing in overall size and robusticity.

Node 13

This is the base of the *Homo* lineage and the immediate ancestor to *H. habilis*. We see the emergence of a moderately developed frontal sulcus, temporal squama pneumatization is reduced, the postglenoid process is small and fused to the tympanic, the anterior zygomatic is inserted above the M^1, and the dental complex and TMJ are closer in terms of overall distance. The nasal clivus is intermediate in length and the mandibular mental foramen hollow is absent. Following on from this, we see the reemergence of *H. habilis*, which is defined by a large number of apomorphic features. These include variability in the presence/absence of an asterionic notch, laterally developed compound temporal crests, laterally inflated mastoid, and the large auditory meatus is aligned to porion. An articular tubercle may or may not be present. The cranial base has become more narrow, the foramen magnum has drifted anteriorly, frontal sinus is intermediately developed, there is a decrease in subnasal height, the anterior palate is shallow, a nasal keel is present, anterior nasal pillars are variable, canine fossa is reduced in depth, and the incisive canal is undeveloped. The condition in *H. habilis* appears to be related to differential neck muscle development and an increase in lower facial prognathism.

Node 14

The last common ancestor to *H. ergaster* and *H. sapiens* has a variable glenoid fossa depth. The vaginal process of the tympanic is large. There is a continued increase in cranial capacity, and the frequency of an occipito-marginal sinus has also increased. The orbital fissure is comma shaped. There is a continued reduction in palate prognathism, and the nasal clivus and nasal floor are defined by a smooth transition with overlap and a corresponding increased development of the incisive canal. There is also continued reduction in male canine size and in premolar size, even greater than the reduction in the molars.

Node 15

The last common ancestor of the two *Kenyanthropus* species has a thickening of the supraorbital, with inflated glabella. The tympanic is tubular, the upper face is broad, there is absolute reduction in subnasal prognathism, and the zygomatic insertion is low. The malar has increased in diagonal length. Finally, incisor heteromorphy is absolutely reduced. The earlier species, *K. platyops*, has a small cranial capacity, the anterior zygomatic is inserted anteriorly at the P^4, and the upper molars have decreased in size. The later *K. rudolfensis* has increased cranial capacity and the upper molars have increased in size.

The small cranial capacity observed in the earlier *Kenyanthropus* species is difficult to resolve. It is hard to accept that there could be any adaptive advantage in the reintroduction of a small brain. The problems in determining how to "objectively" weigh such a character is beyond this study. The most parsimonious way to interpret this difficult question is to accept that *Kenyanthropus* and *Australopithecus* originate from a similar base, and whether *Homo* evolved from *Australopithecus* or *Kenyanthropus* remains to be seen. To accept this scheme, we must also believe that the later species of *Kenyanthropus*, *K. rudolfensis*, must have evolved a large brain independently from *Homo*. While these taxa do appear to represent hominins (*sensu* DWC), the phylogenetic complexities of this group remains obscure.

From these analyses, their interpretation, and our discussion, the origins of the later Pleistocene hominins can be observed from the clade containing *Homo habilis* and *H. ergaster*. As we will see in the chapter to follow, it was *H. ergaster* that was the first hominin to disperse out of Africa around 2 Ma. It was from this group that ultimately a number of hominin species were to later evolve, and in some cases different hominin species would come into contact with each other. Ultimately, however, only one species would survive to continue the hominin lineage into the later Pleistocene and Holocene, *H. sapiens*.

INTERLUDE 3
Of Men's Beards and Peacock's Tails

Sex is a big deal. We know, very broadly speaking, why this should be: Without it, the species has no future. But exactly why it takes the form it does in the human species and why there is this huge variety — now those are other questions.

Gibbons don't differ much in size between the sexes, though they differ in their calls and, in some species, in their color. In orangutans, males are bigger than females, but some males are much bigger than others. Mature male gorillas are always hugely bigger than females and have a conspicuous grey back. Male chimpanzees are only slightly bigger than females but have simply enormous testicles. Men are bigger than females, though not vastly so, but have very big penises (all of them — did you know that?); and men and women are just astoundingly different in all sorts of ways. Very odd. Why?

It has to do with social organization, and this in turn has to do with how species make a living. What it's called is *sexual selection*. It's males getting bigger because they have to compete with each other — but it's much more than that.

Natural selection, which Darwin described in 1859 in *On the Origin of Species*, is the simplest mechanism of change in evolution. If one individual is ever so slightly better adapted to its environment than another, it has a slightly better chance of leaving offspring. That's essentially it. But in 1872, in *The Descent of Man, and Selection in Relation to Sex*, Darwin pointed out that sexual preference has similar consequences. Peahens prefer peacocks with bigger tails, so the bigger the tail, the more offspring a peacock has—on average.

Sir Ronald Fisher was one of those who forged the new evolutionary synthesis in the 1930s — the welding together of natural selection and the new science of genetics to produce the first rounded, biologically satisfying model of evolution. Whether it is a complete model has been discussed ever since, but nobody denies that it is a valid model. Fisher was brilliant and (therefore?) opinionated and realized that the course of selection could and should be modeled mathematically. In his 1930 book, *The Genetical Theory of Natural Selection*, he wrote a great deal not only about natural selection, but also about sexual selection. But this promising start did not last; sexual selection, for some reason, fell into disrepute — for 40 years. It was not until 1972, when Bernard Campbell edited a book called *Sexual Selection and the Descent of Man*, that it gradually eased its way back into respectability. The slowly emerging field of evolutionary psychology took it up, and sexual selection is now established as a mainstream field for research and experimentation.

As far as reproduction is concerned, there is a basic asymmetry between males and females, and it is this: Females are the ones that produce the young, one or two (or however many it is) at a time, whereas males simply do the fertilizing. This makes the female the indispensable sex; there have to be lots of females, but in theory there need only be one single male in the entire population. Males are therefore in fierce competition to determine which one of them it shall be. The best way to do this is, to put it bluntly, to find out what females want and to be as like that as they can. Whichever male pleases the females most — that's the one who gets the most matings and leaves the most offspring.

Female peahens like males with big, gaudy tails. Why? Because tails like these handicap the males; they make them more conspicuous to predators, they make it more difficult to fly, and when folded up they drag on the ground and pick up dirt and nasties. If there is a peacock who, despite all this, has a really big and gorgeous tail, it means he has overcome all these problems and thrived. He must be a very fit specimen indeed — go mate with him, all you peahens, be fertilized by his excellent genes and produce wonderfully fit offspring.

The African long-tailed widow bird (*Euplectes progne*) is like peafowl, but half the size and easier to experiment with. The female is a dowdy mottled brown; the male is red and black with a tail one-and-a-half meters long. The male jumps into the air in front of females, displaying its extraordinary tail to them. Malte Andersson in 1982 caught quite a number of males. He cut portions out of some males' tails, glued them into others' tails, and then released them. For controls, he caught yet other males and either released them again unaltered or cut through their tail feathers and simply reglued them. He found that the number of females who nested in different males' territories was exactly as predicted — most in the territories of the males whose tails had been lengthened, less in the unaltered ones, least in those whose tails had been shortened.

In zebra finches (*Poephila guttata*), males have redder beaks than females (on average). The redness is due to pigments called carotenoids. These are not genetically controlled, but obtained through the diet — and they are not just pigments; they are also antioxidants that stimulate the immune system. Some British zoologists (in a study published in April 2003) supplied some males with pure water as usual but gave their brothers — literally, their full brothers — water with carotenoids in it. The ones with the carotenoid supplements had redder beaks and enhanced immune systems — and were more attractive to the females.

Think what this means. We infer that peacocks and male widow-birds must be fit if they can overcome the handicaps of their tails, and we suppose that females must subconsciously infer this too. With the zebra finches, we actually know it: Those with red beaks really are the fit ones. In the wild, if you can get a lot of carotenoids into your diet — if you can get and keep a good territory with the right resources — you will get a redder beak and females will know that you are ever so fit. And get this: A female, if she chooses the right male, can get access to his territory and she too can get a red beak and a fantastic immune system.

Sexual selection is based on natural selection; what it does is to take the indicators of fitness and run with them.

The fruit on which gibbons feed in their treetop habitat in the southeast Asian rainforests is clumped, and the clumps are scattered. There is enough to feed just small groups, provided that the groups are well spaced through the forest. So gibbons are monogamous and territorial — meaning that their social groups consist of one male and one female plus offspring, and this pair is the sole occupant of an area of treetops (about 20–40 hectares is usual). It used to be thought that the breeding male and female were faithful unto death. But we now know that there is divorce (one member of the pair may simply swing off and pair up with a different mate). There is also extra-pair copulation (cheating, not to put too fine a point on it) — a bit like human monogamy really.

In the white-handed gibbon, both members of the pair defend their territory. And the males don't actually fight for mates, so males and females are about the same size and both have long, stabbing canines. In fact, there's very little difference between them at all; the male doesn't even have a scrotum (for the benefit of the curious, they're in tiny individual sacs on either side of the penis). A female has to attract a mate and keep him there, and she sings loudly — a soaring, melodious aria, to which the male adds an insignificant little coda. In silvery and dwarf gibbons, the male doesn't even bother to do that. In other species the relationship is more equal, and in some the male has as elaborate a song as the female, and a scrotum too. In a few species the male and female are different colors. Complicated — and we know too little about most species to say exactly what sort of sexual selection is going on, but in general monogamy means equality.

Orangutans live in some of the same southeast Asian rainforests as gibbons do, but because of their huge size, they have to move much more cautiously and to live solitary lives. But they can open enormous fruits that are beyond the capabilities of gibbons. How huge are orangutans? Females weigh about 40 kg, fully mature males more than twice that — 85–95 kg. We might deduce that there's a lot of overt competition between these males, and we would be right — they fight when they meet, and smaller ones hear the deep booming voices of the bigger ones and keep well out of the way. And these huge males have wide, solid flanges on their cheeks, which may sway back and forth as they move, obviously some sexual adornment for the females' admiration.

And there are undersized males with no flanges. The females are not in the least attracted to them, but these males chase after the females and, frankly, force themselves on them. For a long while it was assumed that these unflanged males were not mature, but it now seems that some of them may stay like that all their lives — it's simply an alternative strategy. So there are two ways of doing things if you're a male orangutan: You can develop into a huge, splendid chap and wait for the females to come to you (and they do), or you can stay small and go out and get the females because they won't come to you (and they don't). DNA studies show that the two ways are equally successful at yielding offspring. This balance of two different ways of operating is called an *Evolutionary Stable Strategy* (ESS).

Gorillas are even larger than orangutans, and males and females are just as different in size (and all males are big — there are no sneaky runts). But they live in Africa, mainly on the ground, and in social groups. They can do this because, although they prefer fruit, if there isn't any fruit available they can make out just fine on the ground herbs that are all around them. There may be one male and several females, and their offspring, in a group. And lo, gorillas, like orangutans, have an ESS: A male, when he matures, may stay in his father's troop if there are spare females in it (and take it over when father becomes senile), or he may leave it and try to steal females from other individuals' troops. Steal them? If a male looks stronger and fitter than the one they're with, the females will join him in a flash.

Chimpanzees weigh 30–60 kg and live in communities where males and females mingle and separate at will. Why are males not much bigger than females — don't the males compete? Yes and no; when a female is in her fertile period she mates with them all, one by one, and it's not they who compete — it's their sperm! Those huge testicles pour out vast quantities of sperm, and the winner takes all. Oh, there are dominant males in the community, and sometimes one will sequester a female and be her sole consort for a while, but sperm competition is the rule.

And us? We are bizarre. Men are bigger than women, but not too much. But men have beards and moustaches, a lot more body hair than women, and broad shoulders and narrow hips. Women have breasts and a buildup of fat on the buttocks, hips, and thighs. Men and women alike are hairless compared even to chimpanzees and have more body fat. Chimpanzee and gorilla mothers develop breasts, but the breasts shrink again when the babies have been weaned; they are not permanent like women's. We live much longer, too. If a female chimpanzee lives into her late forties, she ceases to breed, and she may be dead before then anyway; women ostentatiously, almost ceremoniously, stop being able to breed at about 50, but live on well after that. (Life, we are told, was nasty, brutish, and short in premedical times. But this is a statistical shortness and is just because so many babies died in their first couple of years that it brings down average life expectancy — if people survived early childhood, most of them lived on into their sixties and seventies and more, like we do today.)

So many people have put forward so many ideas for why we are hairless. To cool down and to swim better are two that we often hear. Recently, Mark Pagel and Walter Bodmer suggested it was to rid ourselves of parasites: Ticks and biting flies can be seen, can't hide away, and can be got rid of, and this became attractive to the opposite sex. Hair was retained on the scalp as a sort of sun hat, in the armpits to waft underarm pheromones (sexual odors) into the world, and on the groin — why? Maybe for the same reason, they suggested.

Pagel and Bodmer haven't gone far enough, have they? On a hairless skin, not only can you be parasite free, you can be *seen* to be parasite free. Scabies and ulcers and wounds show up brightly on naked skin. Unblemished skin advertises in the starkest possible way that you are fit.

Different body shapes and different hair patterns signal sexual fitness from a great distance. Excuse us for asking, dear reader, but have you ever been to a nude beach? People are walking way over at the other end of the beach, and you can tell their sex because of their body shape. As they come nearer, you can begin to discern something of the hair distribution. Women have a dark patch *there* and a separate one *there*, but men have a continuous swath from one to other, parting company on either side of the face, reuniting on the chin, straggling a bit as it goes down the chest and belly, but pretty much continuous all the same — or was in preshaving days. But even so, the five o'clock shadow is discernable well before the sun has slipped below the yardarm.

A Cambridge zoologist, C.B. Goodhardt, said all this long ago, in the early 1960s. He gave lectures in which he would explain that sexual differences in hairiness and hair distribution were deliberately exaggerated, as supernormal stimuli, by "living savage races," and would cause a roar of laughter from (most of) the audience by then showing a slide of a Scottish soldier in bearskin and sporran. But he thought that, for this to work, the ancestral skin color would have to be light. Actually, it doesn't. Compared to Caucasoids (both pink Europeans and brown Arabs or Indians), black Africans are exceptionally hairless except on those same places, and the matte hair stands out against the shiny skin. It works anyhow.

Geoffrey Miller (2000), in *The Mating Mind*, has gone on to show how mental characteristics, too, have been sexually selected; and evolutionary psychologists have done experiments to test whether it's true (it is).

We say, "Vive la difference, mate!"

CHAPTER 6

The First African Exodus: The Emergence of Early *Homo* in Europe and Asia

The sabertooth big cat had been dead for only a few minutes and had not yet attracted attention from the roaming scavengers of the sky and savanna plains. Quietly the hominin approached the cat. It was most unusual to be the first on the scene of such a prize, a large cache of fresh meat. Looking around she picked up a number of large volcanic rock fragments and begun to hammer out a number of crude but very sharp flakes. She looked around to make sure that no other carnivores were approaching. Seeing no sign of approaching danger, she quickly cut into the still-warm carcass. Tearing through the body she came across the liver and proceeded to cut into it, pulling it free. Just as she succeeded in doing this, she was startled by a hyena, who began to circle around her. This meant trouble. She knew that close by would be the rest of the pack. She gathered up the meat she had managed to scavenge and quickly left the cat's carcass to the hyena. She had managed at least to retrieve a small portion of meat, some of which she could share with members of her group. It made no sense to traverse the savanna carrying quantities of fresh meat, the smell of which would surely attract any number of predators. She sat down beside a small tree for cover and proceeded to eat the portion of liver she had managed to retrieve.

It was almost three weeks since she had came across the dead big cat. Now she lay dying a terrible death. The group had been pleased with her prize, and that night some meat was added to their diet of tubers, nuts, and other plant material. Within hours of consuming the liver, she had begun vomiting as well as developing an intense headache, which progressively got worse. Within days her joints had begun to ache and she had uncontrollable diarrhea. It was not long before her skin started to peel and blood began oozing from her pores. The members of the group cared for her as best they could. They managed to keep her alive for a few weeks, but the end was near. Some in the group knew that she must have consumed the easily eaten soft flesh; they had seen this condition before in others long ago who had eaten it. Soon she was dead, and the group moved on, leaving her body to the elements.

Now she is known as KNM-ER 1808, and her discovery and the interpretation of the pathology of her skeleton has given us important
clues to early hominin behavior. She represents an early species of our
own genus, *Homo*, known as *Homo ergaster*. She lived and died in East
Africa around 1.5 Ma. Her bones tell us that she was a likely victim of
hypervitaminosis A (Walker *et al.*, 1982; Walker & Shipman, 1996).
A study of KNM-ER 1808 by Alan Walker and colleagues demonstrated
that shortly before death, her bone had become increasingly brittle,
fibrous, and coarse-textured, suggesting an increased breakdown within
the bone-forming cycle. Today this pattern is seen in rare but severe cases
of hypervitaminosis A. The highest concentration of vitamin A is in carnivore liver; because KNM-ER 1808 was the victim of severe vitamin
A toxic poisoning, she most likely had eaten carnivore liver. She appears
to have been a victim of the early introduction of increased meat supple-
mentation into the early hominin diet, a period of trial and error (Walker
et al., 1982). Later hominins would know better and not consume this
toxic part of an animal carcass. Her death also tells us something of her
life.

   Walker realized immediately the significance of his pathological interpretation of KNM-ER 1808 for early hominin behavior — "Someone else
took care of her" (Walker & Shipman, 1996: 134). There is no way she
could have survived alone for long in the African savanna. She would
quickly have succumbed to the roaming carnivores. We know she survived
for some time because it would take at least a few weeks, if not months,
for this pathology to show up in her bones, so someone must have been
feeding her and protecting her from the carnivores, including hyena packs.
Perhaps more importantly, this someone must have had some way to bring

er water, requiring some form of water "container," which requires plan-
ing. It is unlikely that she would be near a watercourse because this
would be a major focus area for carnivores looking for a drink and a feed.
The group dynamics of early *Homo* must have been based on some form
of mutual support.

It is with the emergence of our own genus that we see a significant expan-
ion of the brain and a dietary shift toward an increasing reliance on meat.
The smaller inverted funnel-shaped rib cage of the proto-australopithecines
was adapted to house a large gut and intestines, so as to process large quan-
ities of plant material that made up much of their diet (Aiello & Wheeler,
1995). With the emergence of early *Homo*, the potbelly of the proto-
australopithecines gave way to a more slim physique, for these humans
relied on a more varied diet, including meat; they had no need for a large gut
and intestine, and their body proportions reflect this (Aiello & Wheeler,
1995; see also Bunn, 2002; Schoeninger *et al.*, 2002). Overall the earlier,
more primitive proto-australopithecines were bipedal "great apes." The
same applies to the earliest representative of our genus, *H. habilis*, as well as
the even more specialized hominin *Paranthropus*. It is from around 2 Ma
that we see within the fossil record a major physical and behavioral shift
with the emergence of *Homo ergaster* in East Africa.

The dietary and behavioral shift to an increased focus on meat eating within
the earliest representatives of *Homo* was not a simple matter. It probably
required a major change in group dynamics involving cooperation and coor-
dination of individuals, an increased dependence on tool technologies to
help in meat acquisition and processing (specimens of *H. ergaster* are associated with the primitive Oldowan stone tool technology), and certainly
a major reconfiguration in hominin gut morphology in order to process
associated fat and meat fibers. As discussed in the previous chapter, it is
likely that some proto-australopithecine species incorporated meat into their
diet, but it was probably not an important component of their diet (Cameron,
1993a; Cameron & Groves, 1993). Indeed, the major dietary focus of these
hominids and the hominin species within *Paranthropus* is usually associated
with a dependence on eating fruit with hard nuts and seeds (see discussions
in Kay & Grine, 1988; Schoeninger *et al.*, 2002). We are still debating the
degree to which *Australopithecus* incorporated meat into its diet, though the
recent isotopic studies of Sponheimer and Lee-Thorp (1999) suggest that
Australopithecus (specimens dating to around 2.5 Ma) consumed large

quantities of animal tissues from large grazing animals. They conclude that it was how *Australopithecus* and *Homo* exploited their food sources as much as the exact sources they used, that distinguished them: Stone tools in particular, enabled *Homo* to disarticulate the skeletons and get at the marrow.

If this is true, it appears that while *Australopithecus* had yet to develop the complex behavioral and technological abilities of early *Homo*, they must have shared at least the incipient beginnings of the specialized gut morphology in order to incorporate increasing degrees of meat into their diet.

The primitive hindgut morphology observed in most primates is related to processing simple carbohydrates obtained from fruit and proteins from leaves, whereas carnivores obtain their energy source from meat fats (Strait *et al.*, 1997; Fleagle, 1999). In hindgut fermenters, the proteins associated with folivory go into the stomach, but they do not break down immediately because they are composed of indigestible fibers requiring a large fermentation chamber. The stomach contains a large number of bacterial colonies to break down these fibers over time, and hindgut fermenters have very large colons and stomach chambers (Chivers & Langer, 1994; Fleagle, 1999; Schoeninger *et al.*, 2002). An increased dependence on meat eating, however, also means an increase in nitrogen, which is toxic to the foregut bacteria. And a corresponding decrease in levels of fiber means that there is a significantly increased rate of colonic twisting (Schoeninger *et al.*, 2002). As hominins became more carnivorous, the stomach and colon would have been reduced in size, while the intestines must have been significantly more developed because it is within the foregut (small intestine) that fat digestion occurs.

While the question of why this shift to increased meat eating occurred does not necessarily require complex explanations (in order to take advantage of a rich dietary niche not previously occupied by hominid groups), the question of how it happened is far more difficult to answer. Perhaps the most elegant model proposed is that recently provided by Bunn (2002) and Schoeninger *et al.* (2002). They emphasize that there was a transitional phase, where an increase in the level of tree-fruit pulps would avoid problems of the colonic twisting, while lipid-rich food items obtained from seeds and other nonmeat materials emphasized the increased ability of the lipid-digesting section of the small intestine. At some point there was a further reliance on lipid-rich foods (including meat) to obtain energy requirements, with corresponding decreases in fiber lipids, resulting in a larger foregut and smaller caecum. Associated with this is the increased development of stone tool technologies, such as grinding stones, in order to process those foods that could not be directly consumed given the switch to an increased foregut. These technologies would enable seed coat removal, enhancing access to seed protein by removing digestion inhibitors. The increased tool technology comes with the emergence of *Homo*, which unlike its Pliocene forebears was now able to take advantage of a much broader dietary base (see Bunn, 2002; Schoeninger *et al.*, 2002).

There is no definitive evidence that *Paranthropus* was a habitual tool-user or tool-maker; its food-processing abilities tend to be associated with a primitive pattern of oral preparation, which is emphasized further by its robust facial structure and enormous grindstone-like premolars and molars. Selection pressures appear to have focused on increasing its robust skull and dental complex, so the ancestral member of this robust lineage, *P. walkeri*, while probably sharing the "transitional" phase in gut morphology (passed on from its late Miocene/early Pliocene ancestor), did not revert to meat eating but emphasized other lipid-rich resources, probably a specialized diet of hard fruits, hard seeds, and other abrasive foods. This dietary overspecialization, in the face of the habitat conflict with other, nonrobust hominin groups, must have contributed to the eventual extinction of the robust hominins.

It is clear that by the time *H. ergaster* appears in the East African fossil record, moderate levels of meat eating had been incorporated into the diet. Specimen KNM-ER 1808 provides evidence that this was not always a successful adaptation and that some form of "trial and error" was still being invoked. It was the broadening of its dietary base, and associated behavioral adaptations as well as increased dependence and development of stone and nonstone technologies (again demonstrated by the protection, feeding, and help given to the individual represented by KNM-ER 1808), that enabled members of this species to increase their territorial range and thus to increase access to resources.

While *H. habilis* and *H. ergaster* are both thought to have manufactured the primitive Oldowan tool technology, it may be that *H. ergaster* was the first to become increasingly reliant on it, while *H. habilis* was an infrequent user of stone tools. Or it may be that *H. ergaster* alone was responsible for their manufacture. The South African hominin StW 53 specimen from Sterkfontein Member 5 (dating to around 1.5 Ma), which until recently had been tentatively allocated to *H. habilis*, is associated with Oldowan-like chopping tools (see Brain, 1981). However, Kuman and

Clarke (2000) propose to reallocate this specimen to *Australopithecus*. They argue that it is part of a secondary deposition and thus was not necessarily deposited at the same time as the tools. If this is so, the jury is still out with regard to *H. habilis* as a tool manufacturer. So far, only *H. habilis* and *P. boisei* are known from Olduvai Gorge Bed I, where Oldowan tools are abundant; but there are so few specimens that at any moment *H. ergaster* or some unexpected species may turn up there. While early demes of *H. ergaster* are associated with Oldowan technology (or Mode I technology), Acheulean tools (or Mode II) are first associated with later populations of *H. ergaster* from Konso, Ethiopia, around 1.4 Ma (Figure 6.1). This new tradition is defined by biface instruments, much more elaborate than those of the Oldowan tradition. This new toolkit consisted of hand axes, cleavers, and picks and must have involved considerable forethought and planning, for the tools represent a predetermined design rather than "blades" struck from a core. This demonstrates that new tool traditions — technologies need not be correlated with the appearance of new hominin species.

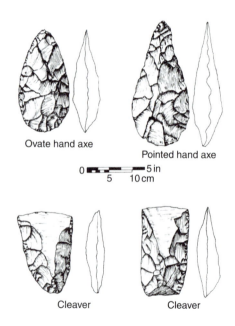

Ovate hand axe

Pointed hand axe

0 ⬛⬛⬛⬛ 5 in
5   10 cm

Cleaver   Cleaver

Figure 6.1 ▶   Typical Acheulean (Mode II) bifacial artifacts.
From Schick and Toth (1993), p. 241.

The Emergence of *H. ergaster*

Homo habilis and *H. ergaster* were sympatric for approximately 400,000 years. This means that either a proto-*ergaster* population split away from an earlier population of *H. habilis* or they share a common ancestor around 2.3 Ma. The first hypothesis appears more likely, for *H. ergaster* does not appear in the fossil record until around 300,000 years after the earliest representatives of *H. habilis*: *H. ergaster* has a temporal range of between 2.0 and 1.5 Ma, while *H. habilis* has a range of between 2.3 and around 1.6 Ma (see Wood & Richmond, 2000). It was *H. ergaster*, however, who survived to move out of Africa, as shown by their presence in Eurasia (Georgia) by at least 1.6 Ma (though, as we will discuss later, these Georgian hominins have recently been allocated to a new species of *Homo*), and the Asian species *H. erectus*, who likely split from Eurasian populations of *H. ergaster* at around the same time. Whether the disappearance of *Homo habilis* from the fossil record was a direct or indirect result of unsuccessful competition between it and *H. ergaster* remains unclear.

While there is still much debate concerning the systematic status of *H. habilis* (some believe it represents a species of *Australopithecus* and call it *A. habilis* [see B.A. Wood & Richmond, 2000]), very few would doubt that the specimens defining *H. ergaster* are indeed representative of our own genus. The most famous and complete specimen of this species is the Nariokotome skeleton from Kenya, dating to around 1.5 mya (see Walker, 1993, 1994; F. Brown & McDougall, 1993). Originally this specimen was allocated to *H. erectus*, though we recognize *H. erectus* as being only an Asian species, and almost all specimens previously allocated as representatives of an African deme of *H. erectus* we allocate to *H. ergaster* (see also Andrews, 1984; B.A. Wood, 1984, 1991; B.A. Wood & Richmond, 2000), the only exception being later immigrants of *H. erectus* back into Africa, such as the Olduvai hominins OH 9 and OH 12 and possibly the 1-million-year-old *Homo* specimens from Danakil (Eritrea) and Bouri (Ethiopia).

Of the specimens allocated to *H. ergaster*, the best preserved and documented are the mandibular-type specimen KNM-ER 992, the mandible KNM-ER 820, the mandible and associated fragments ER 730, the skulls KNM-ER 3733 and KNM-ER 3883, the pathological skeleton ER 1808, and the juvenile skeleton KNM-WT 15000 (see B.A. Wood, 1991; Walker, 1993; B.A. Wood & Richmond, 2000) (Figure 6.2). Compared to its proposed ancestor (*H. habilis*), *H. ergaster* is differentiated by reduced

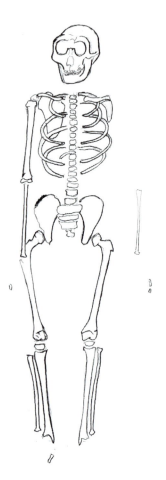

Figure 6.2 ▶   *Homo ergaster* specimen KNM-WT 15000 from Nariokotome, Kenya. This species is the first in the fossil record to show modern human-like body proportions, for even *H. habilis* is defined by the chimpanzee-like proportions of longer forelimbs (arms as well as trunk) but shortened hind limbs (legs).

size of its dental complex relative to body size, increased cranial expansion associated with increasing brain size, reduced supraorbital torus and frontal sulcus development, reduced postorbital constriction (associated with increased cranial base flexure and reduced temporalis muscle development), and the emergence of frontal keeling, which it shares with *H. erectus* from Asia, though *H. ergaster* is not characterized by a number of more specialized features observed in *H. erectus* (Cameron *et al.*, in press; see also, partly, Walker & Leakey, 1993a). As indicated by WT

15000, the postcranial skeleton is very much derived in the modern human condition, different from the primitive australopithecine-like condition of *H. habilis* (Walker & Leakey, 1993b; B.A. Wood & Richmond, 2000). Its rib cage is not funnel shaped, its pelvis is narrower, and in body proportions, especially in its limbs, it is very much like modern humans, with longer lower limbs, indicating a full striding gait. Overall, *H. ergaster* had a locomotor pattern that was very similar to later humans; long-range full terrestrial bipedality (Ruff & Walker, 1993; Walker & Ruff, 1993; B.A. Wood & Collard, 1999; B.A. Wood & Richmond, 2000).

Not only is the Nariokotome specimen of international significance because of its excellent preservation and the number of parts preserved, but the analysis of this skeleton has truly changed our views of early hominin evolution. This specimen is an adolescent boy, between 9 and 11 years old, who obviously would have kept growing in height and stature, because his epiphyses have yet to fuse, clearly indicating that growth would have continued if the youth had not died (Walker & Leakey, 1993b). Estimates of its adult height by Ruff and Walker (1993) place it at around 185 cm (6′1″), with a body weight of around 68 kg (150 lbs), which is truly surprising, not just because of the relatively small stature of earlier hominids and hominins, including *H. habilis*; this adolescent would also be taller than most modern human adults. Indeed, Walker (1993) places this and other *H. ergaster* specimens within the top 17% of modern human populations in terms of height. All evidence suggests that this was no anomaly, but a relatively normal growth pattern for the species (see Ruff & Walker, 1993; Tattersall & Schwartz, 2000).

As suggested by Walker and Ruff (1993), the analysis of body stature and height indicates that *H. ergaster* was the first hominin to adopt a more modern human-like body plan; an adaptation that can be associated with the tropical conditions and most likely a loss of body hair, resulting in increased sweating, which would help keep the body cool (Walker, 1993). The increased body height would enlarge the surface area, which would help further regulate body cooling. The pigment of the skin was probably also dark, to help prevent the formation of skin cancers due to a lack of body hair. Why this adaptation occurred is problematic, because *H. habilis*, with its more primitive australopithecine-like condition, was occupying the same region in time. A major climatic shock occurred around 1.8 Ma, resulting in a further reduction in forests. But the earliest *H. ergaster* specimens predate this by at least 200,000 years, and the speciation event is probably at least 200,000 years earlier than that. Climate and habitat deterioration, however, must have its beginnings in an earlier time, and this phase of climate change can be traced back to at least 2.5 Ma (Potts, 1996).

The development of the derived morphological pattern of *H. ergaster* may be associated with its more efficient type of bipedal locomotion. Carrier (1984) suggests that this type of body plan would result in increased running speeds and endurance, which would force prey to avoid them in an inefficient manner, making *H. ergaster* an effective predator who could run down prey without succumbing to overheating (see also Walker, 1993), though this itself does not explain the adaptive pressures associated with the evolution of its more efficient bipedalism. One must be very careful not to fall into circular and ad hoc argument when trying to explain aspects of physical and cultural evolution. Regardless of the reasons for the change, there was a clear demarcation in body plan with the emergence of *H. ergaster.*

That is not to say that in all aspects *H. ergasters* reflect the modern human condition. Given its mean age of 10 years, we would expect, using modern-day human analogies, for the second and third molars in the Nariokotome youth not to have erupted at this stage. In modern human children, the second molar tends to erupt around age 12 and the third molar around age 18. While the third molar had yet to erupt, the second molar in the Nariokotome boy had erupted, suggesting that it matured earlier than modern humans (H.B. Smith, 1993; see also B. Brown & Walker, 1993; Walker, 1993), so he should be given an age of 12-plus according to the dental evidence. But the overwhelming postcranial evidence (patterns of epiphyseal development — closure of the end plates of the long bones onto the shaft) strongly indicates that the age is between 9 and 10 years. Either way there is a discrepancy between dental and skeletal development in *H. ergaster* that is unique, quite different from the pattern observed in later hominins (see H.B. Smith, 1993; Walker, 1993).

MacLarnon (1993) studied in detail the vertebral canal of the Nariokotome skeleton and noticed that this youth has a relatively narrow thoracic canal (Figure 6.3), suggesting that the thoracic spinal nerves were reduced compared to modern humans. This, according to MacLarnon, gives rise to two possible functional explanations. The first is less muscular movement or control of the trunk, which involves both the intercostal and abdominal muscles, implying a slightly less efficient mode of bipedalism than in modern humans. The second explanation is reduced muscular control associated with breathing, which may reflect poorer control of vocalization — speech. MacLarnon (1993) concluded that the spinal cord of the Nariokotome youth, beginning

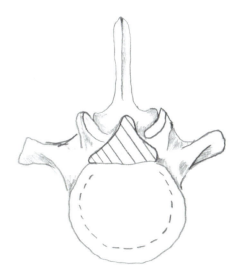

Figure 6.3 ▶ Vertebra from *H. ergaster* specimen KNM-WT 15000 showing the reduced size of the vertebral canal (striped area), indicating that their spinal cord was less developed than in modern humans.
Adapted from MacLarnon (1993).

high up in his neck, displayed fewer nerve fibers than in modern humans; fewer fibers available for innervation of the skeleton and soft tissues means less ability to control muscles associated with vocalization and breathing and less innervation of the involuntary-voluntary action of the diaphragm muscle, which is used by modern humans in the production of speech (see also Walker, 1993). If this is so, it may go some way to refuting the idea that *H. ergaster* could "run down" prey (see earlier), given its reduced control of the diaphragm muscles.

In summary, the emergence of *H. ergaster* in Africa is associated with a major shift in both anatomy and behavior. Its body plan is much closer to that of later hominins than is that of *H. habilis* and earlier hominins/hominids. Though it has its own unique growth pattern, this nonetheless approaches the modern human condition. There is also a shift to increased meat eating and a reliance on technology, including not only an increase in stone tool manufacture, but most likely also "soft tool" manufacture (e.g., water-/food-carrying items made from wood, skins, vegetative material), as suggested by the long-term survival of ER 1808. The story of *H. ergaster*, however, does not stop here; around 1.8 Ma, populations of this species were the first to migrate out of Africa into Europe.

The Original "Out of Africa"

The earliest fossil hominins so far found outside of the African continent are from Dmanisi, in Georgia. Archaeologists digging a medieval site found the first specimen, a mandible, in association with Oldowan-like tools; it was dated to between 1.96 and 1.77 Ma and allocated to *H. erectus* (Gabunia & Vekua, 1995). Recently three skulls have been discovered from the same locality, and most now consider them to represent members of a European deme of *H. ergaster* (Gabunia *et al.*, 2000a; Vekua *et al.*, 2002); revised dates indicate an age of around 1.7 Ma (Gabunia *et al.*, 2000a).

Gabunia *et al.* (2000a) suggest that the Dmanisi hominins share most of their features with *H. ergaster*, including similar supraorbital torus and frontal bone morphology, reduced postorbital constriction, moderate height of the cranial vault, substantial increase in mastoid development, and similar proportions of the facial skeleton. Comparisons of the mandibles, however, including a remarkably large and robust one discovered in 2002, persuaded Gabunia *et al.* (2002) to allocate the Dmanisi sample to a new species, *Homo georgicus* (Figure 6.4). The phylogenetic analyses of Cameron *et al.* (in press) of hominin crania support this, for they do not share a sister-group relationship with either the African *H. ergaster* or Asian *H. erectus*. The Dmanisi hominins have a rather small cranial capacity, with a mean of just 675 ml, which is closer to *H. habilis* at 610 ml than to *H. ergaster* at 851 ml, let alone *H. erectus* at 1151 ml. And, unlike either the African *H. ergaster* or Asian *H. erectus*, they lack frontal keeling.

Figure 6.4 ▶   *Homo georgicus* (or *H. ergaster?*) specimen D2282 from Dmanisi, Georgia.

Vekua *et al.* (2002) suggested that the Dmanisi hominins may share a closer relationship to *H. habilis*. We cannot support this assessment. The recent description of a metatarsal from Dmanisi shows that *Homo georgicus* clearly fits within the more derived *H. ergaster* body plan, not with the australopithecine body plan retained in *H. habilis* (Gabunia *et al.*, 2000b).

Regardless of their eventual taxonomic status, the Dmanisi hominins indicate that early *Homo* had migrated into Eastern Europe by 1.7 Ma. They do not show any features foreshadowing the later hominins that were to occupy western Europe almost 1 million years later — *H. antecessor, H. heidelbergensis, H neanderthalensis*, which have origins from a later migration into Europe from Africa.

This scenario also helps explain a longstanding archaeological "dilemma" concerning the presence of two "geographical toolkits" by 1.5 Ma. As we have discussed, the simple Oldowan chopper toolkit (Mode 1 technology) has a long prehistory, established around 2.5 Ma. But by 1.5 Ma a new tool type appears in Africa — the Acheulean tradition (Mode 2 technology). The new tool type represents a major reconfiguration in tool production — the "biface hand ax culture." While the Acheulean tradition is found throughout Africa, Europe, and the Levantine corridor after 1.5 Ma, with few exceptions it does not appear in eastern Asia, where the Oldowan tradition continues to dominate, though Torre *et al.* (2003) have recently suggested that an incipient form of a Mode 2 technology was being developed in Pakistan.

Given that until recently most accepted that hominins first appeared in Asia around 1.0 Ma, it has been difficult to explain the absence of this culture in Asia, for surely the later immigrants would have brought their Acheulean stone tool tradition with them. Gabunia *et al.* (2000a) suggested that the early appearance of hominins in Dmanisi associated with Oldowan tools indicates that the original migration into Asia occurred before the Acheulean tool tradition was developed. Even were this not the case, the Acheulean does not appear in Africa before the ancestors of *H. erectus* have left, so the dispersal from Africa was driven not by technological innovation but from biological and ecological considerations, including greater reliance on animal protein. A further implication is that there had been only a few early migrations into eastern Asia, and after the initial occupation by these Oldowan tool-manufacturing hominins, the migrations ceased and there was no cultural diffusion between the Acheulean and Oldowan tool cultures (see Larick & Ciochon, 1996; Tattersall & Schwartz, 2000).

There appears to have been a short-term migration into China around 250,000 years ago of an *H. heidelbergensis*-like form, as represented by the Dali and Jinniushan crania and perhaps the Maba cranium (see also Stringer, 1985; Groves, 1989a; Etler & Li, 1994; Tattersall, 1995; Wu & Poirier, 1995; Tattersall & Schwartz, 2000). This may help explain the suggestion that some Acheulean-like artifacts have occasionally been identified from some localities in China and South Korea (Yi & Clarke, 1983; Foley & Lahr, 1997). But *H. erectus* and *H. pekinensis* left no extant descendants; their long-unchallenged occupation of southeastern and eastern Asia ended in extinction.

It is now widely accepted that the earliest appearance of *H. erectus* in Java is around 1.8–1.7 Ma, before 1.5 Ma anyway (Larick & Ciochon, 1996; Swisher *et al.*, 1994, 2000). The earliest appearance of *Homo* (*H. erectus*?) in China is the Gongwangling (Lantian) cranium, dated to around 1.2–1.0 Ma (Wu & Poirier, 1995; Klein, 1999) (Figure 6.5). It has also been suggested that an *H. ergaster*-like hominin is present in Longgupo Cave in southern China (Larick & Ciohon, 1996), though the specimen is simply a fragmentary piece of jaw with a premolar and first molar, and Wu Xinzhi has shown, in a very decisive analysis, that it is

Figure 6.5 ▶   Photograph of original Gongwangling (Lantian) cranium from China.

an orangutan-like ape, not a hominin at all (Wu, 2000). The evidence from Lantian (Wu & Poirier, 1995) and stone tool evidence from Hebei (Jia, 1985) both suggest a hominin entry date into China around 1 Ma.

Larick and Ciochon (1996) suggest that middle Pliocene cooling and drying encouraged a South Asian dispersal by proto-human groups as a band of open tropical and subtropical habitats appeared in present-day Arabia, continuing on through into Southeast Asia. With the lowering of the sea level as a result of increased glacial ice sheets, movement between Africa, Arabia, and Asia was made easier and the route considerably shorter with the emergence of land bridges between these regions. For example, rather than having to move north into the Levant and then traverse a northern route into Asia, hominins could cross from present-day Eritrea into southern Arabia (linked by land bridges) and from there into the Indus Valley and then into Southeast Asia. This route is partially supported by the presence of Oldowan tools in northern Pakistan dated to around 1.9 Ma (Dennell et al., 1994; Larick & Ciochon, 1996).

The route leading to eastern Europe reflects a more diverse paleohabitat (Gabunia et al., 2000b). Faunal and botanical remains from the Dmanisi region are suggestive of a moderately dry climate with a fairly extensive open landscape. The northern expansion of early Homo from East Africa required an adaptation to middle latitudes, where upland habitats where marked by a number of mosaic habitat types. It may be that the populations that gave rise to the Dmanisi hominins moved through the Levantine corridor, into Georgia, and then into eastern Asia from a northern direction. The Dmanisi locality itself appears to have been part of a lake margin, adjacent to a forest-steppe formation, which would have contained rich resources, not only in plants and animals, but also raw material for tool manufacture, as indicated by nearby river gravels (see Gabunia et al., 2000b).

All this indicates that the dispersal from Africa to Asia was a complex one, with numerous routes being used. It also indicates that within Eurasia at least, there was more than one major migratory event. While there was a cultural and biological exchange between Eurasia and Africa, this appears not to have been the case between these regions and Southeast Asia, which seems to have become isolated after its initial colonization.

The early H. erectus specimens from Indonesia are unlikely to represent the ancestral population of this species, but rather reflect a recent arrival from Eurasia (that is, H. erectus is not endemic to eastern Asia) (Figures 6.6–6.8). The presence of an H. erectus-like specimen from Ceprano in present-day

(a)

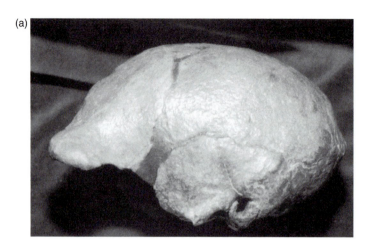

(b)

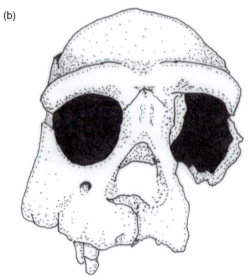

Figure 6.6 ▶   (a) Photograph of original *Homo erectus* specimen Sangiran II from Indonesia. (b) *Homo erectus* Sangiran specimen XVII.

Italy, dated to around 900,000 years ago, indicates a more complex process (see Ascenzi *et al.*, 2000; Clarke, 2000; Schwartz & Tattersall, 2002). It indicates to us that the likely place of origin for this species was somewhere in Eurasia, with a later Eurasian *H. erectus* population moving into western Europe around 1 Ma, before becoming extinct. Indeed, the problematic Olduvai hominins OH 9 from Upper Bed II (dated to 1.2 Ma) and OH 12 and

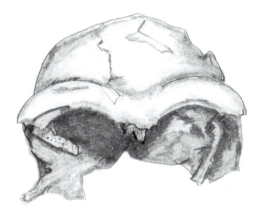

Figure 6.7 ►   The *Homo erectus* calvaria from Ceprano Italy.

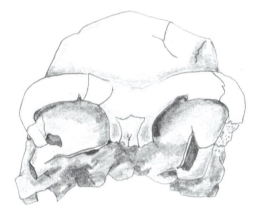

Figure 6.8 ►   The remarkable *Homo erectus*-like specimen OH 9 from Olduvai Gorge, in Tanzania.

other fragmentary specimens from Beds III/IV (dated to around 700,000 years ago), which have been considered by some to represent "classic" specimens of *H. erectus* (Clarke, 2000; see also partly Maier & Nkini, 1984; Rightmire, 1985, 1990; Ascenzi *et al.*, 2000), may represent a southern expansion of this species back into Africa from Eurasia. Indeed, Clarke (2000) suggests that the Ceprano and OH 9 specimens are morphologically identical; and both specimens are morphologically closer to the earlier samples of *H. erectus* than to the later Asian endemic *H. erectus* deme from Ngandong, which are defined by increased neuro-orbital disjunction (see later). While the Ceprano specimen has recently been allocated to its own species, *Homo cepranensis* (Mallegni *et al.*, 2003), we disagree. We maintain that it represents a European deme of *H. erectus*, which was likely associated with dispersals "Into Africa." As demonstrated by Vrba (1985, 1999), faunal migration and territorial expansion during the Plio-/Pleistocene did not only mean an "Out of Africa" exodus, but also an "Into Africa" dispersal: There were several large mammal and rodent migrations back into Africa from Eurasia during this time, and there is no reason why such migrants did not include hominins (see Cameron, in press a). We should thus not be surprised to find later additional nonendemic African specimens of *H. erectus* in East Africa at least.

Further fossil evidence for a possible "Into Africa" migration by Eurasian demes of *H. erectus* have recently been recovered from localities in Eritrea and Ethiopia, both dating to around 1 Ma. These are the *H. erectus*-like fossils from Danakil and Bouri (Abbate *et al.*, 1998; Asfaw *et al.*, 2002). The cranium from Danakil exhibits features said to be distinctive of *H. erectus*, including the greatest cranial breadth across the supramastoids' crests, massive supraorbital torus, and an opisthocranion that coincides with the inion (Abbate *et al.*, 1998). A similar *H. erectus*-like morphological pattern is present in the cranium from Daka (Bouri, Middle Awash), which is believed to be contemporary with the Danakil skull. The Daka hominins are said to be aligned morphologically with OH 9 and to share many derived characters with Asian demes of *H. erectus* (Asfaw *et al.*, 2002).

With the arrival of *H. erectus* in Asia, we can see a pattern of evolutionary stasis. Asian hominins are marked by few speciation events, compared with the numerous ones that occur in Africa and Eurasia. The only speciation event so far recognized in eastern Asia is the evolution of a more derived species in China, *H. pekinensis* from the famous Zhoukoudian Cave close to Beijing and from a few other sites in China (Figure 6.9). Specimens of *H. pekinensis* are currently dated to around 400,000 years ago, with some specimens possibly dating to as far back as 800,000 years ago (Shen *et al.*, 2002; see also Goldberg *et al.*, 2001). This later Chinese species is likely to have originated from an earlier *H. erectus* deme, either from Eurasia or perhaps from a migrant population from Indonesia. While the African and European species *H. heidelbergensis* is also present in China, as represented by the Dali specimen dated to 200,000 years ago and the Jinniushan specimen dated to around 250,000 years ago (Wu & Poirier, 1995) (Figure 6.10), these appear to have been part of a short-term

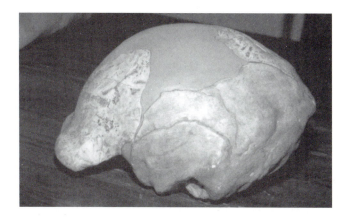

Figure 6.9 ▶ Zhoukoudian specimen ZHK 1966 from China.

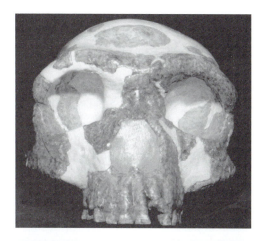

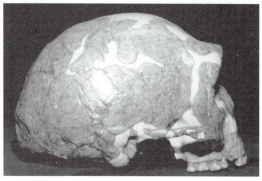

Figure 6.10 ▶ Jinniushan cranium from China.

migratory event from the west and not an Asian speciation event as such. There is also some evidence for Acheulean-like tools in northern China, though they are few and far between (see Foley & Lahr, 1997). This late migration of European hominins into Asia may explain the Narmada cranium (from Central India), which has clear European affinities and is associated with Acheulean tools (see M.A. de Lumley & Sonakia, 1985; Cameron *et al.*, in press) (Figure 6.11); or Narmada may represent another, later, more Neanderthal-like incursion.

Within *H. erectus* we can also see an ongoing trend of evolution from the earliest specimen from Sangiran (1.5 Ma) to the later specimens from Ngandong (40,000 years ago?) (Figure 6.12), a survival in a little-changed form in a way unique from all other hominins. *Homo erectus*, over its long history, underwent increased neuro-orbital disjunction, while African and Eurasian *Homo* developed the opposite condition, increased neuro-orbital convergence. These two morphological trends can be linked to differential patterns of brain morphology, which will affect not only facial hafting to the braincase, but also degrees of postorbital constriction and differential patterns of anterior cranial base angulation (see Weidenreich, 1943; partly D.E. Lieberman, 1995, 2000; Cameron, in press a, b).

While *H. erectus* and *H. pekinensis* both show signs of increased encephalization, their overall pattern of brain development is different from that observed in later African and Eurasian *Homo* in that their frontal lobes are located in a more inferoposterior position so that the frontal is low, with increased supraorbital development and postorbital constriction. Overall the brain case is pushed back from the face (neuro-orbital disjunction). In later western *Homo* the trend is for the forehead to become higher, as the frontal lobes become situated directly above the orbits, which results not only in a reduced supraorbital, but also a high cranium and reduced postorbital constriction. These features are correlated with increased cranial base flexion, associated with the forward and superior migration of the frontal lobes (neuro-orbital convergence). The impact of increased cranial base flexion, or extension, has significant impacts on facial morphology, especially degrees of prognathism (D.E. Lieberman, 1995, 2000; Cameron, in press a, b).

Also associated with this are a number of other features including a significant reduction in mastoid size. This indicates a differential pattern of head and neck musculature, as the mastoid is the major attachment site for the sternocleidomastoideus muscle, a major neck muscle, which controls head rotation amongst other things. *Homo erectus* and *H. pekinensis* are also defined by the development of an angular torus, which suggests

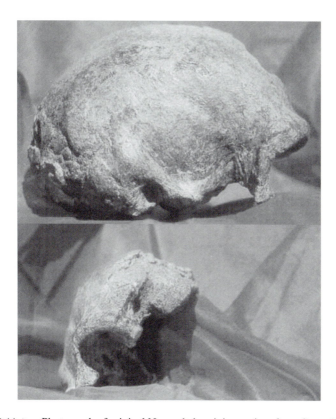

Figure 6.11 ► Photograph of original Narmada hominin cranium from Central India. Kindly supplied by Dr. Rajeev Patnaik, Punjab University.

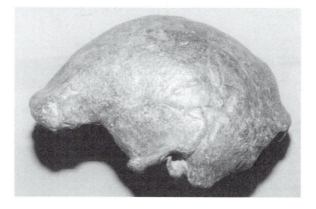

Figure 6.12 ► Ngandong (Solo) V specimen from Indonesia.

increased temporalis muscle size/mass. The temporalis muscle is a major muscle of mastication. There is further differentiation of the neck and masticatory musculature between *H. erectus* and *H. pekinensis*: the Chinese species has increased postorbital constriction (increased temporalis development?) and the connection of the supramastoid and mastoid crests, suggesting increased development of associated muscles, at the expense of the sternocleidomastoideus, or a repositioning of this muscle. One surprising difference between these two Asian species is that while *H. pekinensis* has increased postorbital constriction, suggesting increased anterior cranial base extension, its supraorbital torus is less developed than in *H. erectus* (Cameron *et al.*, in press).

Two major patterns of speciation are evident in the emergence of early *Homo* upon leaving Africa. In Eurasia there is a pattern of rapid cladogenesis, with the rise and fall of numerous hominin species. After the initial colonization of Asia, however, the pattern is one of anagenesis, with *H. erectus* slowly adapting to existing conditions, and only one known endemic speciation event, with the emergence of *H. pekinensis* in China from an earlier *H. erectus* population.

Speciation in Africa can be attributed to increasing climatic and geological upheavals, which would result in the isolation of numerous human and other animal groups (Kingdon, 2003). By 1.8 Ma the formation of large glacial ice sheets had reached its maximum in the northern hemisphere, which resulted in cooler and drier climates, associated with oscillations between more forested and more open habitats. It is also at this time that we see in East Africa renewed activity in the formation of the Rift Valley system, with valley floor spreading associated with high levels of tectonic disruption and volcanic activity. The Rift Valley system would block drainage systems, forming great scarps, and with volcanic eruptions spewing out lava and ash. The previous homogeneous habitat was split into a number of mosaic and restricted habitat zones (see Foley, 1987; Potts, 1996). Within these habitat zones numerous proto-human populations (and other fauna) adapted to their differing ecological settings; in many cases, given the differing associated resources, many groups may have been isolated not so much as a result of physical barriers, but in terms of preferred environments, such as forest as opposed to savanna (Kingdon, 2003).

Over time, geographic isolation and/or isolation based on habitat preference would propel populations down many differing evolutionary pathways. In some cases, however, populations from differing regions might
evolve in parallel as they adapted to the same conditions in a similar
anatomical and behavioral way. In some cases, the result would be extinction of a hominin group, in other cases the absorption of one group into
another; in yet other cases the result would be speciation.

The human paleontological record is making it increasingly clear that,
rather than a simplistic model of just one or two hominin species being
present at any one time, numerous species, some sympatric, emerged in
Africa during the Plio-/Pleistocene, most often followed by the extinction
of one or more of them. This should be no surprise if we view hominins as
just like any other large mammal. For at this same time, we witness in the
fossil record a diversity of experiments within other mammal groups, with
rapid speciation events and extinctions — hominins were just joining in
on the act (Vrba, 1985, 1999; Foley, 1987; Potts, 1996). The same pattern
of rapid and ongoing speciation is also observed in the earlier Miocene
hominids of Africa and Eurasia, for they also had climatic and habitat hur-
dles to jump, which resulted in considerable hominid species diversity
(Cameron, in press a; see also Harrison, 2002; Begun, 2002; Kelley, 2002;
S. Ward & Duren, 2002).

The territorial expansion into Eurasia would also fit this model; as we
have already discussed, the migration to the Levantine corridor and southeastern Europe (Dmanisi) would meet differing mosaic habitats. It is within
Eurasia that we think the origin of *H. erectus* occurs, with its migration into
East and Southeast Asia proper. Here, there is evidence for tropical forests
as well as some grasslands. But what appears to be different is that there is
no major disruption in these habitat zones through time, and populations of
H. erectus were not confronted with major periods of climatic or habitat disruption, unlike the conditions in Africa and Europe, which were marked by
continuous periods of such disruption. The only major difference in habitat
in Asia documented to date is that at Zhoukoutian, which is associated with
a more open woodland habitat, which may go some way to explaining the
emergence of *H. pekinensis* as well as the short-lived migration event of a
H. heidelbergensis population into this region around 200,000 years ago.

Finally, it is interesting to speculate that the survival of *H. erectus* in
Southeast Asia can be attributed to their isolation from other hominin
species. Accepting that populations of *H. erectus* survived at Ngandong as

recently as just 40,000 years ago (Swisher *et al.*, 2000), we have some insight into their extinction. This last appearance datum for *H. erectus* correlates with the slightly earlier appearance of modern humans in this region. There is a similar correlation in Europe, where the endemic populations of *H. neanderthalensis* become extinct with the appearance of modern humans in Europe at around the same time, though at this point let's not get ahead of ourselves.

CHAPTER 7

Human Evolution in the
Middle Pleistocene

Normally the death of an individual was not of major concern, at least in terms of disposing of the body. The group would merely move on, leaving the body where it lay. Now, however, it was the depths of an ice age winter and the body of the old female could not remain within the cave. This was not the first time that this group had come across such a problem. This was, after all, their winter cave, and it was also the most frequent time when members of the clan succumbed to the elements. Per usual practice, the males started to drag the body to the back of the cave, until they reached the deep shaft. They then proceeded to push the body over the lip of the opening. The body fell into the darkness below, and seconds later a thud could be heard as it hit the bottom of the sinkhole. But by then the men had already headed back to the main chamber of the cave to rejoin the group.

In the early 1980s a major discovery was made in the Acueva Mayor Cueva del Silo complex in northern Spain. At the bottom of a deep vertical shaft was a large collection of hominin bones representing at least 30 individuals, adult males and females as well as children. Also found were the remains of a number of cave bears (Arsuaga *et al.*, 1997; Arsuaga, 2002; Agusti & Antón, 2002). The fossil hominins are now usually allocated to the species *H. heidelbergensis*, a species that appears to have its origins in Africa. But as we will see, they may stand at the base of a different species that dominated Europe for 200,000 years.

Most of the fossil materials to be discussed in this chapter have until recently been referred to as "archaic *Homo sapiens*." This classification was used as a type of "dustbin" category for middle Pleistocene specimens that could not easily be allocated to *H. erectus, H. neanderthalensis* or *H. sapiens*. The problem was that they display a mosaic of features such as could also be seen in these other, morphologically and geographically diverse species. With the introduction of phylogenetic systematics and the recognition of primitive and derived features, we can now begin to identify a number of lineages within this group, each defined by derived features (see Chapter 5). As such, we now recognize three species within what was formerly termed "archaic *H. sapiens*." Most recently the species *Homo antecessor* has been named for one group from Spain (Bermúdez de Castro *et al.*, 1997), while the species *H. heidelbergensis* has been resurrected for a number of specimens from Africa and Europe (see Groves, 1989a). Here, we also recognize, at least for the time being, a third species, *H. steinheimensis*, which was endemic to Europe, though we have severe doubts as to whether it is anything more than a name of convenience (see partly Howell, 1998; Tattersall & Schwartz, 2000). The phylogenetic relationship of these three species to the later *H. sapiens* (the modern human lineage) and *H. neanderthalensis* (the Neanderthal lineage) is the subject of this chapter.

Homo antecessor

The earliest non-*ergaster/georgicus/erectus* hominins in Europe are the specimens from Gran Dolina Cave, Level TD-6, in the Atapuerca Hills of northern Spain, allocated to their own species, *Homo antecessor* (Bermúdez de Castro *et al.*, 1997) and dating to around 780,000 years ago (Parés & Pérez-González, 1995; Falguères, 1999) (Figure 7.1). Bermúdez de Castro *et al.* (1997) argue that the new species evolved from *H. ergaster* and represents the likely stem species that gave rise to the Neanderthal lineage and to modern humans. Given the fragmentary nature of the specimens from Gran Dolina, however, their specific distinction from the later European and African *H. heidelbergensis* remains debatable. The supposedly distinctive features of this *H. antecessor*, including increased cranial capacity, reduced facial prognathism, well-developed canine fossa, and angled inferior cheekbone, are also characteristic of

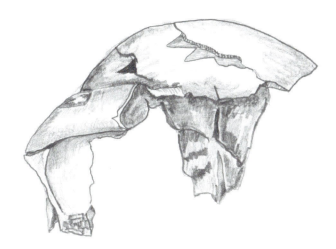

Figure 7.1 ▶ *Homo antecessor* cranial fragment (ATD6-15) from Atapuerca: Gran Dolina.

H. heidelbergensis (F.H. Smith, 2002). Primitive features that *H. antecessor* shares with *H. ergaster* and *H. erectus* include the presence of male lower canine and premolar cingula, and asymmetry in the crowns of the lower third premolars (B.A. Wood & Richmond, 2000).

Postcranially, *H. antecessor* was relatively gracile, most similar to *H. ergaster* and modern *H. sapiens*, contrasting with most of its successors, including *H. heidelbergensis* and *H. neanderthalensis*. As such, it was evidently not adapted to cold conditions, but like *H. ergaster* and *H. georgicus,* it was adapted to a relatively warm, temperate climate (see partly B.A. Wood & Richmond, 2000).

The allocation of these specimens to either *H. antecessor* or *H. heidelbergensis* is not merely an academic exercise; it has major implications for the way we interpret later human evolution. For example, if the Gran Dolina hominins represent *H. heidelbergensis*, then this species may have origins in Europe as opposed to Africa, or it might represent a hypothetical earlier African deme of *H. heidelbergensis* that had migrated into Europe. Most people currently recognize as the earliest representative of *H. heidelbergensis* the large-brained Bodo skull from Ethiopia, which has a cranial capacity of around 1300 cc (Rightmire, 1996) and is currently dated to 600,000 years ago (Clark *et al.*, 1994), which is almost 200,000 years later than the hominins from Gran Dolina. If the Gran Dolina material does, however, represent a new species, did *H. heidelbergensis* evolve from this European population, thus implying a migration back into Africa?

The material culture associated with these early European hominins has been assigned to the Mode 1 technology observed in Africa (Carbonell *et al.*, 1995), though this type of technology has also been associated with *H. erectus* in Asia (Foley & Lahr, 1997) and *H. ergaster* (*H. georgicus*?) in Georgia (Gabunia *et al.*, 2000a). This system of lithic classification was originally defined by Clark (1977) and is based on a pebble tool industry (Oldowan) commonly associated with simple flakes struck off pebbles resulting in choppers and flakes (also see Foley & Lahr, 1997). Thus these hominins are not associated with the derived Mode 2 technology that characterizes the later African demes of *H. ergaster*, not to mention *H. heidelbergensis*.

While it is currently not possible to ascertain the phylogenetic significance of *H. antecessor*, we can see that the pattern of human evolution from their first appearance in Europe is not as straightforward as previously thought, especially if we consider the implied *H. erectus* migration into Europe, as suggested by the Italian Ceprano *H. erectus* specimen (see previous chapter) at around the same time as the Gran Dolina hominins in Spain make their appearance, not to mention the earlier migration into far southeastern Europe of *H. ergaster*. While these species are anatomically distinctive, they all maintained an earlier inherited Mode 1 technology from their likely African ancestor(s).

The Rise of *Homo heidelbergensis*

The earliest widely accepted appearance of *H. heidelbergensis* is around 600,000 years ago, as represented by the Bodo skull from Ethiopia (Figure 7.2). There appears to be a continuum of this African deme, to at least 260,000 years ago (Grün *et al.*, 1996), when the South African Florisbad specimen begins to show real changes, or perhaps to even later if one accepts that the Jebel Irhoud specimens from Morocco represent members of this species (Howell, 1998; F.H. Smith, 2002) — which we decidedly do not! (to be discussed in Chapter 9). The best-preserved African specimen of this species is the Kabwe 1 cranium (Broken Hill) from Zambia (Figure 7.3). While an absolute age for this cranium cannot be ascertained, the best evidence at present suggests it might date to the

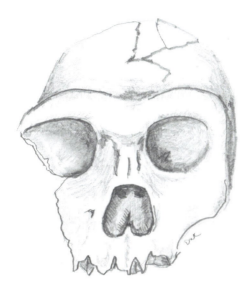

Figure 7.2 ▶   The African *Homo heidelbergensis* specimen from Bodo, Middle Awash, Ethiopia.

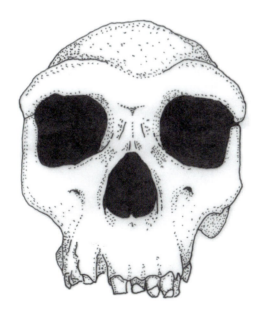

Figure 7.3 ▶   The African *Homo heidelbergensis* specimen Kabwe 1 (Broken Hill) from Zambia.

later middle Pleistocene, perhaps around 300,000 years ago (Rightmire, 1990; Stringer, 2000a). Kabwe has a similar cranial capacity to Bodo at around 1285 cc (Holloway, 1981; Aiello & Dean, 1990), and they share quite a number of other similarities. Kabwe 2 is a maxillary fragment and from the same site come an innominate and limb bones. Other well-preserved African examples of *H. heidelbergensis* include the Saldanha calvaria, from the Western Cape; the rather gracile Ndutu cranium from Tanzania, about 400,000 years old; three mandibles and a parietal from Tighenif (Ternifine), Algeria, which may be as much as 700,000 years old; mandibles from the Cave of Hearths at Makapansgat (South Africa), Kapthurin (Kenya), and Rabat and Casablanca (Morocco); and a gracile facial skeleton from the Thomas Cave, near Casablanca. These extend our knowledge of the species' range of variation, and show that not all specimens are as robust as Kabwe 1 or Bodo — presumably a marked degree of sexual dimorphism is involved (Schwartz & Tattersall, 2003).

H. heidelbergensis is defined by a relatively large brain compared to its predecessors, a receding forehead that is placed posterior to the well-developed supraorbital torus, reduced facial prognathism, a large face puffed out as a result of its inflated frontal and maxillary sinus system, a robust mandible but no retromolar space (as seen in Neanderthals) with relatively small teeth, and no chin. In terms of its preserved postcranial anatomy, it is nearly completely modern in appearance, although more robustly built (Stringer, 2000a; Tattersall & Schwartz, 2000). Almost but not quite like *H. ergaster*, it had a robust pelvis with an accentuated iliac pillar (acetabulo-cristal buttress), and a femur with a thick cortex, its shaft narrowing to its minimum breadth just above the condyles.

The earliest undoubted appearance of *H. heidelbergensis* in Europe is the type specimen, a mandible from Mauer, near Heidelberg, Germany. This key specimen, found in 1907, is thought to date to around 500,000 years ago (Stringer, 2000a). While the mandibular body is very robust, like that observed in earlier species of *Homo*, its molar teeth are relatively small, approaching that of later hominins. A near contemporary with Mauer is the robust tibial fragment from Boxgrove in England (Pitts & Roberts, 1998). It may be significant that both these two specimens, the earliest of the species in Europe, date to 100,000 years after the earliest appearance of this species in Africa (if Bodo is the earliest, but to 200,000 years if the Tighenif fossils are really 700,000 years old).

Following on from these specimens are those from Arago in southern France, dated to around 450,000 years ago (Yokoyama & Nguyen, 1981), and Petralona in northern Greece, dated to perhaps 250,000–150,000 years ago (Grün, 1996). The fragmentary remains from Bilzingsleben in Germany, dated to around 400,000–300,000 years ago, and Vértesszöllös in Hungary, dated to around 210,000–185,000 years ago, are often said to resemble Petralona. If so, there is no bar to considering them representative of *H. heidelbergensis* (Stringer *et al.*, 1979; Stringer, 2000a; Schwartz & Tattersall, 2002). As we suggested in the previous chapter, there also appears to have been a migration of this species into Asia, as represented by the Dali and Jinniushan specimens.

The specimens from La Caune de Arago, Tautavel, in southern France, include isolated teeth and cranial, mandibular, as well as some fragmentary postcranial specimens (Schwartz & Tattersall, 2002). In cranial capacity, Arago 21 (Figure 7.4) is slightly smaller at 1150 cc than the earlier African specimens. Its craniofacial anatomy is very similar to the condition already described for the African deme of this species (see earlier), though the frontal sinus system is less developed and the supraorbital features are a separate, distinct torus above each orbit, as opposed to a shelflike structure, which is similar to the condition that was later to develop in Neanderthals (Stringer, 2000a). The two mandibles, like those of Tighenif in Algeria, span quite a range of sizes, supporting the idea that

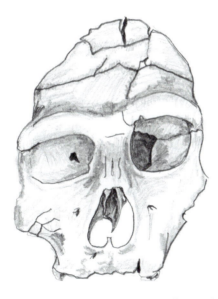

Figure 7.4 ▶   The badly distorted Arago 21 *Homo heidelbergensis* specimen from southern France.

the species was quite sexually dimorphic. The original describers of th
Arago specimens referred to them as alternatively "anteneanderthals" o
an advanced form of *H. erectus*, and they argued that they ultimatel
resulted in the evolution of the neanderthal lineage (H. de Lumley & d
Lumley, 1971). Most today, however, accept these specimens as represent
ing a European population of *H. heidelbergensis* (Hublin, 1985; Stringer
1985; Tattersall, 1986, 1995; Rightmire, 1990; Stringer & Gamble, 1993
Howell, 1998; Schwartz & Tattersall, 2002).

Like the Arago specimens, Petralona falls well within the range of varia
tion of *H. heidelbergensis* (Figure 7.5). In terms of its craniofacia
anatomy, it is very similar to those specimens already discussed wit
derived features, including increased cranial capacity of around 1230 c
(Stringer, 1984; Aiello & Dean, 1990), no angular torus, a receding frontal
reduced mid-facial prognathism, and inflated facial features (Tattersall
1995). Like Arago, the supraorbital in Petralona is not shelflike; rather, i
has separate tori above each orbit. Other features are clearly primitive
including its thick, angular occipital bone with its centrally strong transverse torus (Stringer & Gamble, 1993). Petralona and Vérteszöllös seem t
represent late surviving members of this species. We will argue later that ar

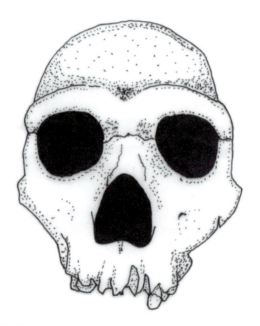

Figure 7.5 ▶   The European Petralona *Homo heidelbergensis* from Greece.

African deme of this species gave rise to early *H. sapiens*, while a European deme of this same species, which we allocate to the "Steinheim group," eventually gave rise to *H. neanderthalensis*.

If this is the correct phylogeny, some may ask what use there is in having a separate species for their common ancestor. Why not put the European fossils into *H. neanderthalensis* and the African ones into *H. sapiens*? The answer is simply a matter of how to divide a probable ancestor–descendant lineage: The most objective way is not to make assumptions about who gave rise to what, but to recognize a separate species at the point where there are fossils that begin to exhibit, in however incipient a form, the uniquely derived character states of the descendant species — the point, in other words, where the phylogenetic connection can for the first time be clearly detected. This is the essence of the Composite Species Concept (Kornet, 1993). We tend to agree with Stringer (1983, 1985) that specimens like Petralona, Arago, Mauer, Kabwe, and Bodo show no specialized Neanderthal features, nor do Kabwe, Ndutu, and Bodo show specialized *sapien* features; so we retain them in a generalized stem species.

In Africa, of course, *H. heidelbergensis* lived in a tropical climate. While some of the European specimens are associated with a cooler-climate fauna, none, as yet, seems to have inhabited a periglacial environment, unlike the Neanderthals. It is quite a thought that each time the ice sheets swept down into Europe, the *H. heidelbergensis* population there collapsed, and the species became confined to Africa; and each time the climate warmed up again, Europe was repopulated from Africa. Only later did a new species develop that, for the very first time, was capable of surviving the extreme cold. When exactly this happened — when the proto-Neanderthals finally emerged — is something we will discuss later.

Homo heidelbergensis is associated with a Mode 2 technology, which includes a large biface industry, including large flakes or cores shaped on both sides to produce "hand axes" (often referred to as Acheulean). This tradition has its origins with the later populations of *H. ergaster* in Africa around 1.5 million years ago. The earlier populations of *H. ergaster*, both the early ones of Koobi Fora and the contemporary deme in Georgia (if they can be considered representative of this species), maintained the more primitive Mode 1 technology (Oldowan). In terms of material culture, then, there is a clear continuum between the African ancestor, as defined by the late *H. ergaster*, and its descendant *H. heidelbergensis*, which occupied Africa and Europe around 1 million years later (see Foley & Lahr, 1997).

There is evidence from Terra Amata, located within the city limits of Nice (dated to around 400,000 years ago), that *H. heidelbergensis* at least had the capacity to construct shelters. This site has long been considered as a seasonal occupation site, where humans built free-standing shelters, though there is still considerable debate regarding their status as real shelters (see Gamble, 1986, 1999). These "huts" (if they are indeed such) were very large, measuring around 8 × 4 meters, and the construction is supposed to have been saplings embedded into the ground and bent toward the middle, where they were tied off. Inside the shelters is evidence of hearths, with considerable numbers of broken animal bones, showing that a diverse type of animal prey was consumed (Gamble, 1999; Tattersall, 1999).

We briefly mentioned in the previous chapter that some specimens from the middle Pleistocene of China reflect a morphology similar to that of *H. heidelbergensis* (see Etler & Li, 1994; Stringer & McKie, 1996; B.A. Wood & Richmond, 2000; Stringer, 2000a). This is particularly true of the two southern Chinese Yunxian crania (dated to perhaps 350,000 years ago but perhaps a lot earlier), which are said to share a close phylogenetic relationship with the African specimens Kabwe, Florisbad, and Irhoud. But Groves has seen the Dali and Jinniushan crania (both dated to around 250,000), and they do certainly closely resemble their European and African contemporaries. When did *H. heidelbergensis* arrive in China, and how long did the species persist there? The date of Yunxian is uncertain, and the middle Pleistocene record of China is even spottier than those of Africa and Europe, so at present we cannot say. We can only observe that, at least at the time of Bodo (the earliest acceptable *H. heidelbergensis* in Africa or anywhere), *Homo pekinensis* was still in occupation of at least the northern part of China.

Mode 2 technologies are well known from India, but eastern Asia is usually considered to have a long, unbroken history of simple Mode 1 technology (Cameron *et al.*, in press). In some areas an Asian biface technology has been recognized (see Schick & Toth, 1993; Foley & Lahr, 1997), though the relationship of these biface technologies to those of Europe and Africa remains obscure, and further description and stratigraphic confirmation of these industries is required. It is even possible that these industries are separate, independent technologies. A number of hypotheses have been put forward to explain the absence of a Mode 2 technology in Asia. The first is that after the initial colonization of Asia by *H. erectus* there was no significant contact between Asia and the rest of the Old World. That is, the original colonizers left Africa/Europe before the development of a Mode 2 technology. The two technologies, that is, vere species specific. Another suggestion is that when human populations rrived in eastern Asia, they became more reliant on the more readily available and accessible forests of bamboo; conversely, there may not have been a readily available and suitable raw lithic material to continue with a Mode 2 technology. Thus the stone tools of Mode 2 technology were subtituted by tools of bamboo. And maybe, if the middle Pleistocene human populations had at most a limited capacity for language, passing the required skills on to the next generation was based on continuous practice and practical demonstration; and failure to keep up the mode, whether because of a lack of suitable raw material or for some other reason, quickly ed to a breakdown in the continuum of Mode 2 technology (see Schick & Toth, 1993).

The Earliest Members of the Neanderthal Lineage?

Another group of mainly European specimens, which some believe represent the basal stock from which the Neanderthal ultimately evolved, have been allocated to a species *H. steinheimensis*; for the moment we (partly following Howell, 1998) will use this name. This species is best represented by its type specimen, the Steinheim skull near Stuttgart in Germany (Figure 7.6), the Sima de los Huesos specimens from Atapuerca in northern Spain, the Swanscombe cranium from England, and the Narmada cranium from India, most of them considered to date between 300,000 and 230,000 years ago (Howell, 1998; Schwartz and Tattersall, 2002; Klein, 1999; Cameron *et al.*, in press), although Stringer and Hublin have recently made a strong case that Swanscombe is actually much older, around 400,000 years old, as may be Atapuerca. This species is contemporary in time and space with the later surviving African and southern European demes of *H. heidelbergensis*. As we will see in the next chapter, the molecular evidence from the Neanderthals suggests that *H. steinheimensis* is close to the probable time of divergence of the Neanderthal lineage from those that ultimately evolved into the earliest modern humans — around 500,000–700,000 years ago.

In its morphology, *H. steinheimensis* clearly displays incipient Neanderthal features, including, to varying degrees, the configuration of the supraorbital tori, the large size of the nasal opening, the medial projection from the side walls of the nasal cavity, developed occipital torus and suprainiac depression, and a long cranium with a slightly more elevated frontal bone

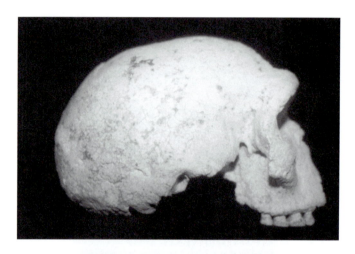

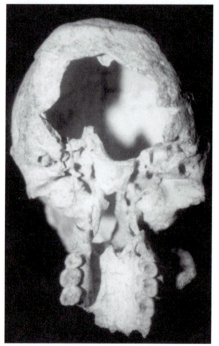

Figure 7.6 ▶   European *Homo "steinheimensis"* specimen from Steinheim, Germany.

associated with neuro-orbital convergence). While some of these features may be seen in specimens of *H. heidelbergensis* (at least in their incipient form), it is this combination that is important; it indicates that *H. steinheimensis* stands at the root of the Neanderthal lineage. Indeed, proposed similarities between some *H. heidelbergensis* specimens (especially Petralona) and the later "classic Neanderthals" can at least be partly explained by their common expansion of their frontal and maxillary sinus systems; the phylogenetic significance of this remains obscure, and it is seen in some African specimens too (such as Bodo). The shared presence of these features in *H. heidelbergensis*, *H. steinheimensis* and *H. neanderthalensis* may represent either a functional and/or developmental convergence or the persistence of a primitive feature that was variably developed in *H. heidelbergensis* but became emphasized in *H. steinheimensis* and even more so in *H. neanderthalensis*. The Atapuerca evidence shows for the first time the robust postcrania that are indicative of an adaptation to a cold climate.

In some aspects of their cranial morphology, however, *H. steinheimensis* is still primitive, and they do not display the Neanderthal "en bombe" shape (Hublin, 1998). Hublin (1998) and Stringer (1998) argued that many of the Neanderthal features probably began as uncommon polymorphic variants, which gradually increased in frequency through time as a result of selection pressures. This ultimately resulted in the "classic" Neanderthal morphological condition, but they are already incipiently developed in *H. steinheimensis*.

The complete but distorted Steinheim cranium was found in 1933 within a gravel pit. The estimated cranial capacity is 1100 cc (Aiello & Dean, 1990), and the specimen is currently dated to around 225,000 years ago (see Schwartz & Tattersall, 2002). Its phylogenetic relationship to *H. erectus*, *H. neanderthalensis*, and *H. sapiens* has long been debated, though the recognition that it has a number of derived Neanderthal-like features suggests that it represents a likely population from which the Neanderthal lineage evolved. Of particular significance is the medial projection within the lateral nasal aperture, which Schwartz and Tattersall (2002) and Tattersall and Schwartz (2000) consider to be a derived feature of the Neanderthal lineage. It also has a number of other incipient Neanderthal-like features, including its similarity in supraorbital form, a midface that is relatively prognathic (forward projection of the nasal aperture), and an occipital torus and suprainiac depression, though only weakly developed, as well as a weakly developed mastoid process (Stringer & Gamble, 1993; Wolpoff, 1999).

The Sima de los Huesos (Atapuerca) fossils have a cranial capacity ranging from 1125 to 1390 cc (Stringer, 2000a) and are currently dated to 300,000 or more years ago (Arsuaga *et al.*, 1997; Schwartz & Tattersall, 2002). In terms of their projected brain mass relative to body mass, they are around 3.1–3.8 times larger than expected for a mammal of its weight. As we will see in the next chapter, according to body weight, Neanderthals have a brain that is around 4.8 times greater than predicted, while *H. sapiens* have a brain weight 5.3 times greater than predicted (Arsuaga, 2002). They show a number of derived Neanderthal features, including the smoothly rolled, double-arched supraorbital torus, large and prognathic nasal aperture, and the horizontal suture over the mastoid region, as well as increased development of the frontal and maxillary sinus system (Tattersall & Schwartz, 2000). The presence and development of the medial internal nasal margin is variable in the Atapuerca specimens (Figure 7.7). For example, this feature is absent in crania 5 and 6 while being slightly developed in AT-638, AT-772, and AT-1665 and more markedly developed in AT-1100, AT-1111, AT-1197, AT-1198, and AT-1666 (Arsuaga *et al.*, 1997). In addition, Arsuaga (2002) has estimated, according to correlations of pelvic breadth and overall height (determined from preserved femora), that the body weight of these hominins would have been even greater than the estimates for the later Neanderthals;

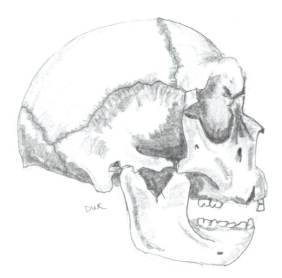

Figure 7.7 ▶ *Homo "steinheimensis"* specimen V from Atapuerca: Sima de los Huesos, Spain.

that is, some individual Sima de los Huesos hominins would have weighed between 198 and perhaps 220 pounds! The combination of these and a number of other features has suggested to Tattersall and Schwartz (2000) that the Atapuerca specimens may in the future need to be allocated to their own species. But we cannot see any real evidence for this, and we accept them as representing *H. steinheimensis*, a species whose validity is itself questionable (see later).

The original Swanscombe occipital was found in 1935, the left parietal was found in 1936, and the right parietal was not found until 1955. But all three specimens can be fitted together and clearly belong to the same individual. Swanscombe has traditionally been viewed as sharing a close relationship with Steinheim (Morant, 1938; Howell, 1960; Stringer *et al.*, 1984; Wolpoff, 1996). Its estimated cranial capacity is 1325 cc (Aiello & Dean, 1990). The suggestion by Santa Luca (1978) and more recently by Stringer and Gamble (1993) and Stringer (2000a) that Swanscombe represents a likely forerunner to the Neanderthals has been endorsed by Schwartz and Tattersall (2002), who have allocated it to *H. steinheimensis*. This allocation is further corroborated by its gracile and double-arched occipital torus, the presence immediately above the torus of a suprainiac fossa, and, finally, the suggestion of a developed juxtamastoid eminence at the occipital margins (Stringer, 2000a). If the analysis of Stringer and Hublin (1999) is cogent and it is 400,000 years old, then it is clearly the oldest example of its species.

Earlier we asked whether there is value in recognizing a species *Homo heidelbergensis*, and we indicated that indeed there is. The question of whether *H. steinheimensis* "exists," in any meaningful sense, is more equivocal. Without a doubt, we are dealing with the stem group of the Neanderthals: All the basic Neanderthal derived traits are present, but in an incipient form. The specimens we have been describing occupy a position intermediate between *H. heidelbergensis*, which has none of these derived states, and true (classic) Neanderthals, in which these states are fully expressed. Under Hublin's "accretion model," the specimens would be placed along an unbroken continuum; in the Composite Species concept (Kornet, 1993), they would be classified as *Homo neanderthalensis*. The latter seems the more sensible option: What we have been describing are early, primitive Neanderthals.

The "Steinheim group," as we will call them, are known mostly from Europe, but there is a surprising exception. The Narmada cranium from central India has an estimated cranial capacity of 1290 cc (Kennedy *et al.*, 1991) and is, according to Cameron *et al.* (in press), 230,000 years old. According to Cameron *et al.* (in press), Narmada shares a numbers of unique features with the Steinheim skull and to a lesser degree with the specimens from Sima de los Huesos. These similarities include a long but more elevated cranium associated with neuro-orbital convergence, similar supraorbital torus development and form, and a similar degree of postorbital constriction. Groves (1989a) saw it as an Indian representative of *Homo heidelbergensis*, relict of the eastward extension of this species that ended up in China; but the presence of definite (proto-) Neanderthal features places that hypothesis in doubt. Indeed, a recent parsimony analysis by Cameron *et al.* (in press) confirms the sister-group relationship between the Narmada and Steinheim specimens. This indicates that the Narmada hominin represents a member of the "Steinheim group" that spread its range to the subcontinent around 250,000 years ago (if not considerably earlier). It may be that the Indian population died out, or maybe it, too, extended its range into China because, as we will show later on, there is quite a definite Classic Neanderthal in the late middle Pleistocene in the far south of China.

While we have no material culture associated with the Sima de los Huesos or Steinheim hominins, the artifacts associated with the Swanscombe cranium represent a classic Mode 2 technology (Acheulean), most similar to that associated with the Boxgrove hominin (Stringer & Gamble, 1993; Gamble, 1999). The near-contemporary Narmada cranium is also associated with an industry consisting of hand axes and numerous cleavers (M.A. de Lumley & Sonakia, 1985; Cameron *et al.*, in press). Thus the material culture of the "Steinheim group," as defined by Swanscombe and Narmada, is associated with a primitive Mode 2 technology.

We have seen that the speciation event that marked the emergence of early *Homo* in Africa is associated with the adaptation of increased meat eating and with corresponding shifts in behavior that are required to obtain the resources from the sort of walking larder that runs away from you — increased territorial range. Perhaps the development of the biface industry (Mode 2 technology) helped to obtain and process these resources. This industry is maintained in Africa for over a million years. It begins with later stages of *H. ergaster* in Africa and persists into *H. heidelbergensis* in both Africa and Europe. The emergent earliest hominins in Europe, whether *H. ergaster* (*H. georgicus*?), *H. erectus*, or *H. antecessor*, however, still used a pebble tool industry (Mode 1 technology), and it is not until the arrival of *H. heidelbergensis* in Europe around 500,000 years ago that a Mode 2 technology appears (Foley & Lahr, 1997). The use of a

Mode 1 toolkit by *H. erectus* has already been discussed. Given the likely westward migration of *H. erectus* into Europe, the late survival of Mode 1 can be explained (see also partly Tattersall & Schwartz, 2000). The maintenance of a Mode 1 technology by *H. antecessor* suggests very strongly an older hominin penetration into Europe from an ancestral group that did not have access to a Mode 2 technology. If we are to believe in a one-to-one association of species and technology mode, then we have argued ourselves into the position that *H. antecessor* was not ancestral to later European populations, but was replaced by *H. heidelbergensis* coming in from Africa. If such associations do not exist, then *H. antecessor* may still have been ancestral to *H. heidelbergensis*, and Mode 2 spread into Europe from Africa without new gene flow.

Mode 2 lasted an amazingly long time without appreciable change. Several people have noticed that hand axes are skillfully made and aesthetically pleasing. Miller (2000) has argued that they were a product of sexual selection — the production of the most elegant examples was an indicator of the producers' intellectual, hence sexual, fitness.

Microscopic analysis of the lithic artefacts from Hoxne in Suffolk (around 400,000 years old) indicate that they were used to process a number of different materials. These all-purpose tools were used to cut meat and work hides, bone, and wood. They were also used to chop up vegetable matter (see Gamble, 1999; Jordan, 1999).

These later middle Pleistocene hominins were not restricted to only using a lithic technology, but clearly also used wood, as evidenced by the Clacton spear found in Essex, which dates to around 450,000 years ago. Whether it is really a spear or a digging stick is still being debated, but either way it is significant because it reminds us that such material, not usually preserved in the archaeological record, should not be discounted — absence of evidence is not evidence of absence. Indeed, additional (presumed) throwing spears that are 2.3 meters in length have been discovered from Schöningen in Germany. There is also evidence of toolkits having been made of stone, including flakes struck on anvils from elephant and rhino long bones, and there is even a bone biface from Castel di Guido in Italy dating to around 450,000 years ago (Gamble, 1999).

While the difference in morphology and between the more "primitive" technology used by the earliest Europeans and the more "advanced" technologies used in Africa can at least be explained by patterns of migration out of (and into) Africa, the later emergence of the "Steinheim group" proto-Neanderthals cannot be associated with any technological innovations: They appear to have used a Mode 2 technology just like *H. heidelbergensis*, and there is no evidence that there was anything different about their behavioral repertoire. Climate change and its associated habitat instability may be solely responsible for the speciation event in the middle Pleistocene of Europe: The more they specialized anatomically, the more they were able to cope with the climatic rigors of ice age Europe and the more of periglacial Europe they could inhabit.

Ever more severe climatic oscillations were starting to dominate Europe, if not Africa. Populations of *H. heidelbergensis* persisted in warmer southern Europe (i.e., Petralona), while contemporary demes of the "Steinheim group" lived in the northern latitudes, within extremely cold conditions. As cold climates spread south, these proto-Neanderthals spread with them: They are found with cold-climate fauna at Atapuerca in northern Spain; when climates ameliorated, *H. heidelbergensis* in its turn was capable of spreading to northern latitudes, such as England (Boxgrove). This reminds us that the distribution of these two species was dictated more by climatic and associated ecological factors than by a simple north/south geographical divide; throughout human evolution our forebears have been, like any other animal, subject to ecological constraints. Our present degree of cultural buffering from environmental extremes is a very recent phenomenon indeed and is still, in fact, far from complete. From time to time, extreme events, such as the prolonged droughts that afflict Australia and West Africa every few decades, or the recurrent monsoon failures of India, or the acid rains and soil salinity that result from botched attempts to further shelter human populations from environmental vicissitudes, remind us how very incomplete is this buffering, even today. Animals we were in the middle Pleistocene, and animals we remain.

Because the African and European *H. heidelbergensis* are similar in all respects, as far as our analysis goes, we should be wary of ascribing large sinus systems and so on to "adaptation to cold conditions." There certainly was no requirement for contemporary African demes to adapt to such conditions. Yet this morphological complex proved an excellent exaptation for those populations that were at the fringes and eventually occupied the northern latitudes of ice age Europe. These northern populations became increasingly isolated physically, culturally, and genetically, eventually giving rise to the proto-Neanderthals. The southern demes, however, continued to exist in the relatively warmer conditions of southern Europe and the temperate to tropical conditions of Africa.

There is some support for such a scheme in the geological and paleon-
ological record, which provides a climatic history of Europe during this
ime (Figure 7.8). For example, while glacial conditions had started to
occur during the Plio-/Pleistocene transition (if not before), a major
change occurred around 750,000 years ago, when the glacial conditions
became longer in duration, while the interglacials were significantly
reduced, and overall climatic fluctuations became more marked. This is
associated with the Brunhes Magnetic Polarity switch to normal. From the

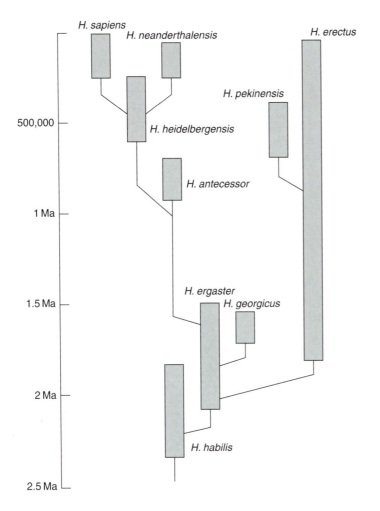

Figure 7.8 ▶ Implied phylogeny of the species we recognize within the genus *Homo*
(see text for details).

Plio-/Pleistocene transition, around 1.8 million years ago, to around 900,000 years ago, a full glacial–interglacial cycle occurred every 40,000 years; between 900,000 and 450,000 years ago, however, it had increased to a 70,000-year cycle; and from 450,000 years ago to the present the cycle is around 100,000 years long (Stringer & Gamble, 1993). One of the longest and coldest patterns of glaciation occurred between 301,000 and 242,000 years ago (Stringer & Gamble, 1993; Gamble, 1999), just at about the time that the "Steinheim group" made its appearance in northern and central Europe and, presumably during the height of the cold phase, spread at least as far as central India. This was quickly followed by another, even colder period between 186,000 and 127,000 years ago (Stringer & Gamble, 1993; Gamble, 1999), corresponding with the emergence of the "classic Neanderthals."

Associated with these unstable cycles of glaciation and interglacial periods in Europe were corresponding fluctuations in the distribution of fauna and flora throughout the continent. During periods of glaciation, most vegetation types became extinct in northern and central Europe, and it is only in southern Europe that continuous pollen sequences exist (Tzedakis & Bennett, 1995; Gamble, 1999; Agusti & Antón, 2002). This was also the case for the many animal groups that were reliant on these food types, which during glacial times must have moved into the refuge of southern Europe. When the interglacial returned, the cold-adapted animal species like mammoth and reindeer moved back to the most northerly regions, as the forests and forest-dependent animals reoccupied central and parts of northern Europe (Geist, 1978; Guthrie, 1984; Gamble, 1986, 1999). It could be that early demes of the "Steinheim group" became more reliant on the cold-adapted faunal groups and followed their migration into the northern regions during interglacial periods, while the contemporary and more southern populations of *H. heidelbergensis* were more reliant on the forest-adapted mammal groups *and* vegetative material to the south. As discussed in the previous chapter, *H. heidelbergensis* made it into China, for a brief time at least, as illustrated by the specimens from Dali and Jinniushan. Thus *H. heidelbergensis* may have remained a warm-climate species, and as such, more of a generalist in dietary requirements; while the "Steinheim group" adapted more and more to an ice-bound Europe and by necessity became less reliant on a vegetable diet, with more and more meat consumption as it became increasingly adapted to its local conditions, and more and more specialized. What this specialization led to is the subject of the next chapter.

INTERLUDE   4
The Geography of Humanity

Where do you live, Australian reader? Sydney, Melbourne, Canberra? Take a trip to Perth, and look around you. The animals are different, aren't they? Not extremely different, perhaps: There are grey kangaroos, but they are western grays, *Macropus fuliginosus*, a different species from the eastern gray (*Macropus giganteus*) that lives in the eastern states. The rosellas are different too. Where are the koalas? There aren't any. But there are those little red things with black and white stripes, numbats — none of them in the eastern states. But the plants are far more different — in fact, the southwest of Australia is famous for its bizarre and gorgeous array.

And go farther afield. Ujung Kulon National Park, in West Java, Indonesia, is a wonderful rainforest paradise, but what you will see there are peacocks, wild boar, monkeys, the wild cattle called banteng, and, if you are very lucky indeed, the Javan rhinoceros — not the ringtail possums, tree kangaroos, and rifle birds you see in a tropical rainforest at the same latitude in Australia. The animals in the Bwindi National Park, in Uganda, at the same latitude, are different again. Those in the Manu National Park in Peru are different yet again.

Although medieval Europeans were vaguely familiar with the animals and plants of North Africa and the Levant, it was not until Columbus's voyages in 1492 that it dawned on them that living organisms are not the same everywhere: There was a whole separate creation on the other side of the Atlantic! As European exploration progressed, it became obvious that there were not just one or two separate creations, but many. How to fit this into the story of Noah's Ark was a headache that remained until Darwin and his contemporaries — particularly the man who had very nearly been Darwin's nemesis but who became first his twin innovator and then his staunchest supporter, Alfred Russell Wallace.

Wallace and his contemporary, P.L. Sclater, and their successors mapped out and documented the distributions of terrestrial animals and plants across the globe and divided the land surface into faunal and floral regions. There are four major faunal realms, called Holarctic, Neotropical (or Neogaean), Paleotropical (or Afro-Tethyan), and Australian (or Notogaean). Each of these is divided and subdivided. There are six major floral kingdoms; four are the same as the faunal realms — Holarctic (or Boreal), Neotropical, Paleotropical, and Australian — but there are two others, whose past or present existence in the faunal sphere is a matter for discussion — Capensic (or South African) and Antarctic. It may simply be that animals are more mobile than plants and that any faunal element corresponding to the remarkable fynbos of the Western Cape, in South Africa, has long since been swamped, and that there are too few land animals capable of surviving the rigors of Antarctica and the subantarctic regions to form a faunal region of their own. These two differences apart, there is a pleasing correspondence between the faunal and floral realms — in outline but not in detail.

The study of the distribution of animals is called *zoogeography*, that of plants, *phytogeography*. Together, they make up the field of *biogeography*. We say that animals and plants that are restricted to particular places are *endemic* to those places, and for obvious reasons the idea of endemicity is central to biogeography.

There is more to biogeography than just noting with interest that kangaroos and eucalypts are found *here* and elephants and baobabs are found *there*. The hard part is finding out why. In the main, it is a compromise between past and present continental patterns, landforms, seaways, soils, and climate, spiced up by the abilities of animals and plants to move by themselves.

The Neotropical realm — South and Central America — has three layers of mammals: (1) The "old endemics" are the ones that are nothing like any mammals found anywhere else, like an order of placental mammals called Xenarthra, containing sloths, anteaters, and armadillos, and two orders of marsupials called Didelphimorphia and Paucituberculata — marsupials they may be, but these opossums and opossum rats are vastly different from the more familiar kangaroos, wombats, and "possums" of Australia. (2) The "young endemics" are the ones that belong to orders found elsewhere but that are represented in the Neotropics by special groups. There are two of these: the platyrrhines, or New World monkeys (marmosets, capuchins, howler and spider monkeys), and the caviomorphs, or New World rodents (tree porcupines, guinea pigs, coypus, and capybaras). (3) The "immigrants" are those that have very close relatives elsewhere: jaguars, pumas and ocelots; zorros, culpeos, and maned wolves; pampas deer, marsh deer, and brockets; and peccaries and tapirs.

It happens that we know a good deal about what South America was like in the past. Let us ignore the Age of Dinosaurs and begin after the asteroid hit, at the very beginning of the Age of Mammals, the Paleocene epoch, from 65 to 55 million years ago. South America is an island, separated from North America by a seaway at least as wide as that separating it from Africa. There, in fossil deposits of the period, are the old endemics: Xenarthra, the Didelphimorphia, and the Paucituberculata and a further array of sabertoothed marsupials and hoofed placentals, which are now dead and gone. Then the curtain closes, and we see no more for 30 million years.

Quite suddenly, the curtain opens again, on a Late Oligocene landscape, about 27 million years ago. The Paleocene collection are still there, but they have been joined by platyrrhines and caviomorphs. How they got there is deeply mysterious. As far as we know, South America had been unbrokenly an island — maybe the ancestors of the young endemics floated across the South Atlantic (which was only half its present width) from Africa on huge rafts of detached vegetation. We don't know. But there they suddenly are, and there they still are today.

At 5 million years ago, at the beginning of the Pliocene, it all changed dramatically, irreparably. Central America happened, a land bridge linking South America to North America. Now, North America had always been part of the rest of the world, with rest-of-the-world mammals. As soon as the land bridge arose, they jostled to get across it and try their luck against the long-isolated inhabitants of the long-isolated continent to the south. The sabertoothed marsupials and hoofed placentals crumbled before them. The other endemics survived, and many prospered; two, an armadillo and an opossum, even went the other way and are still spreading their range in North America.

And here the effect of present-day climate comes in, because South America always had a major tropical component, and Central America is tropical too. So the tropical vegetation spread into Central America, and even as far as the southeastern and southwestern coastal areas of Mexico. With them spread spider monkeys, howler monkeys, sloths, and other endemics.

Let us remind ourselves that human beings did not come down into the Neotropics along with the jaguar. They came later, in a much deadlier wave. What the Pliocene immigrants did for some of the old endemics, *Homo sapiens* is doing today to a far greater degree in sweeping the whole lot away — old endemics, new endemics, and immigrants alike.

It is astonishing that before Jonathan Kingdon, almost no one had thought of human evolution as having a biogeographical element (though we discovered just prior to publication that Juan Luis Arsuaga, in his book *The Neanderthal's Necklace*, published in 2001, also discusses this concept in depth). Kingdon's book, *Lowly Origins*, published in 2003, is going to make a lot of paleoanthropologists say, as Huxley did when he read Darwin's *Origin of Species*, "How extraordinarily stupid of me not to have thought of that myself!" Because, of course, humans and other apes are a Paleotropical group, and humans, chimpanzees, and gorillas belong to a specific part of it: the Afrotropical region. Today the Afrotropical region covers sub-Saharan Africa and coastal southern Arabia and appears to merge into the palearctic region (the Eurasian part of the Holarctic realm) along the Red Sea coast and into Israel. Here, African elephants, African buffaloes, hippos, black and white rhinos, servals, black-backed and side-striped jackals, spotted and brown hyenas, Cape hunting dogs, baboons, guenons, mangabeys, colobus monkeys, giraffes, and hordes and hordes of species of antelopes have their only home. Until recently, it seems that the lesser kudu and perhaps the giraffe extended into southwestern Arabia; the hamadryas baboon still does. Until recently the bubaline hartebeest lived in Israel; the Cape hyrax still does.

And hominines — gorilla, chimpanzee, and human — are Afrotropical. They evolved and diversified in the region, after splitting from the ancestors of their closest relative, the orangutan, in the Oriental region (the Indian and Southeast Asian part of the Paleotropical realm). It should not be too surprising, then, that the entire drama of proto-human diversification should have taken place in the Afrotropical region.

Gorillas (obligatorily) and chimpanzees (preferentially) are rainforest species; presumably, then, the earliest proto-humans were too. Where? The great rainforest belt of Central and West Africa, which pulsated back and forth as drier climate alternated with wetter, was already spoken for by gorillas and chimpanzees. But as Jonathan Kingdon points out, there is a neglected forest belt down the eastern seaboard of Africa. The eastern forests today are dissected fragments, but under a wetter climate regime they would be unified into a giant bloc, with fingers pushing inland alongside the major rivers — Juba, Tana, Galana, Pangani, Ruvuma, Zambezi, Limpopo, etc. Here is the best bet for a region where proto-humans separated from proto-chimpanzees, and here the more open nature of the forests — probably never real rainforests — selected for a ground-living ape more specialized than either of its closest relatives, which themselves are 50% or more terrestrial. And the gallery forests along the big rivers are the inland route opening out into the woodland and savanna habitat where we know the australopithecines best.

Look at a map: Israel, Lebanon, Syria, Turkey, and then we reach the Republic of Georgia. Dmanisi is 1000 km north of Israel, but in a warm phase, some of the Afrotropical fauna reached that far north. The fauna of Dmanisi, 1.7 million years ago, is dominated by Palearctic mammals like deer, bears, and marmots, but there are some Afrotropical elements, like ostriches and brown hyaenas. Like Israel today, the Georgian fauna then was mixed: perhaps, like Israel, Palearctic in the mountains, gazing down on Afrotropicals in the scorching lowlands below. *Homo georgicus* was perhaps at the extremity of its faunal region — but not beyond it. What we still don't know is whether the range of the species (or species group) contracted into Africa when the climate changed, or whether extremity living had enabled it to adapt to new zones and to push on east into the Oriental region.

We can ask the same question much much later. In the latest middle Pleistocene, newly evolved *Homo sapiens* extended its range into Israel (Qafzeh, Skhul) along with Afrotropical fauna when the climate there warmed up, and it gave place to *Homo neanderthalensis* and Palearctic fauna (Tabun, Kebara, Amud) when the climate cooled down again. Eventually, as we know, the range of *Homo sapiens* did not recede back into Africa but pushed on into Europe. But was it able, 115,000 years ago, to push eastward? Is this when China and Southeast Asia were first populated by our own species, or did that event have to await a later push out of Africa? The fossil record of eastern Asia is not good enough to tell us.

In China the boundary between the Paleotropical and Holarctic realms is wide open. The Qinling Range, running east–west across the center, is the nearest thing to a biogeographical boundary, but it is readily breached, and in the past, without a doubt, the boundary fluctuated according to climate. Presumably, early humans fluctuated with it, but at some point, a widening ecological tolerance enabled them to stay when the next cooling occurred, just as it did farther west, in Europe.

The most remarkable faunal boundary in the world bisects Indonesia. Between Borneo and Sulawesi runs Wallace's Line. West of it lies the Oriental region, with its tigers and leopards, civets and mongoose, squirrels, colugos, tree-shrews, and the Asian versions of elephants, rhinos, tapirs, and deer. East of it, in Sulawesi, live the dwarf buffalo known as anoa, the grotesque piglike babirusa, the extraordinary shrew mice, and the northwesternmost marsupials (two species groups of cuscus). Some 800 km east of that is another line, Lydekker's Line, east of which, in New Guinea, begins the real Australian fauna. Between Wallace's and Lydekker's lines is the region of Wallacea, but nobody knows what to do with it — is it Oriental or Australian? In fact, only Sulawesi has a substantial mammal fauna of its own. But as you go east through the Moluccas, what fauna there is becomes more and more Australian.

It is Wallace's Line that has attracted all the attention from an anthropological point of view because, after all, the main drift of people was from west to east across it: There was a time that western Indonesia (Sundaland) was occupied by human beings but New Guinea and Australia were not. Geologically, Sulawesi has been separate from Sundaland for quite a while — maybe since the Late Miocene — and the Makassar Strait, which separates it from Borneo, is very deep and, unlike the islands of Sundaland (and for that matter, New Guinea and Australia on the other side of Lydekker's Line), the two would not have been joined at times of low sea level.

When and how did people first cross Wallace's line?

Great excitement was caused in the 1990s when Dutch, Australian, and Indonesian teams found human artifacts on the island of Flores, dating to nearly 800,000 years ago — because Flores lies east of Wallace's Line! As a matter of fact, Wallace's Line is much less substantial to the south than it is farther north; it is reputed to run between Bali and Lombok. It is true that Bali is part of Sundaland and was connected to Java and thence to Sumatra, Borneo, and the Asian mainland when the sea level was low, whereas the narrow straits between Bali and Lombok are very deep. But it turns out that Lombok shares more of its mammals with Bali than we once thought. It looks as much like a climatic difference as a strictly biogeographical one, and the Oriental fauna dies away successively from west to east as the climate gets drier: West Java, East Java, Bali, Lombok, Sumbawa. Maybe the Lombok Strait was deepened by tectonic activity fairly recently in this highly tectonically active part of the world? Still, east of Sumbawa is another narrow but deep strait, and then we get Flores and its offshore islands. And Flores does have a little mammalian fauna of its own (consisting almost entirely of funny rats, including a wonderful giant one, *Papagomys armandvillei*, half a meter long, excluding the tail). So there is reason to think that humans, more than three-quarters of a million years ago, wandered unusually far east, crossing the sea as they did so. Does this mean that they had boats? We don't know. But the only humans in Southeast Asia at the time, as far as we knew, were *Homo erectus*. So our beetle-browed cousin was not such a slouch after all. But stand by for further, stunning evidence from Flores …

There is not a scrap of evidence that anyone got farther east or southeast until much later. The sea gaps between Timor and Australia and New Guinea, even between Flores and Timor, are really, really substantial. To cross them — that was *Homo sapiens*' work.

CHAPTER 8

"The Grisly Folk": The Emergence of the Neanderthals

Most of the group had left the cave earlier that day. Only one elderly male was left behind. In his youth he had been attacked by a saber-tooth when part of a hunting group; with the help of his clan he had just managed to survive, his right arm torn off, his right leg crippled, his left eye put out as he fell heavily in the attack. He was now a cripple and depended on the generosity of his group to survive.

It was midafternoon when the valley and the surrounding region were rocked by a massive earthquake. The man in the cave could not escape before a large portion of the roof collapsed on him, killing him instantly. Members of the group had immediately rushed back to the cave from their hunting and foraging, only to find the cave entrance almost sealed off by rock debris. Their guttural yells into the cave were met with no reply. Within a few hours they had managed to open up the entrance so that at least they could enter. The old man could be partially seen, his body crushed by the rock debris that had until that afternoon been part of the cave ceiling. Members of the group mourned the loss of their kinsman. They removed the rock debris covering the old man and dug a shallow pit. They placed his body in the grave, turning him on his side, placing his arms across his chest, and fully extending his legs as though in a resting position. They then covered him with smaller pieces of the cave debris. They gathered whatever possessions they could find and moved out of the cave. While it was possible to clear away much of the debris and resettle the cave, all knew that the man's spirit would not enable them to stay. They needed to find another refuge. The leader of the group knew of another suitable cave only a few days distant. They set off down the valley.

In 1957, the skeleton of this elderly, crippled Neanderthal man was excavated from Shanidar, the cave in northern Iraq that had entombed him. This one cave was eventually to yield the remains of nine Neanderthals, some of them apparently buried — one, it is claimed, with a final gift of flowers placed in the grave. This area was just as unstable geologically in the recent past as it is today, and the roof fall was probably the result of an earthquake or major tremor. The significance of the Shanidar excavations and their later interpretation will be discussed presently.

In the previous chapter we examined the emergence of hominins within Europe and the likely role that *H. heidelbergensis* and the "Steinheim group" played in the later evolution of the Neanderthals. It is from around 200,000 to 25,000 years ago that we see the rise and fall of the "classic" Neanderthals that for a time dominated Europe, while slightly more gracile Neanderthal specimens have been recovered from eastern Europe (Croatia and elsewhere) and western Asia (Israel, Iraq, Uzbekistan). While *Homo neanderthalensis* flourished, populations of modern humans (*H. sapiens*) also appeared in western Asia and later in Europe; in numerous cases the two species coexisted, or interdigitated, for a short time at least.

The first fossil hominin to be recognized as not representing a modern human was the specimen from Feldhofer Cave in the Neander Valley, Germany (1856), found and recognized just three years before the publication of Charles Darwin's *On the Origin of Species*. This was not the first discovery of a Neanderthal specimen; unrecognized specimens had previously been discovered in Engis Cave in Belgium between 1829 and 1830 and in Forbes Cave in Gibraltar in 1848 (Stringer, 2000b). Not all accepted the significance of the Neander discovery. For example, some suggested that the robust nature of the specimen, with its thick supraorbital torus, merely represented an idiot who squinted a great deal (thus the developed brow ridges), while others argued that the "bowlegged" nature of the associated leg bone indicated that the individual was a Cossack who had chased Napoleon's army back into western Europe from his disastrous invasion of

Russia in 1812. The most staunch critique of the Feldhofer remains (or any other Neanderthal remains for that matter) as representing an early form of human was the man acknowledged as the father of paleopathology, Professor Virchow of Berlin University, who throughout the second half of the 19th century would characterize all of the unique morphological features of the Neanderthals as "pathological in nature," even when there was clearly no evidence for this (see Trinkaus & Shipman, 1993; Shreeve, 1995; Jordan, 1999). Some in England agreed with Virchow's argument. One commentator stated the following (quoted from Jordan 1999:16):

> It may have been one of those wild men, half-crazed, half-idiotic, cruel and strong, who are always more or less to be found on the outskirts of barbarous tribes, and who now and then appear in civilized communities to be consigned perhaps to the penitentiary or gallows, when their murderous propensities manifest themselves.

The continuing discovery of similar fossils throughout Europe, including those from La Naulette and Spy in Belgium as well as from La Chapelle-aux-Saints (Figure 8.1), La Ferrassie, and La Quina in France showed that such interpretations could not stand up to the ever-increasing evidence, especially when these fossils were found with a distinct material culture as well as remains of extinct fauna. Many of the early researchers were now beginning to place these finds within a geological and archaeological context. In 1889 the first Neanderthals were being excavated from Krapina, in Croatia, and this site would soon yield the remains of two or three dozen Neanderthals. How could anyone now, including Virchow and his followers, still seriously argue that all of these remains represented a family or clan of pathological idiots! By the turn of the century, most accepted that the Neanderthals were a distinct nonpathological population, clearly different from ourselves. The argument now became whether they were our direct ancestors, or whether they represent a primitive human lineage that had become extinct.

The earliest fossil evidence for the emergence of full *H. neanderthalensis* is the French mandible from Montmaurin, which has the characteristic retromolar space typical of Neanderthals and is dated to between 130,000 and 190,000 years ago. The recently redated skulls from Saccopastore in Italy come in next, at between 120,000 and 130,000 years old. The terminal point for the Neanderthals, at around 27 Ka (thousand years ago), is the leg bone fragment and lower jaw bone from Zafarraya in Spain. Most European

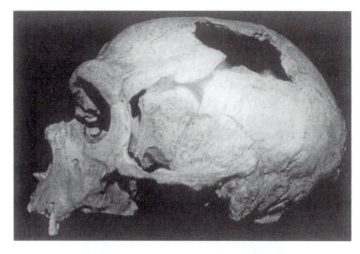

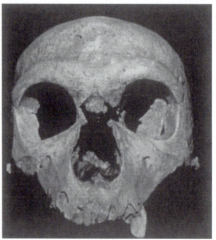

Figure 8.1 ▶   The classic La Chapelle-aux-Saints Neanderthal specimen from France.

and western Asian Neanderthal specimens fall between 40,000 and 65,000
years ago; these include the specimens from France, La Chappelle-aux-
Saints (47–56 Ka), La Quina (35–65 Ka), La Ferrassie (<60 Ka)
Régourdou (<60 Ka), Le Moustier (40–42 Ka); and Peche de L'Azé
(45–55 Ka); from Italy, Grotta Guattari (50–60 Ka); Archi (<60 Ka)
and Saccopastore (about 120 ka) (see Figure 8.2); from Belgium, Spy
(<60 Ka) and Engis (<60 Ka); from Germany, the Feldhofer Cave in
the Neander Valley itself (40 Ka); from Gibraltar, Forbes Cave
(50 Ka); from Croatia, Vindija Cave (42 Ka, but some specimens may

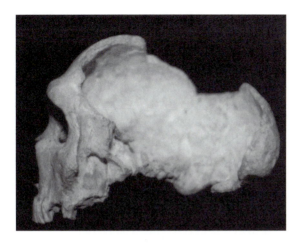

Figure 8.2 ▶ One of the Italian Neanderthal specimens from Saccopastore.

actually be as young as 28,000 years); from Hungary, Subalyuk (60 Ka?); from its farthest eastern record, Uzbekistan, Teshik-Tash (70 Ka); from Iraq, Shanidar Cave (<50 Ka); from Syria, Dederiyeh (perhaps as much as 75 Ka); and from Israel, Kebara and Amud (<60 Ka). The dating of the Tabun Neanderthals, from the Mount Carmel foothills in Israel, remains problematic, but most consider them to be from between 70,000 and perhaps 120,000 years ago; if so, they currently represent the earliest penetration of western Asia by Neanderthal populations (see Trinkaus, 1983; Grün & Stringer, 1991; Grün et al., 1991; Stringer, 1998; Tchernov, 1998; Bar-Yosef, 1998; Tattersall & Schwartz, 2000; Schwartz & Tattersall, 2002, 2003; F.H. Smith, 2002).

The anatomical features of *H. neanderthalensis* are quite distinct from those of both the pre-Neanderthal populations of the middle Pleistocene (taking into account primitive retentions) and the later early specimens of our own lineage, *H. sapiens* (Figure 8.3). While they share primitive features with *H. heidelbergensis*, including the elongated cranium, low frontal, and developed supraorbital tori, the shape of the braincase is distinct in the two species: The cranial vault in *H. neanderthalensis* is higher and more rounded, with laterally projecting and rounded parietals, with a rounded and posteriorly projecting occipital bone, that is, an occipital "bun," and with undeveloped mastoids, though a marked mastoid crest. Seen from the back, the Neanderthal braincase has a characteristic cylindrical shape, described as "en bombe." The Neanderthals also had an

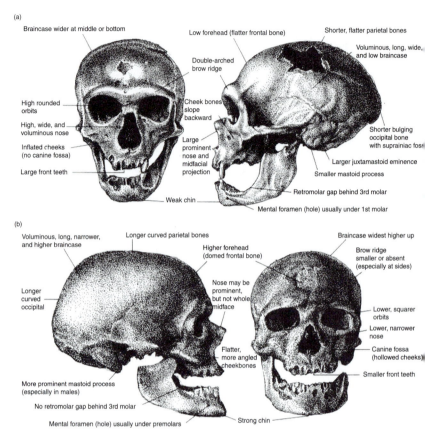

(a)

Braincase wider at middle or bottom

Low forehead (flatter frontal bone)

Shorter, flatter parietal bones

Voluminous, long, wide, and low braincase

Double-arched brow ridge

High rounded orbits

Cheek bones slope backward

High, wide, and voluminous nose

Inflated cheeks (no canine fossa)

Large prominent nose and midfacial projection

Shorter bulging occipital bone with suprainiac fossa

Large front teeth

Larger juxtamastoid eminence

Smaller mastoid process

Weak chin

Retromolar gap behind 3rd molar

Mental foramen (hole) usually under 1st molar

(b)

Voluminous, long, narrower, and higher braincase

Longer curved parietal bones

Braincase widest higher up

Higher forehead (domed frontal bone)

Brow ridge smaller or absent (especially at sides)

Longer curved occipital

Nose may be prominent, but not whole midface

Lower, squarer orbits

Lower, narrower nose

Canine fossa (hollowed cheeks)

Flatter, more angled cheekbones

Smaller front teeth

More prominent mastoid process (especially in males)

No retromolar gap behind 3rd molar

Mental foramen (hole) usually under premolars

Strong chin

Figure 8.3 ▶   (a) The Chapelle-aux-Saints skull, indicating Neanderthal features. (b) The skull of anatomically modern Cro-Magnon 1, with the features characteristic of *Homo sapiens* labeled.

Taken from Stringer and Gamble (1993), pp. 76–77.

enormous cranial capacity, with males averaging around 1600 cc (greater average than that of modern humans), though there is some evi dence that the cranial capacity of females was much less than that o males. The orbits are also round, as opposed to being more rhomboid. An while the supraorbitals in *H. neanderthalensis* are strongly developed, lik those of *H. heidelbergensis*, they are different in shape, with distinct tor above each orbit as opposed to a single torus "shelflike" structure, and th bony thickening curves without a break down the lateral margins of th orbits. The midface is more prognatic and has often been described a though one had got hold of the nose and somehow stretched the midfac forward, and this results in the zygomatics (cheekbones) appearing to be swept backward away from the midface. They have a well-developed nasal aperture, (broad, high, and prominent with marked internal nasal crest), a capacious maxillary sinus system, and a large anterior dental complex (mainly the incisor teeth). The mandible is also distinctive, with its retromolar space (a wide space between the last lower third molar and the ramus), a result of the forward-standing position of the dentition, so that the mental foramen appears to sit farther back, underneath the first molar, and the ascending ramus slopes back.

In body build, the Neanderthals were characteristically different both from their predecessors and from modern humans (Figure 8.4). The postcranial bones are robust and stout, with broad rib cage, long clavicle, wide pelvis, and relatively short and robust limbs that have well-developed muscle markings (see Stringer & Trinkaus, 1981; Trinkaus, 1983; Stringer & Gamble, 1993;

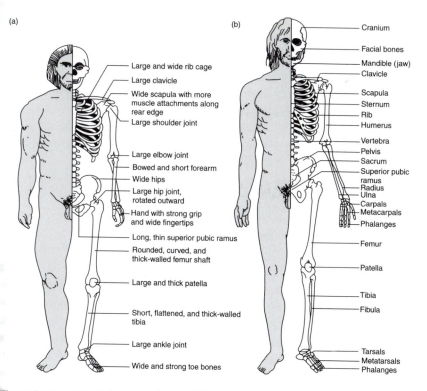

Figure 8.4 ▶ The skeleton of a "classic" Neanderthal (a) indicating typical features compared against the skeleton of an anatomically modern *Homo sapiens* (b). Taken from Stringer and Gamble (1993), p. 79.

Howell, 1998; B.A. Wood & Richmond, 2000; Tattersall & Schwartz, 2000 Agusti & Antón, 2002; Schwartz & Tattersall, 2002). Indeed, Ruff *et al* (1997) and Arsuaga (2002), using an association between the breadth of th hip bones (biiliac breadth) and overall estimated height, suggest that the aver age Neanderthal would have weighed around 168 pounds (75 kg), whil many males would have exceeded 175 pounds (80 kg). Applying brai weight to body weight suggests that the Neanderthal brain was reduced a compared to *H. sapiens* because they have a projected brain that is 4.8 time the size expected for a mammal of its bulk, while *H. sapiens* has a brain tha is around 5.3 times its expected weight (see Ruff *et al.*, 1997; Arsuaga 2002). The hand anatomy was strikingly modified: *H. neanderthalensis* ha a powerful grip, with a broad, long palm, short fingers, and deep grooves fo the interosseous muscles.

Whether *H. neanderthalensis* was cable of complex speech remains a endless, ongoing topic of debate. We currently have not the slightest ide whether Neanderthals processed the necessary neural adaptations for com plex speech, for these simply do not fossilize (B.A. Wood & Richmond 2000). Based on a study of Neanderthal mandibular form and basicrania morphology, Philip Lieberman (1989, 1991) argued that the Neanderthal were unlikely to have had a fully developed vocal tract and thus had limite abilities for speech. Arensburg (1989), in his study of the Kebara Neandertha hyoid bone, however, has refuted much of Lieberman's argument. Som additional inferences have been made from the fossil record in relation t the size of the hypoglossal and vertebral canals, though whether they ar developed enough for the necessary innervation of the tongue and breath ing to control movements related to speech remains problematic (Aiello & Dean, 1990; Kay *et al.*, 1998; DeGusta *et al.*, 1999; MacLarnon & Hewitt 1999; B.A. Wood & Richmond, 2000).

Much has been said about the unique morphology of *H. neanderthalensis* and the functional adaptation of the body plan has been discussed a nauseam. Their robust body structure suggests that the upper limbs wer likely adapted to heavy foraging, such as spear thrusting, while the lowe limbs are those of long-range bipeds (B.A. Wood & Richmond, 2000) The large size of the anterior dentition and their observed wear pattern (much more heavily worn on the labial surfaces) suggest that they wer used as "dental tools" in food preparation and perhaps as a vice, helping t grip mammal hides in their teeth while skinning them with stone tool (Trinkaus, 1983; F.H. Smith, 1983).

Many have considered the rise of classic *H. neanderthalensis* from western Europe to be associated with an adaptation to cold climates, while the slightly less robust Neanderthals from eastern Europe and western Asia might suggest tolerance of the slightly more temperate conditions in these areas. We think that the suggested cold adaptation of the "classic" neanderthals was an exaptation, prefigured in some of the morphologies of *H. heidelbergensis* and enabling the earliest representatives of the lineage to become increasingly adapted to the cold while not restricting them to the periglacial habitats of Ice Age Europe. We must remember, too, that not all morphological features are the result of functional adaptation, but can be related to genetic drift (random change), which must have impacted to some degree the small and isolated populations. Neanderthal children from a very early age display typically robust Neanderthal features, and they are already quite different from modern humans of the same age; the typical Neanderthal features continued to develop in slow progression from gracile juvenile to robust adult (see Trinkaus, 1983, 1986; Stringer *et al.*, 1990). Tattersall and Schwartz (2000:207) insisted that the most remarkable thing about the Neanderthals was their extraordinary adaptability. With this we can only agree. The Neanderthal populations within Europe and western Asia do not bracket a consistently cold time period, and in Syria, Israel, Iraq, and southeastern Europe they lived in a very tolerably warm, productive climate — temperate at worst! — and only those from the younger European sites lived under the harsh Ice Age regimes to which our minds try to adapt them. And yet we must remember that they did not penetrate the tropics, because someone else was, however slightly, better adapted to the tropics; and to this crucial point we will return.

So how would we likely view the Neanderthals today if we saw a family of them walking down the main street of a busy city? Carleton Coon in the 1960s suggested that if we dressed up a Neanderthal man in a suit and gave him a shave and a shower, he would look a little different, but overall would not stand out in the crowd. We take the opposite stance and note, with a sad acknowledgment of the persistence of human racial prejudice and xenophobia, the stance of an unknown commenter who suggested that if a Cro-Magnon got on the same train and sat down next to you, you would probably change seats, but if a Neanderthal sat down next to you, you would change trains. The Neanderthals were very different from ourselves. Not one whit less intelligent, imaginative, dextrous, or adaptable, probably—merely different. As emphasized by Arsuaga (2002: 67):

> The Neanderthals were not simply primitive versions of ourselves ... [They] continued to evolve and ... developed their own distinctive characteristics, just as we did elsewhere. The Neanderthals were not living fossils. They did not belong to the past, and they were not anachronistic. In their particular epoch, they were just as "modern" as our ancestors, the Cro-Magnons, were. The two species were simply different.

With the coming of the last interglacial (oxygen isotope Stage 5) in northern Europe around 130,000 years ago, we see the recolonization of high latitudes by the warmth-loving plant and animal communities that had previously been driven south. Pollen sequences show brief episodes of birch and pine, on to elm and oak, then alder, hazel, yew, and hornbeam, and finally returning to the beginning of the colder conditions with pine, spruce, and silver fir (see Mellars, 1996). It is also at the start of the previous interglacial that Europe was dominated by warm fauna, including forest rhino, hippo, straight-tusked elephant, lion, and hyena, even in northern Europe (Turner & Antón, 1997; Jordan, 1999; Agusti & Antón, 2002).

From 115,000 to around 75,000 years ago there was a succession of cold and slightly more temperate pulses, and then at 75,000 years ago there was a plunge into another long glacial period (Stage 4). It was during this period that glaciers covered most of present-day Scandinavia, and the British Isles and tundra/prairies, or "polar deserts," covered much of northern and central Europe. There was a brief return to the glacial and temporal pulses (Stage 3), but around 32,000 years ago the cold set in once more (Stage 2); at 21,000–17,000 years ago there was a return to maximum glacial conditions (the last glacial maximum, LGM), with glaciers covering much of present-day Poland, northern Germany, Great Britain, and Ireland (see Mellars, 1996; Agusti & Antón, 2002). The shorter interstadial periods during this time were in most cases either too brief or not sufficiently marked to show up in paleontological and geological climatic records, and probably also too brief for trees and associated faunal groups to migrate into northern Europe from their southern refuge areas (Mellars, 1996). That is not to say that these were necessarily hard times. These tundra/prairies would provide an abundance of herd game, which could probably be more easily hunted without the requirements of stalking and chasing individual deer and boars in the thick forests of the south. Some of these more temperate pulses around 30,000–40,000 years ago encouraged the migration of *H. sapiens* into

Europe. Then the Neanderthals lost out to the newcomers while populations of *H. sapiens* maintained and expanded their grip on these more favorable habitats. But that is another story, to which we shall return in the last section.

Molecular evidence has now offered strong support for a long separation of the Neanderthals from *H. sapiens*. In 1997 mtDNA was successfully extracted from the Neanderthal specimen from Feldhofer Cave in the Neander Valley, Germany (see Kahn & Gibbons, 1997; Krings *et al.*, 1997; R. Ward & Stringer, 1997). Only small fragments of Neanderthal mtDNA were recovered. After copying these strands using the polymerase chain reaction technique, researchers were able to identify 378 base pairs. Comparison of these with samples of recent humans showed that they differ in around 27 base pairs (around 7%), while the variability within the modern human sample averages only 8 base pairs (2%). Subsequently, mtDNA has been extracted from two other Neanderthals: specimens from Vindija, in Croatia (Krings *et al.*, 2000), and from Mezmaiskaya, in the Russian Caucasus (Ovchinnikov *et al.*, 2000), proved as similar to the Feldofer samples as three random *H. sapiens* samples are to each other. It must be acknowledged that Gutiérrez *et al.* (2002) have raised questions about these results, both as to the analytical methodologies and the possibility of postmortem changes; yet the amplified sequence of a 12,000-year-old skull from Cheddar Gorge in southwest England showed only two deviations from people living in the area today. As such, modern humans and those living over 12,000 years ago are almost identical in their genetic structure, perhaps allaying some of the problems raised by Gutiérrez *et al.* (2002) and, in the process, further underlining the distinctiveness of the Neanderthals (Sykes, 2001). Likewise, the mtDNA of 8000- to 12,000-year-old Australian skeletons from Kow Swamp, in northern Victoria, as analyzed by Adcock *et al.* (2001), fits easily into the modern aboriginal Australian range of variation (we will comment in a later chapter about the Mungo DNA, which has been claimed to be different). But it is as well to remember that, as yet, no DNA has been extracted from any Cro-Magnon specimen — the *H. sapiens* that succeeded Neanderthals in Europe — so we cannot exclude the possibility that they, too, might have been "different."

Be that as it may, Neanderthals and modern humans are perfectly distinct. Modern humans arose recently in Africa as a distinct species and replaced the Neanderthals, who contributed few if any genes to modern human populations. In other words, there was little or no gene flow between them (see Kahn & Gibbons, 1997; Krings *et al.*, 1997, 2000 R. Ward & Stringer, 1997; Klein, 1999), although occasional hybridizatio may indeed have occurred — see later! Thus there is nothing within th molecular data to support the idea that modern humans evolved from a earlier Neanderthal population. Indeed, using a "molecular clock," the las common ancestor between the modern human and Neanderthal lineage lived around 500,000–700,000 years ago. This is at around the same tim that *H. heidelbergensis* appears in the fossil record.

Material Culture

Homo neanderthalensis made stone tools that were considerably mor sophisticated than those observed in either *H. heidelbergensis* or their pre sumed ancestors from the "Steinheim group": the Mousterian industry (Figure 8.5). This toolkit is usually defined as part of a Mode 3 Technology which makes its first appearance in the archeological record aroun 250,000 years ago (Clark, 1977; Foley & Lahr, 1997; Agusti & Antón, 2002)

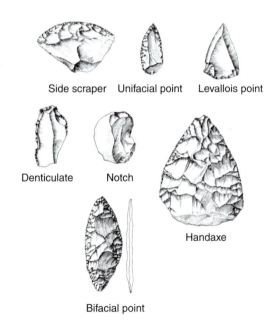

Side scraper   Unifacial point   Levallois point

Denticulate   Notch

Handaxe

Bifacial point

Figure 8.5 ▶   Typical Mousterian (Mode III) stone artifacts.
From Schick and Toth (1993), p. 290.

It must be stressed, however, that the Mousterian technology is not associated only with Neanderthals; some of the earliest representatives of *H. sapiens* have also been found with a Mousterian assemblage, such as Jebel Qafzeh in Israel, while conversely some Neanderthals are found with more advanced toolkits quite like those usually associated with modern *H. sapiens* (Cro-Magnon), such as a Châtelperronian toolkit associated with the St. Césaire Neanderthal in France (see Bar-Yosef, 2000; Mellars, 2000; Pilbeam & Bar-Yosef, 2000).

The Mousterian complex is basically a stone core technology, though it was so carefully fashioned that a number of single blows would detach a ready-made series of finished tools to a predefined pattern in a diversity of forms, including various blade techniques. Unlike their predecessors, the Neanderthals fashioned separate tools for specific purposes, as opposed to multipurpose tools. These tools were retouched to provide a continuous cutting edge around the edges of the artifact. The most commonly recognized lithic artifact is the small, triangular *Levallois point*, which may have been hafted to a spear (see Tattersall & Schwartz, 2000). Evidence of this has been found in spears and spear shafts, whose ends appear to have originally held some form of stone spearhead point, and, as discussed earlier, spears with a fire-hardened point have also been excavated, associated with mammoths. It has been demonstrated that at least six clearly separate stages, all requiring considerable planning abilities from the initial stages of the flaking sequences, are required to produce this technology, suggestive of considerable planning abilities (see Mellars, 1996).

Depending on the availability of raw materials, tools were often resharpened and reused, resulting in numerous forms and sizes as part of this process (Marks & Volkman, 1983; Stringer & Gamble, 1993). Studies of the cutting edges of Mousterian tools (associated with polished edges) have shown that some form of woodworking was being done, possibly mostly the fashioning of spear shafts (Tattersall & Schwartz, 2000). There is little evidence, however, for the use of nonlithic tools within their toolkit, for there is little or no strong evidence for tools made of bone that can be directly attributed to Neanderthals. Thus, unlike their predecessors, *H. heidelbergensis* and *H. steinheimensis*, the Neanderthals, for whatever reason, appear to have ceased using bone as a raw material for tool manufacture.

There is little convincing archeological evidence for symbolic behavior, meaning the production of decorative or artistic items. Much of the suggested evidence that Neanderthals produced such artifacts has been based on what were interpreted as holes intentionally bored within shells or pieces of bone. Most of these bore-holes, however, are now considered to be simply the results of natural damage, caused by chemical erosion of the bones or by carnivore activity (Mellars, 1989; see also Davidson & Noble, 1989, 1993; Cameron, 1993b; Stringer & Gamble, 1993). The presence of ochre and other coloring agents within Neanderthal living sites, however, is not disputed and suggests that pigments were applied to the body or other items, such as clothing and wooden artifacts. Indeed, there is evidence that pigments were applied to stone tools and pieces of bone (see Mellars, 1996); this suggests that symbolic art may not be the exclusive domain of *H. sapiens*.

Neanderthals have also long been associated with a "cave bear cult." This was originally suggested as a result of excavations of the Drachenloch Cave in the Swiss Alps between 1917 and 1923. It was said that Neanderthals collected and placed a large number of cave bear skulls and long bones on natural shelves of rock or slabs of rock piled up to make artificial shelves within the cave (see Jordan, 1999; Tattersall, 1999). While no Neanderthal remains were found in the cave, Mousterian artifacts were discovered. Most today agree, however, that the evidence form Drachenloch, far from being proof of a cave bear cult, is rather the result of shoddy excavation supervision. The slabs were the result of single large blocks that fell from the cave ceiling and later split along bedding planes by frost action. The accumulation of cave bear remains appears to have been a result of a long-term occupation of the cave by cave bears over hundreds if not thousands of years and thus represents the natural remains of the bears' den, with individuals dying in the cave and their remains being shuffled around by later bear occupants as they formed their own sleeping areas within the cave (see Tattersall, 1999).

Implied Social Dynamics

Most researchers today believe that Neanderthals deliberately and intentionally buried their dead, which suggests compassion for members of their clan, though Gargett (1989, 1996, 1999) has argued against intentional burials, based on spatial taphonomic studies (see also Noble & Davidson, 1996). The most famous case to date for an intentional Neanderthal burial is from Shanidar Cave in northern Iraq. Between 1953 and 1960, Ralph Solecki excavated the Shanidar deposits and recovered the remains of nine Neanderthal individuals, among which one group dated to around 60,000 years ago and another group to between 70,000 and 80,000 years ago (Trinkaus, 1983; Tattersall & Schwartz, 2000). Some individuals found in the upper level of the excavations appear to have been killed by a collapse of the cave roof (Shanidar 1, 3, and 5), for they were excavated immediately below rockfall debris (Trinkaus, 1983). Others, however, appear to have been intentionally buried, the most famous being Shanidar 4, found in the lower levels of the excavation. This individual was lying on his left side, with the right arm across the body and the legs partially flexed. In addition to this, pollen was said to be found in association with the burial. Shanidar 4 soon became famous as representing a "flower burial," or, if we are to agree with the title of Solecki's well-known book, they represent *The First Flower People* (1971). Recently, however, it has been suggested that the flower pollen may have been inadvertently introduced to the site by workmen while excavating Shanidar in the 1950s!

Evidence of recognized kinship and perhaps a concept of loss is suggested by an examination of the "old man," Shanidar 1. This individual, as excavated, was lying on his back, turned slightly onto his right side, with his arms across his chest and his legs fully extended. While he appears to have died as a result of trauma associated with the roof fall, the arrangement of Shanidar 1's body suggests that after the disaster, the body was deliberately covered with layers of small pieces of limestone, by members of his social group in an attempt to bury him (Trinkaus, 1983). The major significance of Shanidar, however, is his long-term physical condition before death. Prior to death he had sustained numerous injuries to his right frontal and left orbit (which suggests he was probably blind in his left eye at least), and had suffered from a massive injury to his right side that resulted in arthritic degenerations of the right knee and ankle as well as extreme withering of the right clavicle, scapula, and humerus, with fractures and a possible amputation of the distal humerus (Trinkaus, 1983). These injuries occurred long before the roof fall because substantial bone healing had occurred, proving that the injuries were not recent and thus not a result of the collapse of the cave roof that killed him. The implication is that this individual must have been cared for by his group in order to have survived to a ripe old age (remembering he would have survived even longer if he had not been killed by the roof fall). This need not be particularly surprising, for something similar had happened even as far back as *H. ergaster*, as indicated by the pathological condition of KNM-ER 1808, who had obviously been cared for by individuals from her group for a considerable period of time.

Neanderthal burials are not restricted to western Asia. In Europe, the La Chapelle-aux-Saints individual had been buried in a rectangular pit at the base of the Mousterian deposits. A similar burial pit occurs at Le Moustier, containing the remains of a Neanderthal teenager. At Regourdou, also in France, a Neanderthal was excavated from a burial pit that was lined with a bed of stones and covered over with a cairn containing a mixture of animal bones and artifacts, though these artifacts may merely be part of the backfill. Burials have also been found at La Ferrassie in France (Figure 8.6). Even more interesting is that these burials appear to represent the world's first cemetery. In all, seven Neanderthals were buried, almost all of whom were oriented in an east–west direction, the only exception being an infant burial in a north–south orientation at the back of the cave wall. There were two adults, the rest being children, and the age and sex distribution is suggestive of a "family plot" (Mellars, 1996; see also Stringer & Gamble, 1993; Jordan, 1999). Whether material culture and faunal remains associated with such burials can be considered grave goods is still unresolved. As suggested by Chase and Dibble (1987), most of the "deliberate grave goods" can probably be attributed to an inadvertent incorporation into the grave at the time of burial, which would have been dug through Mousterian layers, and the artifacts might have been incorporated into the grave as part of the backfilling operation.

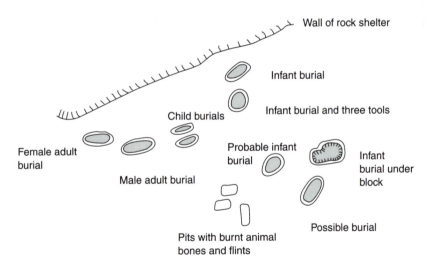

Figure 8.6 ▶   The Neanderthal "cemetery" at La Ferrassie. North is at the top of the figure. Not to scale.

Adapted from Jordan, 1999.

Some have suggested likely cannibalistic bahavior associated with ome Neanderthal remains. While there is some evidence of defleshing n specimens from Moula-Guercy in France (Defleur *et al.*, 1999; 3.A. Wood & Richmond, 2000), this in itself is not evidence of cannibalsm, for some burial rituals in the recent past also involved defleshing but 1ot cannibalism. This is especially true when such burials are associated vith secondary burial practices (T.D. White, 1992). The often-quoted ritual acrifice and/or cannibalistic behavior as suggested by the damage at the base of the Neanderthal skull from Grotta Guattari (which was also hought to have been placed within a ring of stones) has more recently been interpreted as the result of hyena damage and taphonomic process ather than the result of human action (Stiner, 1991, 1994). There is ncreasing evidence, however for cannibalistic behavior at Krapina Croatia) because much of the Neanderthal material is burnt, with long bones being smashed open at marrow-rich points and the base of the skulls enlarged, suggestive of removing the brain — though again we cannot completely discount postmortem defleshing or nonhuman activity Gamble, 1986; Jordan, 1999) (Figure 8.7).

Even given the lack of evidence relating to grave goods associated with Neanderthal burials, clearly there is some form of symbolic behavior tied to the act of burying the dead. The burials cannot merely be seen as a way that Neanderthals disposed of the dead — if so, there are far easier and less labor-intensive ways of doing so. The act of burial is most likely

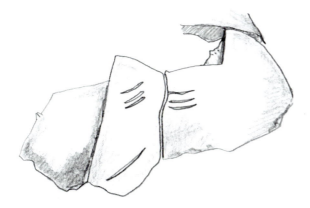

Figure 8.7 ▶   One of the Krapina Neanderthal specimens from Croatia, with defined cut marks on the skull. We, as many, believe that this does not necessarily equate with cannibalism but, rather, likely represents part of a postmortem burial practice.

associated with some form of emotional attachment to the individual by members of the group and a recognition of their importance and status. The act of burial is symbolic of their loss.

And how long did they live? There are two individuals dubbed "old men" — Shanidar 1 and La Chapelle-aux-Saints. Trinkaus doubts that they really were "old" and puts them in their late 30s; only to La Ferrassie 1 does he attribute an age definitely greater than 40. Yet we wonder if this was right. Among mammals, potential longevity is strongly correlated with brain size — and because Neanderthal brain size was of the same order as that of modern humans (even a little larger), there is some theoretical reason to expect that they could have reached ages of 70 or 80, as we do. Aging adult skeletons is problematic; perhaps "impossible" would be a better word (Bocquet-Appel & Masset, 1982). All one can say is that breakdowns of functioning parts accumulate with increasing age. The "old man" of La Chapelle was riddled with arthritis; even La Ferrassie 1 had a touch of it (Cave & Straus, 1957). High infant mortality pulls down mean life expectancy; even in modern populations whose mean life expectancy is, say, 35, there are still plenty of dotards in their 80s. To us, everything seems to point to a similar pattern for the Neanderthals.

As we have already observed in the study of Neanderthal anatomy, genetics, and material culture, Neanderthals are distinctive not only from their predecessors but also from modern humans (also see next chapter). This also applies to patterns of land use: They appear to have developed a "Neanderthal niche" distinct from all other hominins.

Land use patterns by Neanderthals are becoming increasingly well documented. While the sample from western Asia is not large enough to make any definitive conclusions, the numerous documented occupation sites in France and Spain enable us to make some general statements, which probably applied to Neanderthal populations in general (see R. White, 1983; Gamble, 1986, 1999; Stiner, 1991, 1994; Mellars, 1996). The current synthesis of Neanderthal land use patterns is that (1) cave and rock shelter sites, in southwestern France at least, share several topographic and environmental features, including well-sheltered locations in positions offering extensive and wide-ranging views over adjacent valley habitats (and these valley habitats were ecologically diverse), and almost all have easy access to abundant and high-quality raw materials; (2) these sites are located in such a fashion as to serve as central places from which diverse conomic and technical activities could be conducted (see Stiner, 1991, 994; Mellars, 1996).

As far as open-air sites are concerned, the consensus seems to be that the majority of the richer ones are predominately on higher and more exposed ocations, usually on major plateaus. They are also commonly close to active springs, lakes, or streams. Whether these sites represent substantial periods of ongoing occupation by large groups or longer-term revisiting over short periods by smaller groups remains debatable (Mellars, 1996). Turq (1989, 1992), however, has demonstrated that lithic artifacts excavated from caves or rock shelters in southwestern France are made from purely local raw materials in 78% of cases, while in open-air sites, local raw material accounts for 94% of the material used. This suggests that most occupations at open-air sites within this region were restricted either to very brief periods of time or to the exploitation of foods and other resources within a short distance from the sites (Mellars, 1996:268).

There is little doubt that Neanderthals knew their environment intimately and that they organized their lives around the changing seasons, which required knowledge of herd patterns and complex hunting and gathering strategies. It is also increasingly clear that they did not flinch from taking on mammoths, bison, and possibly even cave bears, with nothing more than spears and clubs. A spear with a fire-hardened tip, dating to around 130,000 years ago, was recently excavated from a site in Germany. Perhaps what is even more interesting is that it was found lodged between the ribs of a mammoth. There is also clear evidence for mass slaughter, as documented at La Cotte de St. Brelade, Jersey, where large mammals appear to have been stampeded over the edge of the cliffs and then subsequently butchered and dragged into the cave (Gamble, 1999; Jordan, 1999).

As well as being competent hunters of large, medium, and small game, Neanderthals turned to foraging strategies using coastal caves and aquatic resources, including shellfish and tortoises. There is also evidence that in some cases, they scavenged animal carcasses from other predators (Stiner, 1994). The major difference in habitat exploitation between Neanderthals and modern humans is not so much in the types of food that were consumed, but in the way they gathered these resources (different hunting strategies) as well as the animal body parts consumed. For example, Stiner (1991, 1994), in her study of Neanderthal sites in Italy, has concluded that Neanderthals focused on younger individuals within animal groups when hunting, and faunal remains tend to be dominated by head parts, or they contain more or less balanced frequencies of head and limb bones. Head-dominated assemblages are also biased toward old-aged prey, while those containing more fleshy parts have a tendency to represent a broader age sample, reflecting more closely extant herd structures. A similar pattern of faunal assemblage has emerged from the French site at Combe Grenal, which has a long occupation sequence, over 70,000 years. A large number of horses' heads have been recovered, which yield little meat. Conversely, young or sickly reindeer are also common, and the more fleshy upper leg parts dominate the assemblage.

Neanderthals are also thought to have relied more on scavenging behavior at specific times; so hunting and scavenging resources were not pooled together, but occupy distinct phases of occupation, or at different sites, a behavior that is different from that observed in modern humans (Stiner 1994). In addition, those sites considered to be scavenging sites, which are dominated by low-yielding meaty parts, such as heads, are also said to have flints that have been subjected to more retouch and are often heavier in weight, which would be required to help process the scavenged material. These lithic artefacts would also be required to help process the plant food that was undoubtedly needed to supplement the meager meat ration from these occupation sites (see Stiner, 1994; Jordan, 1999).

Stiner (1994) also argued that the Mousterian differs from the upper Paleolithic in the geographic scale of habitat exploitation and that while both Neanderthal and modern early humans were mobile groups, the Neanderthals appear to have been differentiated by their responses to resource opportunities, which were governed more by local or immediate exigencies (Stiner, 1994). As such, the major difference between *H. neanderthalensis* and *H. sapiens* is not so much the development of a new behavioral repertoire, but a shift in emphasis of existing universal hominid behavior (see also Cameron, 1993a; Cameron & Groves, 1993).

The Fate of *Homo neanderthalensis*

All the available paleoanthropological, archeological, and molecular evidence strongly supports the displacement of more "primitive" endemic human populations by modern humans who recently arrived from Africa. In the case of the Neanderthals at least, it appears to have been displacement by rapid extinction.

The correlation in western and central Europe between the disappearance of Neanderthals from both the paleontological and archeological records, around 27,000 years ago, and the appearance of modern humans, Cro-Magnons, around 30,000 years ago (with some evidence of earlier ɔlonization around 40,000 years ago), strongly supports the rapid-extinction ıodel (see Stringer & Gamble, 1993; Stringer & McKie, 1996; Klein, 999; Tattersall & Schwartz, 2000). While it may be argued that the associ-tion between the Saint-Césaire Neanderthal (Figure 8.8) and an apparent ʰhâtelperronian tradition (see earlier) supports continuity between Middle

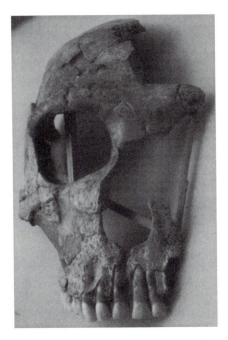

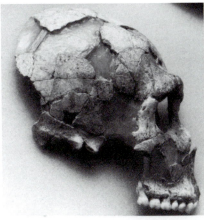

Figure 8.8 ▶   The classic Saint-Césaire Neanderthal specimen from France.

and Upper Paleolithic populations, it is also possible that this was a local Mousterian cultural adaptation. Indeed, the Châtelperronian industry associated with the Saint-Césaire Neanderthal is recognized as closely resembling the Mousterian tradition Type B, thus likely representing a local variant of a Mousterian technology (Klein, 1999).

Other evidence does suggest, however, that the two groups interacted a little. Is it really a coincidence that Neanderthals began developing bladelike variants of the Mousterian at precisely the time when *H. sapiens* was spreading among them? At Arcy-sur-Cure, a late (<33,000 years ago) Neanderthal site in southern France, there is even, for the very first time, decorative art — animal teeth pierced as if for a necklace and ivory rings — as well as bone and ivory tools (Hublin *et al.*, 1996). Did they obtain these objects from Cro-Magnons by trade, or did they make them themselves by copying their new neighbors? Overall, however, there appears to have been limited contact between these groups. As suggested by the paleodemographer Jean-Pierre Bocquet-Appel (in Arsuaga, 2002), the Neanderthals never attained a high population density, and far-flung populations were obliged to exchange individuals in order to keep up a viable population. This resulted in a spreading of biological (and cultural) resources very thinly. But with the emergence of *H. sapiens*, we witness the appearance of ever-increasing groups. As these large and dense population clusters became reproductively viable and economically self-sufficient, they also became more biologically and culturally closed. Or to put it another way, we may be witnessing the emergence of a form of biological and cultural elitism.

Finally, we return to the theme of interbreeding. From time to time, quite a number of fossils have been claimed to be hybrids, or "Neanderthals showing evidence of gene flow," or even "modern humans showing evidence of gene flow." But only one of these claims carries conviction: the skeleton of a 4-year-old child, about 25,000 years old, from Lagar Velho in Portugal (Duarte *et al.*, 1999). Even given its extreme youth, the skeleton shows characteristics that are unmistakeably those of a mixture: Like modern humans it has small teeth (with especially small front teeth) and a modern pelvis, but like Neanderthals the limb bones are robust and short, and there is a telltale large juxtamastoid eminence on the temporal bone. The lower jaw by itself shows mixed characters — it combines a Neanderthal-like retreating mandibular symphysis with a modern pointed chin (note, the chin in modern humans results from the recession of the superior incisor alveolar border of the mandible due to the retraction of the dental complex, which "leaves behind — inferiorly" a chin; thus a chin is not an outgrowth, but exactly the opposite condition [see Enlow & Hans, 1996]). This does not mean that we bear Neanderthal genes — they have long since disappeared — but it does mean that possibly some of the last Neanderthals, the last of their kind, had joined Cro-Magnon hordes and interbred with them.

The populations of *H. neanderthalensis* from western Asia appear *after* the emergence of modern *H. sapiens* in this region and eastern Africa. For example, the relatively modern-looking populations from Herto in Ethiopia date to around 160,000 years ago, and the Skhul and Qafzeh hominins date to perhaps 120,000 years ago, while the Neanderthals from Kebara, Amud, and Shanidar at least date to less than 70,000 years ago (see Trinkaus, 1983; Stringer, 1998; Tchernov, 1998; T.D. White *et al.*, 2003). Of significance is that the modern-looking humans from Skhul and Qafzeh are associated with typically African fauna and represent a likely northern migration from Africa into the coastal zones of the Levant. The Neanderthals, however, are associated with typical Eurasian fauna and, unlike the modern humans from Skhul and Qafzeh, were not restricted to the coastal plains but even occupied parts of present-day Syria and Iraq. It is likely that the modern humans located around the coastal regions migrated or expanded their range into this region from the south during the more mild and temperate periods of 130,000–70,000 years ago. With the change in climatic conditions from 70,000 years ago, however, it is likely that Neanderthal populations migrated or expanded their range into western Asia from the north, displacing the modern human populations (see Stringer & Gamble, 1993; Stringer, 1998; Klein, 1999).

It is only with the emergence of *Homo sapiens* that humans have been in a position to significantly change their environment, rather than being changed by it. Before this, the human was just another large mammal, with its own role to play within the large-mammal community system. As such, hominins were open to the same selective pressures that impacted other mammals, and they were certainly part of any dispersal pattern that affected these mammalian communities. Turner (1984), Vrba (1985), and Tchernov (1998) have each demonstrated that the presence of hominins within the Plio-/Pleistocene fossil record have been shown to correlate strongly with the presence of principle mammalian fauna, showing that they were an integral component of their mammalian communities. The increasing southward and eastward expansion of glacial conditions resulted in a retreat from these northern regions of the existing floral and faunal groups. This southeast migration also included herds of red deer, wild horse, and roe deer. It is also likely that at least some Neanderthal populations dependent on these

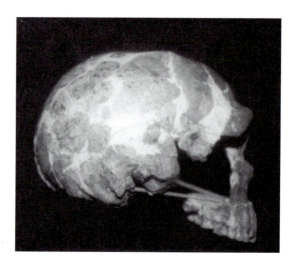

Figure 8.9 ▶   The Tabun Neanderthal from Mount Carmel, Israel.

resources also shifted southeastward, eventually settling in western Asia, as convincingly argued by Tchernov (1988:80, 87):

> It is probably true that any hominid dispersal was a natural part of any emigration until the late upper Palaeolithic period. Given that hominids play an integral part in the ecological events, hominid dispersals and faunal changes are actually part of the same phenomenon.... [T]he new environmental configuration across Europe challenged the foraging technologies, the social structure, and the spatial organization of Mousterian populations, forcing them to move into the Mediterranean lands.

It is likely that such dispersal patterns were intermittent, as shown by the early penetration into this region by the Tabun Neanderthals (Figure 8.9), who may be the only Neanderthal population to have preceded *H. sapiens* in western Asia (Stringer, 1998), with later Neanderthal populations moving into the region around 70,000 years ago. The disappearance of *H. neanderthalensis* from this region is similar to that outlined for Europe, that is, rapid extinction. From around 40,000 years ago we see in this region a more sophisticated toolkit, similar to those documented in Europe from around 30,000 years ago. We also see for the first time remains of modern *H. sapiens*, dating to around 40,000 years ago (see Stringer & Gamble, 1993; Klein, 1999; Tattersall, 1999).

Until recently, the easternmost Neanderthal appeared to be the child burial at Teshik-Tash, in Uzbekistan. Did the high Tianshan Range limit

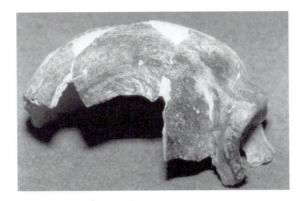

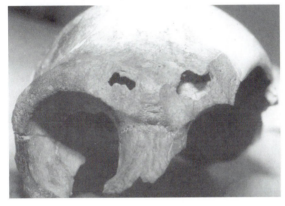

Figure 8.10 ▶   The Chinese Neanderthal-like specimen form Maba.

them to the east? Perhaps not, because Mousterian surface scatters are
known from Mongolia, too. Now comes a startling finding: One of us
[Groves (in press)] shows that the Maba calotte, from Guangxi in southeastern China, is a Neanderthal (Figure 8.10)! The conclusion seems
inescapable: the en bombe vault, the rounded orbits and supraorbital torus,
the huge frontal sinus, and the protruding nasal skeleton. By what route
did they get there, these Chinese Neanderthals, and when? Are they more
like members of the "Steinheim group," descended from the Narmada
people? The Neanderthals are the best known of all our fossil kin, but we
still have a great deal to learn about them.

In conclusion, all available evidence suggests that *H. neanderthalensis*
originated from an indigenous European ancestor, which we associate

with the "Steinheim group." Paleontological evidence from Atapuerca and the recent molecular evidence both suggest that the Neanderthal lineage is significantly older than previously thought, dating back to around 500,000 or perhaps to 700,000 years ago. The geographic and genetic isolation of a European Ice Age population resulted in its morphological differentiation, which over time resulted in speciation. We also suggest that initially the "cold-adapted" morphology of *H. neanderthalensis* was an exaptation, which developed further as a result of selection pressures, which eventually enabled them to occupy their distinct ecological niche, though it in no way restricted them to this type of habitat (still, this same Neanderthal niche may have inhibited occupation by other, less-adapted non-Neanderthal populations).

As we will see in the next chapter, modern *H. sapiens* originated in Africa around 200,000 years ago. It is also from around this time that we see an initial short-term territorial expansion into western Asia by early modern humans, associated with a similar migration by other African faunal groups during the more temperate periods from around 120,000–70,000 years ago. Conversely, around 70,000 years ago we see the return of colder climates and the colonization, or recolonization if Tabun is as much as 120,000 years old, of large areas in western Asia by *H. neanderthalensis*, with perhaps a remigration from Europe around 50,000 years ago. It is from around 40,000 years ago that we see within the paleontological and archeological record of both western Asia and Europe the arrival of truly modern humans into these regions. By 30,000 years ago, modern *H. sapiens* had firmly established themselves in these regions, and the last known Neanderthal refuge, Zafarraya Cave in the far south of Spain, in the extreme southwest of Europe, dating to 27,000 years ago, witnesses their extinction.

CHAPTER 9

The Second African Exodus: The Emergence of Modern Humans

The two human groups had lived side by side in the adjoining valleys for as long as anyone could remember. Relations between the groups had always been based on avoidance. While the light-skinned, robust, stocky people largely restricted themselves to the surrounding valley systems to the north, the dark-skinned, taller, more gracile people controlled a much larger region to the south and often moved through the landscape. Each of these groups, however, had a home base that was close by. The rough terrain between these two valley systems helped in the desire for each group to avoid the other.

The region to the south, however, was eventually not large enough to support the increasing members of the dark-skinned clan. The only avenue for territorial expansion and access to new resources was the land to the north. The climate, too, was warming up. The gazelles, hartebeest, and wild asses favored by the dark skins were increasing, and the wild goats, deer, and horses hunted by the light skins were becoming scarcer. Soon confrontations between the two groups became more regular and the region to the north was increasingly being occupied by members of the dark-skinned clan. Unlike the light-skinned people, the dark-skinned clan had long ago made friends with the wolves that shared their landscape and their rocky ledges. The wolves helped them hunt by bailing up their prey, and to the wolves they tossed food, gradually taming them. The wolves barked to sound the alarm, played with the kids when the adults were off hunting and gathering by day, and huddled close to keep them warm at night.

The robusts had no choice but to seek land farther to the north, though they soon came into conflict with another dark-skinned clan moving in from the northeast. The light-skinned band was in decline. The social structure of the group had always been based on a largely sedentary structure, with a longstanding home base. Their need to keep moving every few years was taxing on the group; soon they were to disappear from the landscape forever.

The Evidence from Africa and the Levant

The earliest evidence for the emergence of truly modern humans is from Africa, with the discoveries of *H. sapiens* from Herto in the Middle Awash of Ethiopia (T.D. White *et al.*, 2003) (Figure 9.1). These fossil specimens and stone artifacts have been dated to between 160,000 and 154,000 years ago by precise age determinations based on the argon isotope method (Clark *et al.*, 2003). They are truly significant because they predate the classic Neanderthals and lack any of the Neanderthal derived features (see previous chapter). Like some previous (and later) hominin populations, there is evidence of postmortem modification to the bones, that is, cut marks. These

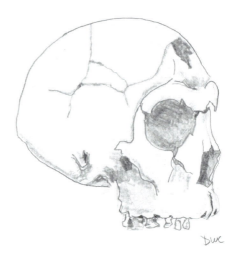

Figure 9.1 ▶ The Herto BOU-VP-16/1 adult cranium.

Adapted from T.D. White *et al.*, 2003.

 are best considered to be mortuary practices rather than cannibalism T.D. White *et al.*, 2003; see also Stringer, 2003). Indeed, the morphological condition of these specimens is intermediate between the condition observed in the African deme of *H. heidelbergensis* (defined by Bodo) and modern *H. sapiens* of today (T.D. White *et al.*, 2003; see also Stringer 2003). As stated by Stringer (2003:693–694):

> Despite the presence of some primitive features, there seems to be enough morphological evidence to regard the Herto material as the oldest definite record of what we currently think of as modern *H. sapiens*. The fact that the geological age of these fossils is close to some estimates obtained by genetic analyses for the origin of modern human variation only heightens their importance.

The stone artifacts associated with the Herto fossils are represented mostly by a Levallois-like tradition, including flake blades and points as well as some biface "hand axes" (Clark *et al.*, 2003). The archeology of the Herto localities suggests that these hominins occupied the margin of a freshwater lake. There is evidence that they were involved in the butchery of large mammal carcasses, especially hippopotami, including a broad ontogenetic age span from newborn calves to adults (Clark *et al.*, 2003).

The next earliest evidence for modern humans from this region are the discoveries from Singa, in the Sudan. Additional specimens soon followed, with discoveries at Dire-Dawa in Ethiopia, Jebel Irhoud and Dar-Soltan, both in Morocco, Border Cave in Kwazulu-Natal, Omo-Kibish (Omo 1) in Ethiopia, and Klasies River Mouth in the southern Cape. None of these fossils, except perhaps Irhoud, is older than 150,000 years, and most are less than 100,000 years old (B.A. Wood & Richmond, 2000).

The most recent attempt to define the morphological condition of modern *H. sapiens* has been that put forward by D.E. Lieberman (1995, 1998) and D.E. Lieberman *et al.*, (2002). They suggest that modern humans can be defined by their globular braincase, a high and vertical frontal, substantial reduction in the supraorbital torus, a developed canine fossa, and a pronounced chin; overall the skull is relatively gracile in appearance. This condition can be attributed to reduced facial prognathism, associated with increased cranial base flexion, resulting in maximum reduction in neuro-orbital disjunction (see also D.E. Lieberman, 2000). Further to this, D.E. Lieberman *et al.* (2002) identified distinct developmental processes of the skull (through ontogenetic studies) that are unique (autapomorphies) to modern *H. sapiens*. They confirm that anatomically modern humans are distinct in two main structural autapomorphies from *H. neanderthalensis* (and other "archaics"); these are in increased facial retraction and neurocranial globularity. These patterns are associated with a combination of shifts in cranial base angle, cranial fossae length and width, and facial length which themselves are likely in response to increased size of the temporal and frontal lobes of the brain as well as their repositioning. Postcranially modern humans are also clearly distinct from *H. neanderthalensis*, with elongated distal limbs relative to the trunk, with a trunk and pelvis that are more narrow, and with low body mass relative to stature (see B.A. Wood & Richmond, 2000).

The Omo-Kibish skeleton (Omo 1), which has been dated to 130,000 years ago (Stringer, 1998), is very modern in appearance, even though the bone table is relatively thick (Figure 9.2). While the face is missing, except for isolated fragments, the supraorbital tori are modern in appearance, the partially preserved zygomatic is reduced in size, and, finally, the mandibular fragment has a distinct chin, all confirming its status as representative of *H. sapiens*. The postcranial anatomy is clearly that of a tall and well-built male individual (Stringer & Gamble, 1993).

Klasies River Mouth, dating back to the last interglacial (between 120,000 and 90,000 years ago) is of interest not only because of the early appearance of modern humans, but also because it has been suggested that the cave dwellers practiced cannibalism, similar to that observed in the late Neanderthals from Krapina in Croatia. The fossils from this cave

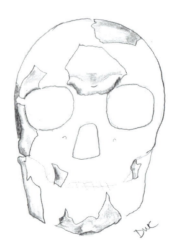

Figure 9.2 ▶   Reconstruction of Omo 1 from Ethiopia, based on preserved parts (shaded).

system are clearly modern in appearance but with a mosaic of features: A frontal specimen has barely developed tori, of modern aspect (Stringer & Gamble, 1993), but of the half-dozen mandibles, the largest (which is above most modern humans in size) has the most prominent, modern chin, while the smallest (which is as small as any modern woman) has virtually no chin at all. The fragmentary nature of the material, the evidence of numerous impact fractures, the numerous cut marks on bone, the lack of long bones, and the evidence that the material had been burnt are what has suggested to some that these remains are part of a cannibal feast (see Deacon & Deacon, 1999), though it cannot be discounted that they are merely part of a secondary burial practice.

The remains from Jebel Irhoud in Morocco (Figure 9.3) are around 150,000 years old, perhaps a little more, and are defined by a combination of modern *H. sapiens* features with a few more primitive traits. For example, while the associated femur is robust and the teeth relatively large, the back of the skull of Irhoud 1 has a primitive appearance. It has a moderately developed supraorbital region, though it is clearly approaching the modern condition in its facial anatomy, with its broad, nonprognathic face, and its zygomatic is hollow. The mandible is marked by a distinct chin. Irhoud 2 is a calvaria, which is higher, rounder, and thoroughly modern. The Jebel Irhoud remains show affinities with the "proto-Cro-Magnons" from Skhul and Qafzeh (Stringer & Gamble, 1993; Jordan, 1999; Hublin, 2000). A circumorbital fragment from Zuttiyeh, in Israel, is about the same age as Irhoud, supporting our assertion that, from time to time, the Levant was part of the African theater.

The fossil evidence and associated dates from the African archeological record place the emergence of modern *H. sapiens* in the African continent to a minimum of 165,000 years ago. This is in full agreement with molecular studies that also depict modern humans diversifying in Africa around the same time. Molecular studies of mtDNA support the idea of recent African origins, and as discussed in the last chapter, the molecular studies of Neanderthal mtDNA also show a divergence from a modern human sample far outside the range observed in modern humans. Indeed, recent studies of nuclear DNA further refine the proposed dates of divergence from an African ancestor. As in studies of mtDNA, patterns of nuclear DNA distribution are significantly more variable in Africa compared to the rest of the world, and it has been possible to estimate the time of the first appearance of some of these patterns in Africa at between 200,000 and 100,000 years ago (Jordan, 1999; Klein, 1999; Wells, 2002; T.D. White *et al.*, 2003; Oppenheimer 2003). The extremely limited genetic distance between modern *H. sapiens* populations in the rest of the world today strongly supports a recent and common ancestry from Africa.

Where in Africa did *Homo sapiens* originate, and from what species? As we have seen, the earliest remains are from both the far north and the far south and in between. Before these earliest specimens there is a hiatus in

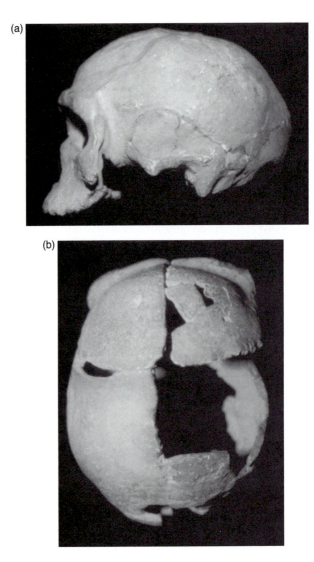

Figure 9.3 ▶   Near-modern *H. sapiens* Jebel Irhoud specimens from Morocco, Africa: (a) Specimen 1, (b) Specimen 2.

ime, and then we have the Florisbad skull (South Africa) at 259,000 years go (Figure 9.4) and the Guomde skull (Kenya) at 272,000–279,000 years go — both of them intermediate between *Homo heidelbergensis* and *H. sapiens*. It looks as if the African record, though less complete, is

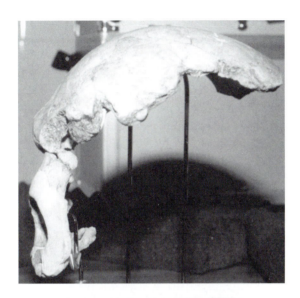

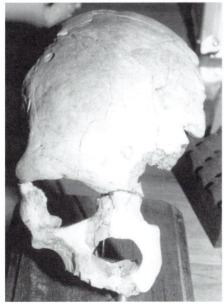

Figure 9.4 ▶   The near-modern *H. sapiens* from Florisbad, South Africa.

complementing the European record: two *H. heidelbergensis* populations diverging, and each rapidly specializing in its own way, resulting in two sister species, individualizing at about the same time. This, of course, gets us no nearer to an actual place of origin. Or maybe there was none — maybe the "sapientizing" traits arose and spread almost instantaneously imperceptibly, so that all African populations appeared to evolve together

The emergence of modern *H. sapiens* is more often than not considered to be associated with the emergence of symbolic behavior, including art and decorative objects, which itself, it is argued, reflects for the first time evidence for complex and elaborate speech (see Noble & Davidson, 1996). It is also with the emergence of *H. sapiens* that we witness in the archeological record a substantially refined toolkit consisting of finely worked flint and bone tools, including needles and fishhooks — the first time we have evidence for sewn clothing or for fishing. There is, however evidence from Africa of a disjuncture between the modern morphology of *H. sapiens* and the emergence of modern behavior: The remains from Klasies River Mouth, for example, are found with an African variant of the Mousterian technology (see Tattersall & Schwartz, 2000; B.A. Wood & Richmond, 2000).

Compared to the later explosion in Europe, there is little evidence for the widespread development of artistic or symbolic behavior within the earliest modern humans of Africa. We can, however, see an incipient behavioral shift toward this direction with incised ostrich eggshell fragments dated to around 50,000 years ago (Tattersall & Schwartz, 2000). Indeed, there is evidence of the development of an advanced tool technology at a number of early sites; thus by 80,000 years ago we see the development of bone harpoon points from Katanda in Zaire (Democratic Republic of Congo) and the remarkably upper Paleolithic-like industry of Howieson's Poort, which arose, flourished briefly, and then inexplicably disappeared in southern Africa.

We mentioned in the previous chapter that two sites in the Levant contain the remains of early modern *H. sapiens* and are associated with a typical African fauna. We consider this region at this time to be a biogeographic extension of the African continent rather than considering the traces of *Homo sapiens* coming out of Africa en route to points east and north Mughartet es-Skhul, discovered between 1931 and 1932, and Djebel Qafzeh, excavated in 1933 and later between 1965 to 1975, are currently believed to fall between 120,000 and 80,000 years ago (see Stringer & Andrews, 1988; Valladas *et al.*, 1998).

The near moderns from Skhul (Figure 9.5) and Qafzeh have a high and
short braincase, round in profile, with reduced supraorbital tori and significantly reduced facial prognathism. The face is broad but with a low nasal
aperture, and the zygomatics, while not inflated, do not run posteriorly as
in Neanderthals, and the orbits are wide and low. Some primitive features
are retained, including, in the case of Qafzeh 9, lower facial prognathism.
In terms of their body proportions, they are even more like modern
humans, with their elongated lower limbs and more slim physique, as if
adapted for long-distance travel, though tropical populations today have a
lankier build and longer limbs than do those from cooler climates, and the
Skhul/Qafzeh proportions probably simply reflect their tropical origins.

In addition, an interesting three-dimensional morphometric analysis of
Skhul and Qafzeh metacarpal fossil remains (finger bones) by Niewoehner
(2001) indicates that these early modern *H. sapiens* had hands that were
essentially like those of people from the Upper Paleolithic (and like us of
today), unlike those of the Neanderthals, who are marked by a number of
unique metacarpal features. Neanderthals are characterized by increased
muscle mechanical advantages at the base of the thumb and on the ulnar
and radial sides of the wrist, relative to the people from Skhul/Qafzeh and
the Upper Paleolithic. Neanderthals are also distinct because of their
increased development of muscle crests and fingertip widths. Together,
these suggest that Neanderthals and the Skhul/Qafzeh peoples are distinct

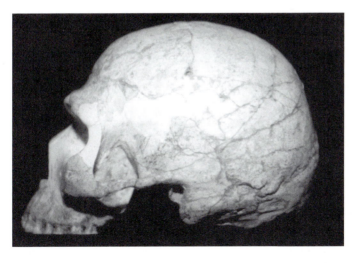

Figure 9.5 ▶   The early *H. sapiens* Skhul V skull from Mount Carmel in Israel.

from each other in the most functionally significant regions of the hand an
that the Skhul/Qafzeh hand remains are morphologically and functionall
within the range of modern humans, unlike Neanderthals. As Niewoehne
(2001: 2983) concludes:

> Given the patterns of between-sample morphological and functional similar-
> ities discovered, . . . the Skhul/Qafzeh hominids were most likely using
> oblique grips and finer finger movements more frequently than were
> Neanderthals. Notably, the skeletal evidence, . . . in the context of late
> Pleistocene patterns of modern human emergence, indicates that significant
> shifts in habitual manipulative behavior were associated with the early emer-
> gence of modern humans. Such behavioral shifts may well have been one of
> the primary components of the subsequent spread of early modern humans.

Perhaps the most interesting thing about the Skuhl/Qafzeh peoples is tha
they shared the Levant with Neanderthals. There are seven well-analyze
human habitation sites known from Israel spanning the late Middl
Pleistocene and the Late Pleistocene: Oumm Qatafa, Zuttiyeh, Tabur
Qafzeh, Skhul, Hayonim, and Kebara. The deposits in the Tabun cave g
right through from about 210,000 to 40,000 years ago. Those in the othe
sites are by no means as long-lasting, but the evidence they yield is entirel
consistent, as shown in a remarkable series of papers by Tchernov (se
Tchernov, 1998, and references therein): During Oxygen Isotope Stage €
when the climate was cool, the Palearctic fauna spread down from Europ
into the Levant. At 130,000 years ago, when the climate warmed up sharpl
at the beginning of Stage 5, the Afrotropical fauna swept in and replaced th
Palearctic. When the climate slowly deteriorated, culminating in the onset o
Stage 4 around 75,000 years ago, the Palearctic fauna came back agai
briefly. Stage 3, beginning about 63,000 years ago, represented a partia
return to mild conditions and Afrotropical fauna again, ending aroun
32,000 years ago with the onset of the last glacial maximum (Stage 2). Th
Palearctic fauna included moles, hamsters, dormice, yellow-necked vole
red deer, wild goats, and horses. The Afrotropical fauna replaced these wit
African shrews, gerbils, the Nile rat, the multimammate mouse, hartebees
ibex, warthog, and wild asses. It was not always a complete replacemen
rather, the two faunas fluctuated in abundance, just as today in Israel th
Palearctic fauna predominates on the highlands of the Carmel, and th
Afrotropical fauna — what remains of it — predominates in the Negev
though it does penetrate the northern parts of the country to a limited degre
And along with the Afrotropical fauna we get *H. sapiens*; along with th

Palearctic fauna we get the Neanderthals. The two species of humans just fluctuated back and forth, like any other animal, until, at 40,000 years ago, *Homo sapiens* began, for whatever reason, spreading into the Neanderthal heartland of Europe. After that there were no more Neanderthals to repopulate the Middle East.

As in Africa, there is a disjunction between biology and culture: Both of these near-modern human populations are associated with a local variant of a Mousterian technology. It is not until around 47,000 years ago, in the Negev at Boker Tachtit, that we see a transitional toolkit that points toward the development of an Upper Paleolithic tool technology (Goren-Inbar & Belfer-Cohen, 1998), defined by increasing frequencies of prismatic cores, with blanks modified into end scrapers, and the development of burins (Bar-Yosef, 1995). Yet more than fifty thousand years earlier, Qafzeh 9 was buried with arms folded and knees bent in a shallow grave, with an infant buried close to his feet. Qafzeh 11, a child, was found with the skull and antlers of a large deer, presumably buried as a complete head (Stringer & Gamble, 1993).

The available paleoanthropological, archeological, and molecular evidence now fully supports the origins of modern humans in Africa. It is likely that the African deme of *H. heidelbergensis*, known from individuals like Kabwe and Bodo, was ancestral to *H. sapiens* as indicated by the "intermediate" anatomical condition observed in the Herto specimens from the Middle Awash, Ethiopia (T.D. White *et al.*, 2003; Stringer, 2003). Probably, as we have seen, *H. heidelbergensis* populations in Europe were ancestral to Neanderthals, though the species survived in parts of southern Europe, especially Greece (as indicated by Petralona) until quite late, while *H. steinheimensis* (if we really want to recognize it as a separate species) succeeded it in central and western Europe. It cannot be discounted, however, that either African or Levantine demes of *H. heidelbergensis* may have given rise to *H. sapiens* in the Levant, which then spread into Africa and, later, Europe, though the molecular evidence does point to an African origin. It is these African and/or Levantine demes of *H. sapiens* that eventually moved into Europe around 40,000 years ago (though 20,000 years earlier they had already settled in Australia — see next chapter), as attested by the presence of the *H. sapiens* child from Ksar Akil in Lebanon, dated to around 37,000 years ago, where an Upper Paleolithic blade lithic industry has also been documented at around 44,000 years ago. This is closely followed by the first documented appearance of this industry in Europe at Bacho Kiro in Bulgaria around 43,000 years ago, before finally making it all the way to Spain by 39,000 years ago (see Tiller, 1998; Gamble, 1999; Jordan, 1999).

The Evidence from Europe

The first discovery of truly modern people was made in 1868, when skeletons were excavated by railway workers at the Cro-Magnon rock shelter at Les Eyzies in France, which gave these people their name, "the Cro-Magnon race" (Figure 9.6). Following this were innumerable discoveries in Europe — in the former Czechoslovakia at Mladec, Predmosti, and Brno; in France at Combe Capelle, Grimaldi, Chancelade, Aurignac, and many other sites; in Germany at Oberkassel; and on and on. Thus, for the first time we have what can be considered a decent population sample, and

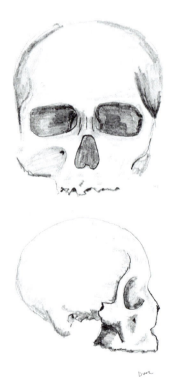

Figure 9.6 ▶   Cro-Magnon specimen 1 from France.

we know recent human prehistory in more detail in Europe than in any other region, at any time.

Only one of the European fossils dates earlier than 32,000 years ago Trinkaus *et al.*, 2003), though the archeological record, as shown earlier, suggests that they had arrived in southern Europe by at least 40,000 years ago. The earliest fossil evidence shows that the first *H. sapiens* to arrive in Europe were not "cold adapted" but were marked by long limbs, with a fully modern cranial and facial morphology. They have high and vertical frontals, with a round braincase, reduced supraorbital tori, a flat facial profile, with reduced zygomatic development, a relatively small dental complex, and a distinctive chin. Unexpectedly, some intraspecies changes occurred even within Europe. The earliest Europeans were tall, though robust, with some brow ridge development and big jaws. The changes toward smaller, more gracile people can be traced through the Late Pleistocene into Recent times. Overall, however, they clearly reflect an African anatomical condition. It would appear that their success within glacial Europe was not based on their physique (which was at least partially the case for the Neanderthals) but rather on a superior material culture, greater cooperative hunting skills, and perhaps a more effective form of language. And perhaps even the beginnings of symbiosis with the ancestors of the dog (Newby, 1997).

The fully fledged Aurignacian industry, which appears in Europe, is different from the earlier tool traditions associated with modern *H. sapiens* as observed in Africa and the Levant. This later technology consists of long, narrow blades struck successively from carefully shaped cylindrical stone cores, the most innovative aspect being the widespread use of bone and antler to fashion additional tools (Figure 9.7). The material culture of the Aurignacian is diverse and is extremely refined in appearance. Nothing like it has been seen in the archeological record before this time. In addition, we also see for the first time undisputed evidence of art and objects of symbolism. And the period is also characterized by frequent use of red ochre and stone-circled hearths, as well as the use of rocks for warmth making (Bar-Yosef, 1995). Clearly a behavioral shift has occurred that reflects our own view of the world and how we interpret it.

The earliest appearance to date of artifacts associated with artistic/symbolic meaning are the caves from Vogelherd in Germany, where a whole series of animal figurines carved from mammoth ivory has been recovered, dating to around 34,000 years ago. The most famous evidence for their artistic expression and possibly associated ritual is the fantastic cave art from France and Spain, which was being produced at Chauvet cave

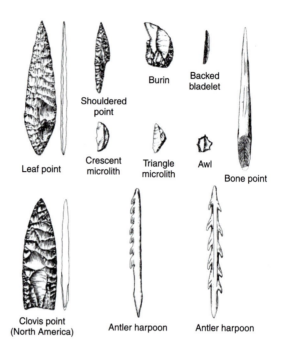

Figure 9.7 ▶   European Mode IV stone artifacts.
From Schick and Toth (1993, p. 297).

in France as early as 32,000 years ago, through to the masterpieces a
Lascaux, now thought to date to around 17,000 years ago, which provide :
form of Ice Age narrative. There is also evidence of music, as shown by th
discovery of bone flutes from the French Pyrenees site of Isturitz, dated t
around 32,000 years ago, with additional flutes and beads from Pair-non
Pair cave in France, dating to around 27,000 years ago, and a flute mad
from a swan's radius recently discovered from Geissenklösterle in souther
Germany, dating to as early as 36,000 years ago (Gamble, 1999).

The Evidence from Asia

Unfortunately, we are still waiting for detailed reporting of much of the evidence relating to the emergence of *H. sapiens* in Asia. Most of the evidenc
of *H. sapiens* in northeast and southeast Asia dates to less than 20,000 year
ago, though a cranium and maxilla have been recovered from Ziyang tha
have been dated to around 40,000 years ago (Wu & Poirier, 1995). Also, ar

occipital bone from Shanxi Province, in China, has been dated to 28,000 years ago (Wu & Zhang, 1983; Kamminga & Wright, 1988; Wu & Poirier, 1995). Indeed, the specimens from the Upper Cave at Zhoukoudian in China, which had previously been considered Late Pleistocene, are now thought to be early Holocene, around 11,000 years old. A special status applies to the Liujiang specimen from the far south of China (Figure 9.8), whose dating has swung back and forth: Formerly thought to be well back in the Late Pleistocene, it was considered by Kamminga and Wright (1988) to represent a Holocene, possibly as late as Neolithic, burial into late Pleistocene deposits. But its dating and stratigraphy have recently been reconsidered, and it has apparently been reconfirmed as at least 67,000 B.P. (Shen *et al.*, 2002).

In southeast Asia, the oldest *H. sapiens* are burials from northern Vietnam dating to 20,000 years ago (Ciochon & Olsen, 1986). The fragmentary skull from Niah Cave in Borneo was originally thought to date to approximately 40,000 years ago, but the original excavator did not realize that the remains were a burial dug into somewhat (or much) older deposits, so an age for this specimen remains problematic (see Brose & Wolpoff, 1971; Kamminga & Wright, 1988). The Wajak skulls from Java, Indonesia, have been thought to

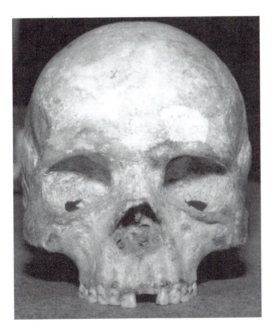

Figure 9.8 ▶ Modern *H. sapiens* specimen from Luijiang, China.

be Late Pleistocene based solely on their robust appearance, but Storm (1995) reports radiocarbon dates of 6,560 B.P. for the human femur and 10,560 for associated fauna (uncalibrated); their purported affinities to indigenous Australians will be discussed in the next chapter.

While there has been some suggestion that there may have been penetration into Europe via Gibraltar because there are some similarities between the North African Aterian and European Acheulean lithic traditions, the human biological evidence does not support any significant exchange between these two populations, and the same applies to possible exchange between endemic Neanderthals and the "pre-Cro-Magnons" of North Africa (Hublin, 2000). The initial movement of modern *H. sapiens* into Europe must surely have been from western Asia, which would doubtless have been facilitated by the expansion of mixed deciduous/coniferous woodland during the warm interstadial into southern Europe, with associated faunal groups, around 40,000 years ago (Gamble, 1999).

As we have described earlier, there is little evidence for precise coexistence between Neanderthals and modern *H. sapiens* in western Asia. Rather, there appear to have been time-successive occupations of the southern Levant (see also Hublin, 2000), and the northward expansion of modern *H. sapiens* into the Levant is associated with environmental changes driving the northern extension of African faunas into the region (Tchernov, 1998; Hublin, 2000). By 40,000 years ago, modern *H. sapiens* had increased their territorial range to include southern, and eventually, central and western Europe.

With the emergence of modern *H. sapiens* in Europe around 40,000 years ago, on the contrary, there is some degree of coexistence between them and Neanderthals. The available paleoanthropological and archeological record suggests a dispersal of modern *H. sapiens* groups from east to west, initially of low demographic density and discontinuous settlements, later with some form of coexistence of central and western Europe by these modern people and Neanderthal populations (see Mellars, 1996; Hublin, 2000). As Mellars (1996) suggests, however, at any one point in time modern *H. sapiens* and Neanderthals were likely to have been confined to separate economic and demographic territories, which largely avoided any direct competition for the use of specific resources at the same time and place.

By 30,000 years ago, the climatic conditions were again deteriorating and glacial conditions were returning to Europe. It may be this event that triggered the final extinction of the surviving Neanderthal populations. For example, with the coming of the new Ice Age, we know that within 3,000 years, between 30,000 and 27,000 B.P., the last Neanderthal populations disappeared from Spain, perhaps slightly later in Portugal if you accept the "hybrid" child, dated to around 25,000 years ago. Conversely, it is from this time that the European and western Asian paleoanthropological and archeological record becomes increasingly dominated by the remains and past activities of modern *H. sapiens*. And the *H. sapiens* that appear in the record just after the demise of the Neanderthals cannot be said to display any persisting Neanderthal-like features (Hublin, 2000). Even before the start of the deteriorating climatic conditions, the newcomers had laid claim to much of Europe, and as far as the Neanderthals were concerned, the competition for resources was already lost. The likely superior hunting skills of *H. sapiens*, perhaps assisted by their proto-dogs, may have also contributed to a decline in Neanderthal birthrates and/or increase in infant mortality rates (or maybe compared to *H. sapiens*, they were already low). And over time, the Neanderthal world was becoming one of increasing populations of moderns and a decreasing population of Neanderthals.

With the final onset of an Ice Age Europe, Neanderthal bands were probably already few and far between and did not represent a viable breeding population. Both *H. sapiens* and *H. neanderthalensis* groups would need to adapt rapidly to the deteriorating conditions, and there had to have been increased competition between them for finite resources. While the Neanderthals may have been able to adapt and survive another cold climatic shift if they alone occupied Europe, the presence of *H. sapiens*, with their competitive edge, pushed *H. neanderthalensis* to extinction (see Mellars, 1996). So the decline of the Neanderthals did not necessarily mean extinction through violence, but extinction through competition, which became even more intense with the onset of the last Ice Age.

CHAPTER 10

The Emergence of Modern Humans in Asia and Australia

The young woman had been cremated earlier that morning, the details hidden from the men by the secret women's ritual. As the cremation was being conducted, some of the men sat around the campfire, close to the lake margin, and discussed the hunting and fishing of the previous day. They were well away from the secret women's business. Some men were showing the children how to make stone tools, knapping away flakes from large stone cores. The smell of cooking fish and kangaroo meat started to filter among the group. Added to the fire were large quantities of freshwater mussels, and a few emu eggs were positioned close by, ready to be heated at the last moment.

The next day, the men went out hunting, while most of the women returned to the cremation site. Some of the younger women stayed behind in the camp to look after the children. The remains of the young woman were collected and her bones smashed into hundreds of pieces. The original cremation pit was dug through and the bones of the young woman interred. The ash from the cremation was then used to help fill the shallow pit. More women's business was performed before the women returned to the campsite late that night.

The discovery in 1968 of the "Mungo Lady" (specimen LM 1) and her excavation in 1969 are significant events in Australian and world prehistory. The remains of this individual pushed back the colonization of Australia to at least 30,000 years ago, which was at that time almost twice

251

as old as previously thought. Of equal significance are the complex cultural practices, clearly demonstrated by the fact that she was cremated. The burial site still represents the earliest evidence of cremation anywhere in the world. Australia is now claimed to have been colonized at least 60,000 years ago (Thorne *et al.*, 1999), though this claim has recently been challenged. But whether the date was 60,000 or closer to 40,000, as Bowler *et al.* (2003) now claim, it was a long, long time ago. Yet what is clear is that these were modern people, not only in appearance but also in behavior.

In the 1930s, an American anthropologist, Joseph Birdsell, traveled around Australia measuring Aboriginal people. He found that in the rainforests of the Atherton Tablelands, in far north Queensland, people are very short and have crisply curled, sometimes "woolly" hair. In the southeast, people are comparatively light-skinned and bulkily built, and the men tend to be hairy and to go bald early. In the tropical north (outside the rainforests), people are very dark and slenderly built. Instead of simply noting these variations and ascribing them to local adaptations, as we would today, Birdsell was convinced that three distinct races, which came in at different times and, in part, replaced each other, populated Australia. First, he argued, came the short people, who survive today in purest form in the Atherton rainforests; he called them Barrineans, after Lake Barrine, near Atherton. Next came the light-skinned people, whom he called the Murrayians because they survived in purest form along the River Murray. Finally came the dark, lanky people; he called them Carpentarians, from the Gulf of Carpentaria. This is Birdsell's Trihybrid hypothesis (Birdsell, 1949, 1967).

Birdsell's model enjoyed some popularity up until the 1960s, but the work of Andrew Abbie, a zoologist from Adelaide, was gradually undermining it. Abbie did not deny the existence of physical differences from place to place, but for him the essential unity of the Australian Aboriginal physique, when compared to any other human group whatever, spoke eloquently of a single origin (Abbie, 1968, 1976).

The modern version of the Birdsell model is Alan Thorne's Dihybrid hypothesis, which dates from the early 1970s and is based not on living people but on fossils. Thorne argued that the original human colonization of Australia had two distinct migration episodes, one coming from China and the other from Indonesia (Thorne, 1984). The so-called "gracile populations" (as represented by many of the Willandra Lakes fossils) are said to have originated from present-day China, while the later "robust populations" (as represented by fossils from Kow Swamp and Coobool Creek), he suggested, originated from present-day Indonesia (Thorne, 1984; Wolpoff *et al.*, 1984). This scheme, of course, is tied to the Multiregional hypothesis, which argues for independent evolution of human groups in different regions of the world through time. It suggests that there is a continuation of anatomical form from *Homo erectus* of Java through to the first Australians (Wolpoff *et al.*, 1984; Thorne & Wolpoff, 1981, 1992; Thorne, 2000).

The alternative hypothesis is that indigenous Australians are represented by one migration occurring anywhere from about 45,000 to 70,000 years ago, with little significant biological input after this initial colonization. This scheme is tied to the "Out of Africa" hypothesis: There was a recent origin of modern humans from Africa, followed by a rapid biological dispersion throughout Europe and Asia, including Australia. These modern populations quickly replaced the original populations in Europe and Asia without any significant inbreeding. The demise of the more "archaic" Old World human populations was based on either a prehistoric genocide (which we consider to be unlikely) or, far more probable, their simply succumbing to competition for the available resources from recently arrived modern humans — although extinction is a perfectly natural biological event, and it is possible that pre-*sapiens* species could simply have become extinct before *H. sapiens* arrived (Stringer, 1984, 1986, 1989, 1998, 1999; Stringer & Andrews, 1988; Groves, 1989a, 1989b, 1994; Howells, 1992; Stringer & McKie, 1997; partly Adcock *et al.*, 2001).

A major problem with previous interpretations of the Australian paleoanthropological record has been the focus on anatomical features, with little attempt to identify the significance of the features being discussed (the few exceptions being Wright, 1976; P. Brown, 1981; Groves, 1989a, 1989b; Habgood, 1989). For example, Macintosh and Larnach (1972, 1976), Thorne (1976), Wolpoff *et al.* (1984), and Thorne and Wolpoff (1981) base much of their case on an implied similarity of the upper face (particularly the supraorbital and frontal regions), arguing for a biological continuity between the Australian "robust" and the Indonesian *H. erectus* populations. Other features, such as "thick" cranial bones, flattened frontals, and enlarged dental complexes, have also been used. Are these features, however, unique to Pleistocene Southeast Asian and Australian populations? Indeed, what is the functional, developmental, and phylogenetic significance of these features anyway?

Before attempting to answer these important questions, we must first briefly describe these two "types," the prehistoric "gracile" and the "robust" Australians.

Interpretations of the Australian Paleoanthropological Evidence

Some of the Mungo individuals represent the oldest Australian human fossils so far discovered. It is one of these, Lake Mungo 3 (LM 3) (Figure 10.1), that has now been dated at either 60,000 years ago (Thorne *et al.*, 1999) or 40,000 years ago (Bowler *et al.*, 2003). They are described as having a high frontal and relatively thin cranial walls. The cranium is spherical in shape, the frontal lacks much of a supraorbital torus, and the face is relatively flat and lies immediately below the frontal (Webb, 1989). In other words, this early population is very modern in appearance.

Apart from LM 1, LM 3, and some other, much less complete, specimens from the Willandra Lakes, the main candidate for the status of "gracile" among the Australian fossils is the skull from Keilor (Melbourne Airport). Keilor, in particular, is said to resemble the cranium from Wajak in Java and some specimens from southern China — Ziyang, dated to around 38,000 years ago, and Liujiang, arguably dated at 67,000 years ago (see previous chapter). The isolated incisor teeth from Liujiang are "shovel-shaped" (Wu & Poirier, 1995), and this incisor morphology has often been argued to represent a regional trait linking Chinese *H. erectus*

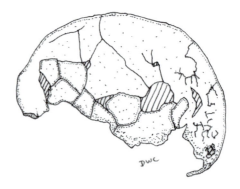

Figure 10.1 ▶ Willandra hominin specimen 3 (Lake Mungo III) from southern New South Wales, Australia.

or *Homo pekinensis*, as we prefer — they differ consistently from the true *Homo erectus* from Java) and modern East Asians (Weidenreich, 1937). This feature, however, is also observed in Neanderthals, *H. heidelbergensis*, and even the presumed ancestral species *Homo ergaster* from Africa (e.g., WT-15000) and *A. africanus*. Thus it is simply a "primitive" feature for the hominins (see also Stringer *et al.*, 1997). Note, too, that incisor shoveling is not strongly developed in Aboriginal Australians, including the Keilor skull, so if this "regional continuity" character were correct, it would be evidence *against* a China–Australia link, not for it!

The so-called "robust" Pleistocene Australians include skulls from Kow Swamp (Figure 10.2), Cohuna, and Coobool; all apparently date from about 9,000–12,000 years ago. They have large jaws and teeth, thick skull bones, prominent continuous brow ridges, and, in particular, flat, receding foreheads. Thorne (1976, 1977, 1984, 2000), Thorne and Wolpoff (1981), Wolpoff *et al.* (1984), and Wolpoff (1989) argue that these "robust" fossils all demonstrate a continuity of morphological features with Javanese *Homo erectus* and share no physical and/or genetic influence from the Pleistocene of Africa. Indeed, much of the entire multiregional scheme has been based on comparisons of early Pleistocene Indonesian *H. erectus* specimens from Sangiran and the Late Pleistocene Australian modern human fossils from Kow Swamp (Thorne, 1976; Thorne & Wolpoff, 1981).

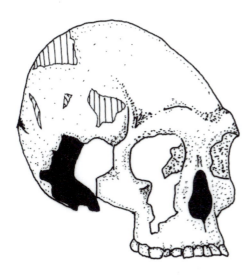

Figure 10.2 ▶ Kow Swamp 1 from Victoria, Australia. See text for details.
Adapted from P. Brown (1981).

What of the other Australian fossils? One from the Willandra Lakes (WLH 50) is indeed very robust, and it has thicker skull bones than any from Kow Swamp (see Webb, 1989; Hawks *et al.*, 2000); its date is 14,000 years ago. Hawks *et al.* (2000) have argued in depth that WLH 50 provides evidence, to them at least, for an *H. erectus*–Australian lineage. It has been argued, however, that the unusual cranial thickness and robusticity observed in WLH 50 are the results of a pathological condition associated with some form of hemoglobinopathy (Webb, 1995). Therefore, while Hawks *et al.* (2000) spend considerable time analyzing WLH 50 in terms of phenotypic and metric features, and while they themselves admit that they cannot rule out that WLH 50 may have suffered a pathological condition, they still maintain the continuity scheme. How seriously can we consider such a relationship when their evolutionary sequence from *H. erectus* to Australia is based on one specimen, especially one that is so problematic?

Another, from Cossack, Western Australia, is very large and appears to have a flat, receding forehead, but it is distorted and compressed, so it is difficult to interpret its actual morphological shape. Nitchie, from western New South Wales, is very much like Keilor but has larger brow ridges. This highlights a problem — the boundary between the two groups is far from clear-cut. Keilor, in fact, is every bit as large as any Kow Swamp, and it has an enormous palate. Lake Mungo 3 has thick skull bones and large teeth but is "gracile" in shape. Both are very different from the tiny LM 1, the only one that is really "gracile" at all.

As far as the dates go, the "graciles" appear earlier: Lake Mungo 1 and 3 are either 40,000 or 60,000 years old; Keilor is very latest Pleistocene (probably 10,000–20,000 years old). Of the "robusts," WLH 50 is 14,000 years old, the Kow Swamp specimens vary from 9,000 to 12,000 years ago, or alternatively 26,000 years ago (Stone and Cupper, 2003), and the Coobool skills vary about the same, while Nitchie (if it truly belongs to this group) is only 6,000 years old. If the Dihybrid model is correct, then the "graciles" arrived first and, the robusts much later.

P. Brown (1981, 1987, 1989, 1992), however, has argued that there is virtually no difference between the two types. A single origin for the first Australians is likely if one takes into account (1) temporal considerations, (2) sexual dimorphism, and (3) cultural practices, such as intentional cranial deformation. We will explain these three factors one by one.

First, there seems to be an increasing robusticity over time. We have already noted that the "graciles" came first. But this could mean either that a new ("robust") population did enter later or that there was gradual change toward greater robusticity, because until about 15,000 years ago the fossil record is very sparse. Antón and Weinstein (1999) were the first to draw attention to this trend in overall increased robusticity during the later Pleistocene fossil hominin record.

Second, there is sexual dimorphism. Webb (1989) adopted the Dihybrid model and at the same time classified them according to sex. Independently, both Pardoe (1991) and Groves (1990) examined these allocations and concluded that, if Webb had been right about sex, almost all his males were "robust" and almost all his females were "gracile." In other words, the criteria one uses to determine sex are the same as those to determine which of the two "types" one is examining!

Finally, there is artificial cranial deformation. This is a result of parents' deliberately or unintentionally altering their children's head shape in infancy; it is said to look more attractive, or it is culturally distinctive, or it may simply result from swaddling or other cradling practices. It has been very widespread throughout the world over time: Within the last few hundred years the Maya, the Aztec, some Plains Indians, some Hawaiians, the people of Espiritu Santo in Vanuatu, offshore islanders in Papua New Guinea, Lebanese Maronites and other Middle Easterners, and some women in the Toulouse region of France, all had artificially deformed crania. Usually these became long and flat, with very flat receding foreheads, though swaddling resulted in the opposite (short and high). P. Brown (1981) showed that the flat frontals observed in the Kow Swamp crania (Figure 10.3) are due to such intentional cranial deformation and so are not natural like those of *H. erectus*. Brown had analyzed crania from Arawe, on the island Melanesia, which are known to have been deliberately deformed, and compared them to crania from the Sepik River Region, known not to be affected by cranial deformation. He demonstrated that many of the "unique" features observed in the Australian "robust" crania are also present in the Arawe deformed specimens, including flattening of the frontal and occipital, an increase in cranial height, an increase in postorbital constriction, a well developed prebregmatic eminence, and a low position of maximum cranial breadth (P. Brown, 1981). Because there is no evidence that *H. erectus* was practicing intentional cranial deformation, this means that any similarity between it and Kow Swamp is purely superficial.

P. Brown (1981, 1987) then went on to examine some of the features used by Thorne and Wolpoff. He suggested that these supposed unique Australasian features are not unique. For example, while the Kow Swamp

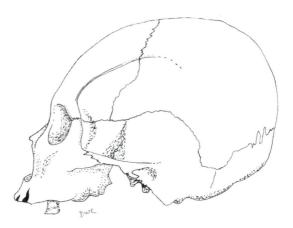

Figure 10.3 ▶   Example of intentional cranial deformation, defined as a sloping forehead and anterior projecting lower face. See text for more details.

Adapted from P. Brown (1981).

crania do show significant cranial thickness, Brown demonstrated that the thickness of the cranium observed in the female LM 1 is within the range observed in the Holocene female Murray Valley sample (P. Brown, 1987), while the male LM 3 specimen falls within the range of not only the male Murray Valley collection, but also the male Kow Swamp sample. Thus, again, the difference in cranial thickness between the robust and the gracile samples is not as distinct as some would have us believe.

It is appreciated, however, that intentional cranial deformation cannot alone explain all of the features said to be synapomorphies between the Indonesian and "robust" Australian fossil samples. The upper face of these groups is marked by well-developed supraorbital torus, a large postcanine dental complex, and large, robust mandibles. These characters will be the focus of the present discussion.

To date, many paleoanthropological studies have focused on phenotypic comparisons of fossil samples, with little or no understanding of the functional and/or developmental processes that lie behind the formation of the morphologies being described and compared. This approach can be misleading, especially for those proposing any clade relationship between populations. For example, the multiregionalists base their model on morphological similarity between groups. This is acceptable if you can show that the morphological features being used are homologies. Such characters are phylogenetically significant because they help us identify past evolutionary relationships. For example, expanded frontal lobes ssociated with the anterosuperior location of the anterior cranial fossa re a phylogenetically significant feature linking all modern humans (see).E. Lieberman, 1995, 2000). Homoplasy is the result of morphological onvergence, parallelism, or reversal and not the result of immediate hared common ancestry, an example being a developed incisive canal in •oth modern humans and the orangutan. Such characters are of no phylo- enetic significance because they have evolved separately.

The "Gracile" and "Robust" Australians of the Pleistocene and Holocene

There is no morphological evidence to support a Chinese origin for the ᵖleistocene Australians (e.g., Mungo and Keilor specimens) or Indonesians e.g., Wajak cranium). And let us remember that the very Keilor specimen hat is said by Thorne (1976) to be an important part of the "gracile" popu- ation is actually shown by Thorne and Wilson (1977) to be within the nodern comparative Murray Valley population in shape, but larger! Indeed, he only truly "gracile" Australian specimen is LM 1. A single specimen falling outside the general range is certainly noteworthy, but must be incon- :lusive. More significantly, however, LM 1 could be a juvenile. When cre- nated, her skull came apart at the sutures, and because of the skull's ʳragility matrix has been left and still covers the crucial basicranial region, ᵥhich would tell us whether she was mature or not (there is a key suture, he basilar suture — strictly speaking, the spheno-occipital synchondrosis — hat fuses at between 18 and 25 years of age). Should she, then, be called 'Mungo girl" and not "Mungo lady"? Maybe, maybe not. But it is very, ᵥery dangerous to base a whole supposed physical (racial, geographical) :ype on such an equivocal specimen.

The origin of the earliest Australians remains problematic. The overall derived morphology of the early Australians is clearly modern, and it best fits in with the arrival of a modern human population, immediately from Indonesia and ultimately "Out of Africa", that had left Africa some 40,000 years earlier. The physical and genetic migration of this modern human African population throughout Eurasia reached Australia by 60,000 or perhaps 45,000 years ago, earlier than any modern human occupation of Europe.

So what of the proposed phylogenetic relationship between the "robust" Australians and the more relict archaic *H. erectus* populations from

Indonesia? It is after all the later "robust" Australian material that is most frequently cited as supporting a multiregional paradigm. The features said to help define an Indonesian and Australian Pleistocene clade can be broken into two distinct matrices. The upper face, including the frontal squama, supraorbital, and facial prognathism, can be explained in terms of developmental processes associated with either (1) neuro-orbital disjunction in *H. erectus* or (2) degrees of cranial flexure and sphenoid lengthening in the "robust" Australians (and other Holocene "robusts" worldwide). The remaining features, including a developed nuchal torus, large postcanine dental complex, and a large robust mandible (all of which characterize all Aboriginal Australians anyway, especially the Pleistocene ones), can be explained in functional terms associated with the masticatory complex, which evolved within the context of adapting to prevailing conditions within Australia.

The upper face of the "robust" late Pleistocene/early Holocene Australians and the Indonesian middle Pleistocene populations has most frequently been used as a derived "complex" uniting these two groups, to the exclusion of all others. This is particularly true of the long, flat frontal and developed supraorbital region (Thorne, 1976; Thorne & Wolpoff, 1981, 1992; Wolpoff *et al.*, 1984; Wolpoff, 1999). These two features, however, are intricately linked in developmental terms, for both cranial and facial developmental processes will readily influence the form of the frontal bone.

The supraorbital region of the Kow Swamp individuals is superficially similar to those observed in the Sangiran specimens, with developed and continuous brow ridges. They are, however, even more similar to Middle Pleistocene African fossils such as Kabwe (Zambia) and European ones such as Petralona (Greece), relative to *H. erectus* (Figure 10.4). Indeed they are well within the range of early and extant *H. sapiens*. The upper facial morphology observed in the "robust" Australians is not a primitive retention but a derived modern human feature, and any similarity is illusory. Given that much of the discussion by the multiregionalists has been based on an examination of the Kow Swamp and Sangiran specimens, especially Sangiran 17 (Thorne & Wolpoff, 1981; Wolpoff *et al.*, 1984), let us review the upper facial morphology of these two groups (Figure 10.5).

Sangiran

The supraorbital is strongly developed and there is a slight inferior depression at its midregion at glabella (Rightmire, 1990). The supraorbital projects anteriorly over the face and is a dominant feature of the face, associated with

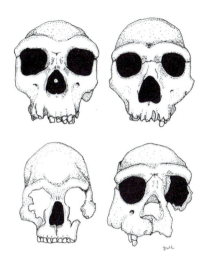

Figure 10.4 ▶ (Top left) *Homo heidelbergensis* specimen from Petralona, Greece. (Top right) *Homo heidelbergensis* specimen from Kabwe, Africa. (Bottom left) Modern human from Kow Swamp, Australia. (Bottom right) *Homo erectus* specimen Sangiran 17 from Indonesia. (See text for details.)

Adapted from Rightmire (1990).

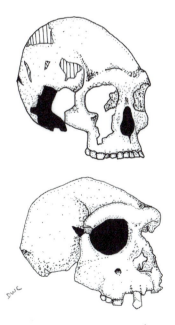

Figure 10.5 ▶ (Top) A modern human from Kow Swamp (KS1) and (bottom) the early *H. erectus* Sangiran 17 specimen from Indonesia.

Adapted from P. Brown (1981).

a developed frontal sinus complex (Thorne & Wolpoff, 1981). The torus deepens toward the lateral part of the orbit. This is also true of the Indonesian Middle Pleistocene (perhaps Upper Pleistocene?) Ngandong and Sambungmacan hominins (Santa Luca, 1980; Delson *et al.*, 2001; Marquez *et al.*, 2001). The frontal has a developed but small supraorbital sulcus (Thorne & Wolpoff, 1981). The supraorbital is only weakly arched. Glabella would have been more developed, though preservation has generally deflated this region (Thorne & Wolpoff, 1981; Rightmire, 1990). The frontal bone is broad and flat, with moderate postorbital constriction, though there is no midsagittal keel (Thorne & Wolpoff, 1981; Rightmire, 1990). The midface is moderately prognathic, relative to the projecting supraorbital torus (Thorne & Wolpoff, 1981; Wolpoff *et al.*, 1984).

Kow Swamp

The supraorbitals, compared to most living humans, are developed, though relative to Indonesian *H. erectus*, they are reduced. Unlike *H. erectus* they tend to be separate arched tori above each orbit, not thicker (deeper) laterally, and the supraorbital region does not project beyond the upper face (see P. Brown, 1981; D.E. Lieberman, 1995). Most Kow Swamp specimens do not have a supraorbital sulcus (Wolpoff *et al.*, 1984; Habgood, 1989), the only exception being Kow Swamp 7 (Wolpoff *et al.*, 1984). Like all modern and more archaic humans, Kow Swamp specimens have a developed frontal sinus complex (primitive hominid feature). Glabella is undeveloped (P. Brown, 1981). As in the Indonesian hominins, the frontal bones are broad and flat, with some degree of postorbital constriction (P. Brown, 1981). The midface is prognathic (Thorne, 1976; Thorne & Wolpoff, 1981), but not to the degree observed in the middle Pleistocene hominins. Indeed, their range of facial prognathism is well within the range of many other terminal Pleistocene non-Australasian "robust" groups (P. Brown, 1992; partially Groves & Thorne, 1999).

The upper face of these hominins is distinct. Over the considerable time that separates these populations (at least 1 million years) there is a marked reduction of robusticity, while still retaining the developed supraorbital region and flat frontal, so defining a general evolutionary trend between the "robust" Australians and the Indonesian *H. erectus* populations, to the exclusion of all others according to Thorne & Wolpoff (1981). (See also Wolpoff *et al.*, 1984; Thorne, 1984; Wolpoff, 1989).

Supraorbital development has been shown not to be under overt strain during mastication; its development is not related to masticatory demands. The original model of supraorbital formation proposed by Weidenreich (1941) has been reinvoked by a number of researchers (Moss & Young, 1960; Biegert, 1963; Shea, 1985, 1993; Ravosa, 1988; Hylander *et al.*, 1991; Hylander & Rovosa, 1992; C. Wood, 1994; D.E. Lieberman, 1998). It is now commonly argued that the development of the supraorbital in most nonhuman primates, including early hominins, is associated with neuro-orbital disjunction. This is when the face is pushed out from the braincase, as in *H. erectus*; the braincase, including the anterior cranial fossae that house the frontal lobes of the brain, is positioned posterior to the face, so the frontal lobes are long and relatively low (Broadfield *et al.*, 2001). This is associated with the moderate degree of postorbital constriction and often a frontal sulcus. In other words, the robust supraorbital of *H. erectus* is the result of frontal bone drift, which is required to bridge the space between the upper face and the more posterior braincase. Thus, the supraorbital torus is a structural "supporting beam" connecting the large and posteriorly orientated braincase to the face (Hylander & Ravosa, 1992; D.E. Lieberman, 2000; Cameron, in press a).

The condition in *H. sapiens* over the last 170,000 years or so is that the anterior cranial fossa has moved anterosuperiorly, and so the frontal lobes are located directly above the orbits, as opposed to the more posterior position of *H. erectus*. This results in a vertically oriented frontal bone, with the face positioned directly beneath, reducing the supraorbital region and the frontal sulcus. This can be referred to as neuro-orbital "convergence." This is clearly the way it is in Kow Swamp individuals (Figure 10.6), even taking into account those individuals whose frontals have been affected by intentional cranial deformation; indeed, individuals that are not influenced much at all by intentional deformation show that the anterior cranial fossae are located directly above the orbits in the typical modern *H. sapiens* fashion, and not posterior to them as in *H. erectus* (see also radiographs published in D.E. Lieberman, 1995, Figure 2). As will be discussed presently, the formation of the supraorbital in these modern human populations is also associated with an increase in sphenoid length and cranial base extension (see D.E. Lieberman, 1998). Therefore, the suggested 'similarity' between these two populations is the result of two very different processes, one biological the other cultural; i.e., they clearly arose independently.

Large supraorbitals are known to occur in a number of recent *H. sapiens*, especially in large robust males with narrow skulls, including some northern Europeans and African Bushmen (D.E. Lieberman, 2000), and in a number

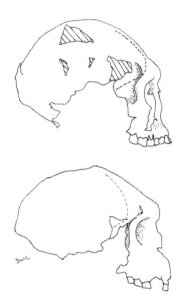

Figure 10.6 ▶   (Top) A modern human from Kow Swamp (KS1) showing the modern human condition of neuro-orbital convergence, as opposed to (bottom) the more archaic condition of neuro-orbital disjunction observed in the early *H. erectus* Sangiran 17 specimen. (See text for details.)

of fossil *H. sapiens* from Africa (Dar es-Soltan), Western Asia (Skhul V, Qafzeh), Europe (Madlec, Predmosti, Cro-Magnon), and East Asia (Zhoukoudian Upper Cave specimen 101) (see Wu, 1961; F.H. Smith & Raymond, 1980; Bräuer, 1984; Stringer *et al.*, 1984; Kamminga & Wright, 1988; Corrucini, 1992; Etler & Li, 1994; Wu & Poirier, 1995; D.E. Lieberman, 2000). The robust supraorbital is a primitive hominin feature and is not evidence for regional continuity (see Groves, 1989b). Its development in these *H. sapiens* is a result of increased anterior cranial base angulation, and thus it is distinct from the neuro-orbital disjunction, which is its origin in *H. erectus* and other earlier hominins. It also explains the increase in midfacial prognathism, which is a by-product of cranial base angulation (see D.E. Lieberman, 1998, 2000).

The degree to which the cranial base is flexed will affect the development of the supraorbital and will directly influence the degree to which the midface is prognathic. A more flexed cranial base will position more of the face beneath the anterior cranial fossa; a more extended cranial base will position more of the face in front of the fossa (D.E. Lieberman, 2000), associated with an increased length of the sphenoid (D.E. Lieberman, 1998).

The sphenoid is the central bone of the cranial base, from which the face grows anteriorly during ontogeny (D.E. Lieberman, 1998); with increased sphenoid length, the upper face will move forward, which in turn results in increased supraorbital development (see Enlow and Hans, 1996). A similar pattern can be observed in some Neanderthal specimens, and it is due to similar processes (D.E. Lieberman, 1998).

Finally, Westaway and Cameron (submitted) have tested the suggested affiliations between the erectine populations of Asia and the "robust" populations of Australia using parsimony analysis (see Chapter 5 for discussion of this method). Because we were testing the proposed relationship of the Ngandong populations (Indonesia) and the Kow Swamp people (Australia), and because the Ngandong fossils are represented by cranial specimens only (i.e., no facial specimens), we used 27 cranial features preserved in the Ngandong fossils. The resulting analyses are shown in Figures 10.7 and 10.8. Figure 10.7 is based on a strict consensus from

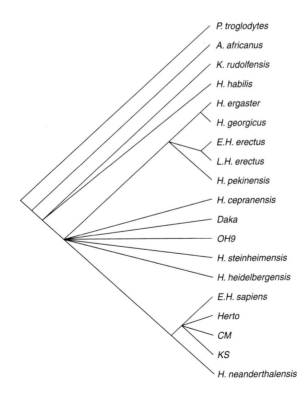

Figure 10.7 ▶   Strict consensus tree of 27 cranial features. (See text for details.)

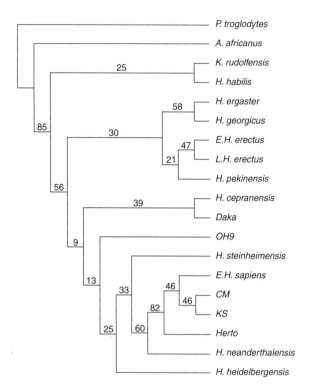

Figure 10.8 ▶   Bootstrap analysis with 1,000 replications of 27 cranial features. (See text for details.)

22 trees, with a tree length of 100 and a rescaled consistency index of 0.279. This analysis clearly defines an erectine clade defined by African, eastern European, and Asian specimens as distinct from all others. The relationship between the Ceprano, Daka, and OH 9 specimens, as well as the populations defined by Steinheim, and the species *H. heidelbergensis* cannot be resolved (perhaps not surprising given that only 27 cranial features are used). Significantly, however, classic specimens of *H. neanderthalensis* have a deeper prehistory than that of the derived *H. sapiens* clade; i.e., early *H. sapiens*, the Herto, Cro-Magnon (CM), and Kow Swamp populations (KS). Indeed, let us remember that early *H. sapiens* and Herto are substantially older in time than the Neanderthals. To further test the significance of these results, a bootstrap analysis was run on these 27 characters, with 1,000 replications. This resulted in greater resolution of proposed relationships, with *H. heidelbergensis* emerging after the older Pleistocene hominins; this is followed by the Steinheim group, and then the

Neanderthals. It is only *after* the emergence of the Neanderthal lineage, however, that we observe the appearance of the modern human condition first defined by Herto, then early *sapiens*, then finally Cro-Magnon and Kow Swamp. Both of these analyses clearly support the emergence of the modern human lineage *after* the emergence of the Neanderthals, which rejects the Multiregional hypothesis.

The remaining features that are said to support a regional continuum between the middle Pleistocene Indonesian hominins and the late Pleistocene Australasians are largely functional in origin. These features include a developed nuchal torus, a large robust mandible, and a large postcanine dental complex (Thorne & Wolpoff, 1981, 1992; Wolpoff *et al.*, 1984; partly P. Brown, 1981). They are associated with the development of the masticatory apparatus and *in vivo* responses to diet. They are closely integrated as a functional complex and should not be isolated as individual features.

As Wolpoff (1971) has shown, a large dental complex is observed in the sub-Saharan Africans, second only to Australians (see also D.E. Lieberman, 1995). These and other modern human groups were in most cases dependent on the mastication of largely unprepared food resources (little preparation before consumption), with the molars pulping and grinding tough food items. Selective pressures would result in an increase in the dental complex, for a biomechanical failure to cope with such stresses would result in early death.

In order to house a large dental complex, one requires a large robust mandible, with its associated musculature. Clearly a small mandibular frame housing a large dental complex will result in masticatory stresses, associated with mandibular wishboning and twisting of the mandible along its long axis, and these will result in disastrous compressive and tensile stress and a breakdown of the masticatory system (see Hylander, 1984, 1988; Aiello & Dean, 1990; McGowan, 1999; Martin *et al.*, 1999).

The development of skeletal form, including the mandible, continually works to provide a state of balance between separate entities as they merge and grow into a "single" functional whole (Enlow & Hans, 1996). A state of balance never really exists because other morphological units are forced to change in response to local "adaptations," which in turn propel changes elsewhere; there is a continual game of "catch-up" between muscle mass and bone mass. Any local change will result in a domino effect throughout the soft tissue and skeletal systems as it adapts to the corresponding changes (Enlow, 1982; Enlow & Hans, 1996; Martin *et al.*, 1999). The major muscles of mastication, the masseter, temporalis, and pterygoids, require significant muscle attachment sites. The same applies to the neck muscles controlling the rotation of the head, the sternocleido-mastoideus, and the rectus capitis posterior muscles, and these will affect the development of the nuchal crest and the mastoids.

The Archeological Evidence

The original Australians are among the earliest known true modern humans. To get to Australia required a number of ocean voyages over considerable distances. The original colonization of this continent represents by far the earliest evidence of oceangoing craft. Indeed, the minimum distance, even during a glacial maximum, would still be around 50 km (Chappell, 1976; R. Jones, 1977; P.J. White & O'Connell, 1982; Bellwood, 1997). Two open sea routes are most likely as the entry points to Australia. The first is the eastern route via Maluku to New Guinea, the other is the southern route from Nusatenggara to Northwest Australia. Either way, it suggests that these people had extensive and sophisticated knowledge of working wood, bamboo, and other related materials (see Klein, 1999). Not only this, but the early Australians are distinguished from the later occupants of Europe in terms of art and burial practice; in both cases in fact, Australia has the earliest evidence for these practices (R. Jones, 1989). For example, at Mandu Mandu Rockshelter, the earliest known evidence of beads (modified cone shells) has been documented at 32,000 years ago, while the female Mungo individual (LM 1) is the oldest cremation known anywhere in the world (see Klein, 1999). By 26,000 years ago, the ancestors of indigenous Australians were using very sophisticated fishing techniques, including gill nets and traps (Balme, 1983).

What this means is not that navigation, art, beadwork, cremation, gill nets, and fish traps were invented in Australia and spread elsewhere. This may be so, but the sea barriers are vast, and contacts must have been few. Instead, we think that these commonalities could mean one of two things. Some of these cultural practices may well have been developed in Australia and in the rest of the world independently: Common humanity has responded to similar challenges in similar ways. But some of the Australia/rest-of-the-world similarities are so complex in all their details hat we can draw only one conclusion: They already existed in the common ancestor, before the colonization of Australia over 60,000 years ago. Language surely arose once, as did the control of fire and religion and music. These are cultural items for which there is no direct evidence, and yet the principle of parsimony suggests very strongly that they are very much earlier in their genesis.

It is likely that the original colonization of Australia was based on coastal movements through the landscape. By 25,000 years ago, however, the inland region had been penetrated (Bowdler, 1997). The occupation of the interior of the continent may have been via the many extensive river systems (see Mulvaney & Kamminga, 1999; and Flood, 1999). The stone tool tradition for the first 30,000–50,000 years is not particularly distinctive and is referred to as the "core tool and scraper tradition." It is not until the mid-Holocene (around 5,000 years ago) that the tool tradition rapidly changes and we get a more refined tradition known as the Australian "small-tool phase" (Figure 10.9). This new technology includes stone knives, daggers, neatly trimmed spear points and barbs, and many other micro all-purpose tools (Mulvaney & Kamminga, 1999).

If there was a separate dual migration into Australia by "robust" and "gracile" populations, then there is certainly no evidence for it within the archeological record. Indeed, overall, the archeological record demonstrates a continuum within and between the earliest inhabitants and the later descendants until the "arrival" of the mid-Holocene small-tool tradition.

The Molecular Evidence

The extraction of ancient mtDNA from a number of Australian Pleistocene human remains (Adcock *et al.*, 2001) is of significance because it enables us for the first time to examine and compare early modern human mtDNA, dating between perhaps 60,000 and 8,000 years ago, with recent modern humans. Mitochondrial DNA samples have been extracted from the early "gracile" specimens from Willandra Lakes (LM 3) and later "robust" populations from Kow Swamp and the "gracile" near-Holocene Willandra Lakes bones. The LM 3 mtDNA is said to be divergent from all modern human populations, including the later Pleistocene Australians. This is at first sight surprising, given that LM 3 undoubtedly represents a modern human. Yet all it really means is that we are seeing an "extinction" over time as a result of drift and natural selection of an ancient form of modern

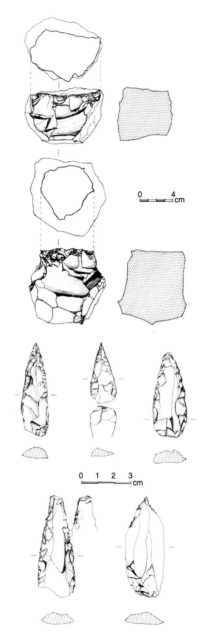

Figure 10.9 ▶   Typical "large core tool tradition" (top) common in Australia until around 5,000 years ago, which was largely replaced by the Australian "small-tool tradition" (bottom).

From P.J. White and O'Connell (1982).

human mtDNA, just as would be expected from the nonrecombining nature of mtDNA (Adcock *et al.*, 2001; Relethford, 2001). For the describers it suggests the extinction of a whole race. This seems to us a rather extravagant interpretation, depending as it does on a one-to-one linking between a mtDNA lineage and a human type. As molecular geneticists have long insisted, a gene tree is not, or not necessarily, to be equated with a species or population tree.

Another significant finding is that the earlier Mungo specimen is closer to modern humans than to the (later) Neanderthal samples, thus confirming the distinctiveness of the Neanderthals from living humans (see also Krings *et al.*, 1997; R. Ward & Stringer, 1997; Stringer, 1999). Perhaps the most relevant finding that enables us to test directly the "Out of Africa" and the Multiregionalist models is that there is no support for a dual migration to Australia. The near-Holocene "graciles" from Willandra Lakes and Kow Swamp show absolutely no difference in their mtDNA, suggesting that the presence of two distinct populations within Australia is most unlikely. Indeed, neither "robust" nor "gracile" populations can be differentiated from living Aboriginal Australians (Adcock *et al.*, 2001; Relethford, 2001).

But there are other wrinkles to the ancient Australian mtDNA story. The first is that, when evolutionary trees, whether of DNA or of whole populations as taxa, are published, one should include some statistics that indicates the support for the particular branches, to show how much credence should be placed on that branch. In their paper on ancient Australian DNA, Adcock *et al.* (2001) gave branch support values for all branches *except* the crucial one. They sequenced chimpanzees as well as humans and gave statistical values for both branches: Humans are not chimpanzees (or vice versa). They also included the data for the Neanderthals and gave the statistical values for the two branches: Good, modern humans are not Neanderthals (or vice versa). They then showed a branch separating Kow Swamp 8 from other *H. sapiens*, but the non-KS 8 branch had only 30% support, so, quite rightly, they concluded that it could not be relied upon: no evidence that KS 8 is outside the modern human range. Then came the branching of LM 3 from the other *H. sapiens*, but curiously no support value was given for the non-LM 3 branch. In other words, no evidence was given that the rest of us *do* form a branch separate from LM 3!

From time to time, DNA, being the mobile chemical it is, results in a lot of mtDNA inserting itself into the nuclear DNA. Inserts like this are called *numts* — there are, we now know, quite a lot of them in the nDNA of many species, including humans. Now, mtDNA evolves very fast. But once it forms into numt, its rate of evolution slows down to the more reduced pace typical of nDNA, so it remains there, a living fossil of what mtDNA used to look like. Adcock *et al.* (2001) found that LM 3's mtDNA strikingly resembles one of the modern human numts. Indeed, Colgan (2001), noting how difficult it is to avoid contamination when sequencing ancient DNA, suggested that, though Greg Adcock had taken the precautions of sequencing his own (and Alan Thorne's) mtDNA to make sure that what he thought was ancient DNA was not his own (or Alan Thorne's) by mistake, he might have inadvertently sequenced one of his own numts! If he had, that would explain a lot. Finally, biologist John Trueman redid the statistical analysis by a different but more "traditional" method and could not replicate Adcock *et al.*'s results anyway.

In summary, the available paleoanthropological evidence does not support a morphological continuum between the Indonesian middle Pleistocene hominins and the late Pleistocene "robust" Australians from Kow Swamp and/or Coobool Creek. Any suggested superficial similarities in the upper face, including supraorbital and midfacial prognathism, are the result of two very different developmental processes, showing that these features are an anatomical analogy as a result of functional convergence (see Figures 10.7 and 10.8).

Given that there is little, if any, biological evidence to support a separate Australasian Pleistocene clade, from where did the original Australians originate? The early dates for the LM 3 individual (at least 40,000–60,000 years ago) support the hypothesis that the much later Australians can easily be accommodated within this initial colonizing population. Any real or imagined differential degree of robusticity between the early and later Australians can be explained by time difference; during the Holocene we see the reverse condition happening, from "robust" to more "gracile" (within only a few thousand years), just as happened in many other, non-Australasian, populations, including Africa (see Wright, 1976; Groves & Thorne, 1999). Cultural influences may also have played a part; the later Kow Swamp people practiced cranial deformation, while the earlier Willandra Lakes population did not. Over a considerable time range, it is not difficult to see adaptive, selective, and cultural forces operating within a relatively isolated population.

The origins of the pre-Mungo Australians, which initially colonized Australia, are clearly an important research issue. Were they part of an

"Out of Africa" population migratory wave (physical and/or genetic), which started around 120,000 years ago? Well-developed neuro-orbital convergence first appears in the African fossil record with specimens like the Omo-Kibish *H. sapiens*, dated to over 130,000 years ago. The more archaic Indonesian middle–late Pleistocene *H. erectus* populations, with their neuro-orbital disjunction, had nothing or very little to do with the biology of the late Pleistocene of Australia.

There is no evidence supporting an early "gracile" migration from China around 60,000 years ago followed later by a "robust" migration from Indonesia around 15,000 years ago. The recent dates for the Ngandong (Solo) *H. erectus* specimens, at around 50,000–30,000 years ago (Swisher *et al.*, 1996), if accepted, postdate the arrival of the modern humans in Australia by at least 10,000 years (probably more). If these dates are correct, then two species of *Homo* occupied this region around 50,000 years ago, which means that the Solo people cannot have been ancestors of people who lived earlier than they did (Swisher *et al.*, 2000), though they could have been survivors of a population that supplied a few migrants to join *H. sapiens* in the voyage to Australia (as we have argued, there is actually no evidence of any such fellow voyagers). Instead, the Solo people were a persisting primitive form, unaffected by the moderns who had to pass through Indonesia to get to Australia. The Ngandong people eventually succumbed to competition pressures from modern *H. sapiens* who settled in the area.

The archeological record supports a single population within Australia from its earliest colonization. It supports a continuum within the cultural practices of the early and later Australians. If there was a sudden significant migratory event after the initial colonization, one would expect to see some cultural differences within the archeological record as evidenced in other parts of the world, but there is nothing to suggest a "cultural" shift until the mid-Holocene.

Finally, Adcock's molecular evidence suggests that some of the original colonizers of Australia had a unique genetic sequence unrepresented in living modern humans, which should not be surprising given that the original inhabitants occupied Australia from at least 40,000–60,000 years ago. The genetic sampling of near-contemporary fossil specimens will be crucial in determining the origins of these earliest Australians. If the "Out of Africa" model is correct, then we can predict that the mtDNA of the Pleistocene Eurasian material should all fall within a similar range and be extremely close to the African material of the same age. These modern human Pleistocene populations should be closer to each other than to the more "archaic" populations they presumably replaced. This has already been partly demonstrated by the molecular evidence from the 12,000-year-old remains of "Cheddar Man," which are almost identical to extant human populations and nothing like mtDNA of the Neander Valley Neanderthal (see Chapter 8). The current molecular evidence shows that there is no molecular support for two populations, one "gracile" and another "robust," within Australia. It supports the presence of a single population.

INTERLUDE 5

Milford Wolpoff in the Garden of Eden

The three great world monotheistic religions — Judaism, Christianity, and Islam — all take the story of Adam and Eve and the Garden of Eden as their myth of origin. Of course, most adherents of these faiths nowadays agree that the story is a myth in the literal sense (it is untrue), and that its purpose is mythical in the anthropological sense (it is full of cultural meaning). Naturally, a few fundamentalists prefer to reject the whole apparatus of modern science and to stick with Adam (a man with an asymmetrical rib cage) and Eve, a talking snake, a deity who needs to keep cool by walking in the shade, and all the rest of it. The Book of Genesis is less bloody and brutal than some other origin myths, such as the battle of the gods and titans in ancient Greece and the slain gods of the Aztecs, and more so than others, such as the cosmic egg of the Dogon. So who's for a bit of science?

So redolent of origin mythology is the Garden of Eden that it has been used as a basis for the preferred model of the scientific origins of modern humans. We know that before 2 million years ago, all our ancestors lived in Africa. But did everyone, through human evolution, come out of Africa? Couldn't Asia have produced an ancestor now and again? No? Not even a little ancestor? Sorry. Dmanisi looks as if it came out of Africa. *Homo erectus* and *Homo pekinensis* have that look about them, too (but separately, or from a common "Out-of-Africa" stem? We don't know). And the Dali–Jinniushan stock seems to have come out of Africa. The Neanderthal lineage — that too came out of Africa.

How about *Homo sapiens*? Bill Howells, one of the founders of modern paleoanthropology, proposed in the early 1970s that our species was the last "Out-of-Africa" product. Gunther Bräuer in the mid-1970s took up the theme and proposed a modification; Chris Stringer at about the same time indicated that he preferred to keep it intact, much as Howells envisaged it. Becky Cann in the late 1980s supported the model with her reconstruction of mitochondrial DNA evolution, and since then, molecular geneticists have almost without exception supported it. Out-of-Africa, or the Garden of Eden model. Africa, not Mesopotamia, was our Eden all the way down.

But there is a different model. In the 1930s and 40s, Franz Weidenreich thought, from his studies of the Zhoukoudian ("Peking Man") and Ngandong ("Solo Man") remains, that he had evidence that the different modern human races had evolved *in situ*. In particular, Zhoukoudian was the ancestor of Mongoloids, Java (Sangiran via Ngandong) the ancestor of Australoids. In 1963, Carleton Coon revived Weidenreich's model, but seemed to be suggesting that the five races, which he recognized, had evolved quite independently from their nonmodern precursors and had crossed the boundary into *Homo sapiens* independently. The model was refined in the mid-1980s by Milford Wolpoff, Wu Xinzhi, and Alan Thorne and dubbed "*regional continuity*" (or the Multiregional hypothesis).

Weidenreich knew rather few fossils, but it did seem impressive that all the Zhoukoudian incisors were shovel-shaped. And where today are shovel-shaped incisors most common? Why, among Mongoloid peoples, in East Asia and, particularly, among North American Indians. And a large triangular bone separated from the rest of the occipital by a horizontal suture occurred in three out of five Zhoukoudian crania, and today it is far more common in Peru than anywhere else, to the extent that it is often called the Inca bone. Incas are Mongoloid — this is less convincing than the shovel-shaped incisors, perhaps, because you have to go so far away from Zhoukoudian to find them. Weidenreich also drew attention to a few other features that occur at Zhoukoudian and are more common in some Mongoloid populations than in other modern peoples.

Then Weidenreich turned to the Javanese *Homo erectus* fossils. Where today do we find such flat, receding foreheads? Why, in Australia and Melanesia. Not precisely Java, but close.

It is difficult now to know what most of Weidenreich's contemporaries thought of his comparisons, but we must remember how very sparse the fossil record was in those days. There were modern and Neanderthal skulls from Europe and those funny Chinese and Javanese ones, and there was gradually increasing acceptance of the australo-pithecines from Africa. So if we took the geographic distribution of fossils at face value, it might appear that our ancestors had first arisen in Africa, then shifted to Asia, and finally moved over into Europe. This seemed a bit unlikely, so perhaps we should wait until we knew more about the distribution of shovel-shaped incisors.

Coon's 1962 book *The Origin of Races* revived the Weidenreich hypothesis — and how! It is a most annoying book to wade through. First, was he really saying that modern human races had evolved from non-*sapiens* ancestors (he called them all *Homo erectus*) quite independently? Mostly he did really seem to be saying this. Yet in places he seemed equally to deny that it was possible. How many modern races did he recognize? In one place he said it didn't matter, but immediately he appeared to backtrack, because through-out the book he spoke of five: Caucasoid (that's Europeans, Arabs, and Indians, the stan-dard definition), Mongoloid (eastern Asians, Polynesians, and Native Americans), Australoid (Aboriginal Australians and Melanesians), Congoid (black Africans) and Capoid (Khoisan — Bushmen and "Hottentots"). He insisted on his five lines of descent, but only in one case did he actually say why; this was the "Peking Man"-to-Mongoloid lineage, the one Weidenreich had already done. Elsewhere his arguments were of the sort that George Orwell had dubbed "bullying arguments," simply iterating and reiterating that it was so (such as about the Kabwe skull: "On the whole this face is mostly Negro" he says baldly on p. 626). Especially confusing was his Capoid lineage: He began it with the Ternifine and other Middle Pleistocene remains from North Africa, so the Capoids had to have picked up their tents and moved south through the Congoids, but evidently without mixing with them, to become Bushmen in southern Africa — all this without letting on why Ternifine and the others were Capoid in the first place.

The anthropological community became very angry with Coon for this book. We are sorry to say that there were actually very few cogent critiques of it. Mostly the anger was directed at the racism that seemed to be implied. Some races (Mongoloid and Caucasoid) had been *Homo sapiens* for much longer than others — he was quite explicit about this — and so, one was left to infer, had had a much longer time in which to get really good at being *sapiens*. There is a simply shocking pair of photos making up Plate XXXII. The caption reads, "The Alpha and Omega of *Homo sapiens*: An Australian aboriginal woman with a cranial capacity of under 1,000 cc. (Topsy, a Tiwi); and a Chinese sage with a brain nearly twice that size (Dr. Li Chi, the renowned archeologist and director of Academia Sinica)". Up till then, racism and antievolutionism had gone hand in hand in places like the Deep South of the United States; suddenly it became acceptable to believe in evolution in a racist context.

Whether Coon had chosen his words carefully or was just being what he might have called a dispassionate scientist, the fact remains that his book was interpreted in a racist sense, and he did not a thing to disabuse anyone.

After this furor, it would have been a brave anthropologist who would revive Weidenreich's model. But a brave anthropologist did: Milford Wolpoff.

Wolpoff is not a racist. He makes it clear that he regards Coon's book as an unfortunate episode in the intellectual lineage from Weidenreich to himself — alright, as a real blot on that otherwise-honorable lineage. He is also one of the most admired paleoanthropologists working in the field today; his book *Paleoanthropology* (first published in 1980 and with a monster 2nd edition in 1999) shows a simply amazing grasp of the material. So we think he's one of the good guys. How then can he be wrong?

His coproposers of the Regional Continuity hypothesis are also good guys. Wu Xinzhi is China's most eminent paleoanthropologist, particularly since the retirement of the remarkable (and indeed sagelike) Wu Rukang. All the significant human fossils of China since Zhoukoudian were discovered or described (or both) by him. Alan Thorne, too, is one of the good guys. It was Alan Thorne who discovered most of the human fossil material in Australia, and he kept public (including Aboriginal) interest in it alive for a quarter of a century.

Today's multiregionalists repudiate Coon. They insist that there was global gene flow as well as genetic continuity over time. So each stage in sapientization was quickly transmitted to all contemporary populations, while at the same time regional characteristics were preserved.

So how can these good guys be so wrong? Somewhere they got stuck. Eventually, a controversy ceases to be based on the balance of the evidence; instead, each side simply looks for evidence to support its long-held view. The multiregionalists seem to be doing that sort of thing. Not the Garden-of-Edenists, no, of course not! We are above that sort of thing!

Alright, the arguments: The similarity of shovel-shaped incisors between *Homo pekinensis* and modern Mongoloids fails because all the earlier stages of the human lineage, not merely the Zhoukoudian people, had shovel-shaped incisors. It's nonshoveled incisors in other modern people that are the "special" thing. And anyway, it's just the frequencies that differ between modern peoples; like all racial characters there's no 0% versus 100%. The Inca bones sporadically recur in other populations, and are again nothing special. As for the "flat" foreheads of modern Australoids and Java *Homo erectus*, the frontal bones of all other middle Pleistocene fossils — *pekinensis* alone excepted – are at least as flat as that. Indeed, the low frontals observed in *H. erectus* are associated with a pattern of neuro-orbital disjunction, while any "flat" pattern observed in modern *H. sapiens* is associated with the exact opposite condition of neuro-orbital convergence and thus refutes any "similarity" in form.

On the positive side, the supraorbital ridges of *Homo heidelbergensis* are different from those of its eastern Asian contemporaries (they arch, and they are deeper over the middle of each orbit, whereas those of *H. erectus* and *H. pekinensis* are straight, even thickened laterally), and they resemble those of brow-ridged modern humans. Eastern Asian "erectines" never have an ossified styloid process, but this is universal in *Homo heidelbergensis* and almost so in modern humans. And so it goes on. And the earliest *Homo sapiens*, wherever they come from, always resemble Australoids in features like brow ridges, frontal flatness, and large jaws and teeth, and modern racial features can be seen to develop quite late in each region.

There is no continuity. Milford Wolpoff finds himself in the Garden of Eden. But he won't eat no apple.

CHAPTER 11

Epilogue

The earliest hominids recognized up to now are the Miocene Kenyapithecinae of Eurasia (*Griphopithecus*) and Africa (*Kenyapithecus*). *Griphopithecus* may have given rise to the later European *Dryopithecus* and possibly to *Oreopithecus* in Italy and *Graecopithecus* in Greece. While *Dryopithecus* and *Oreopithecus* became extinct, *Graecopithecus* may have been part of the large-scale mammal dispersal back into the African continent, where it gave rise to *Samburupithecus* and ultimately the living African hominids — *Gorilla, Pan,* and of course that other ultimately African hominid, *Homo*.

Whether the first hominids were endemic to Africa or not remains debatable. But what is clear is that the earliest hominins — members of our clade — were African. This is true whether or not *Sahelanthropus* was a hominin. We witness in the late Miocene and early Pliocene of Africa an explosion in hominid diversity, and from among this diversification the earliest hominins emerged. In the details, we two authors agree to disagree. The taxonomic scheme of Groves (Table 11.1) indicates a diverse range of species allocated to the genus *Homo*; the scheme presented by Cameron (Table 11.2), however, acknowledges the long-distinct evolutionary trajectories (in terms of phylogeny and adaptation) and on this basis allocates the same species to a number of different genera. The undoubted hominins *Australopithecus* and *Kenyanthropus* are likely to have originated from a *Sahelanthropus*-like (Cameron), *Orrorin*-like (Groves), or Garhi-like ancestor. The east African species previously allocated to *Australopithecus*, including *afarensis*, may have played no direct role in the evolution of

TABLE 11.1 ► Colin's Hominid Taxonomy

Hominidae
   Kenyapithecinae
   Ponginae
   Dryopithecinae
   Homininae
        Gorillini
           Graecopithecus
           Gorilla
               G. gorilla
               G. beringei
      Hominini
         Pan
             P. troglodytes
             P. paniscus
         Orrorin
             O. tugenensis
         Homo
           (stem group)
               H. kadabba
               H. ramidus
           (australopithecine group)
               H. anamensis
               H. bahrelghazali
               H. afarensis
               H. garhi
               H. africanus
           (paranthropine group)
               H. walkeri
               H. boisei
               H. robustus
           (kenyanthropine group)
               H. platyops
               H. rudolfensis
           (habiline group)
               H. habilis
           (erectine group)
               H. ergaster
               H. georgicus
               H. erectus
               H. pekinensis
               H. antecessor
           (sapient group)
               H. heidelbergensis
               H. neanderthalensis
               H. sapiens
   Hominidae indet.
     Sahelanthropus tschadensis
     Lothagam hominid

TABLE 11.2 ▶   Dave's Hominid Taxonomy

Hominidae
    Kenyapithecinae
    Ponginae
    Dryopithecinae
    Oreopithecinae
    Gorillinae
        Graecopithecini
            Graecopithecus
        Gorillini
            Gorilla
                G. gorilla
    Paninae
        Panini
            Pan
                P. troglodytes
                P. paniscus
    Homininae
        Hominini
            Orrorin
                O. tugenensis
            Sahelanthropus
                S. tschadensis
            Garhi deme
            Australopithecus
                A. africanus
            Paranthropus
                P. walkeri
                P. boisei
                P. robustus
            Kenyanthropus
                K. platyops
                K. rudolfensis
            Homo
                H. habilis
                H. ergaster
                H. georgicus
                H. erectus
                H. pekinensis
                H. heidelbergensis
                H. neanderthalensis
                H. sapiens
Hominidae indet.
    Lothagam hominid
    Ardipithecus ramidus
    Praeanthropus afarensis
    Anamensis hominids
    Bahrelghazali hominids

humans (or indeed chimpanzees). Cameron entertains the idea that *Ardipithecus* may eventually turn out to be a kind of proto-chimpanzee or a "third chimpanzee" if you will, notwithstanding its proposed primitive form of bipedal locomotion. For a fossil group to be considered hominin, clearly it must display bipedal characteristics (along with a number of other derived features). But as we have observed with the proto-australopithecines bipedalism alone does not define a hominin.

During the later Pliocene, *Paranthropus* and early true *Homo* emerge from this same basal hominin population. By this time, the later species of *Kenyanthropus* appears, but apparently it could not compete with the more specialized species of *Paranthropus* or the more general and opportunistic behavioral repertoire of early *Homo*. And later the overspecialized species of *Paranthropus* were probably also pushed over the edge into extinction by the appearance of *Homo ergaster*, which likely increased its ecological niche to include and take over the more marginal habitats of *Paranthropus*. There is no convincing evidence to date that species other than *Homo* were involved in the manufacture of stone tools — and these increased their ability to process and consume large quantities of food items that had previously been denied other hominins, including bone marrow. *Paranthropus*, we suppose, was not a habitual tool user or toolmaker, and its food-processing abilities were dependent upon its derived masticatory apparatus rather than tool technology. Up until this point, human evolution followed the course of other mammal groups: New species would arise followed by climatic and environmental change, resulting in the rise of further taxa but also the extinction of some of the older ones. It is with the emergence of *H. ergaster* that we see a physical shift to the human condition, both in body proportions and in more complex behavior, which ultimately resulted in the first dispersal "Out of Africa" around 1.7 million years ago.

Even after the dispersal of *Homo* out of Africa, speciation events were still occurring. We see the emergence of *H. georgicus* in western Asia, and within eastern Asia the rise of *H. erectus* and *H. pekinensis*. The long morphological stasis observed in *H. erectus*, lasting over 1 million years, is likely associated with the long-lasting ecological stability in this region in strong contrast to the dramatic fluctuations observed in China, Africa and especially Europe.

The earliest exodus of *H. ergaster* from Africa must have occurred before the development of the Acheulean industry because *Homo georgicus* and *H. erectus* are associated with a more primitive, Oldowan-like technology while those demes of *H. ergaster* that remained in Africa developed the more advanced Acheulean tradition. Indeed, the Acheulean tool complex would ultimately spread into Europe along with later immigrants from Africa during the middle Pleistocene, *Homo heidelbergensis*. It was not, however, a "one-way street," as witnessed by the presence of something like "classic" Asian *H. erectus* in Italy around 800,000 years ago (as represented by Ceprano) and even earlier into Africa before 1 million years ago (as indicated by OH 9 and Daka).

Around 800,000 years ago, at the same time that a Eurasian deme of *H. erectus* was moving into Italy, a new species of *Homo* emerged in Spain. This was *H. antecessor*, and it was closely followed by the appearance in Africa of *H. heidelbergensis*. We doubt whether the European *antecessor* is ancestral to the African *heidelbergensis*, for two main reasons. First, there are no synapomorphies uniting these taxa. Second, *H. antecessor* is associated with an Oldowan technology, not with the more advanced Acheulean technology that is associated with *H. heidelbergensis*. We think that *H. heidelbergensis* probably inherited this technology from its more likely ancestor, *H. ergaster*. Thus *H. antecessor* may represent a pre-Acheulean-dependent *H. ergaster* penetration of Europe — perhaps *H. georgicus*.

By 500,000 years ago, two distinct lineages were evolving in the west. The first was endemic to Europe, which we call the Steinheim group (basal Neanderthals), though it is not restricted to Europe alone because we believe that the Narmada cranium from central India and Maba from China also represent part of this lineage. The second lineage of *H. heidelbergensis* is endemic to Africa, later moving into Europe and even China (Dali and Jinniushan). Later, African demes of *H. heidelbergensis* gave rise to early *H. sapiens*, the initial split occurring around 300,000 years ago.

The current fossil, molecular, and archeological evidence strongly supports the specific distinction of the Neanderthals from early and modern *H. sapiens*. The Neanderthal lineage first appeared some 500,000 years ago, according to the molecular clock, and the fossil record suggests that the "Steinheim group" reflects a basal Neanderthal anatomical condition. This group exploited the colder conditions of an Ice Age Europe, their survival in this region enabled by the increasing development of their "cold exaptations." The initial split between these two groups was not based on a greater ability to withstand Ice Age conditions (though ultimately this must have pushed them farther apart over time), but occurred for some other reason, perhaps cultural.

The molecular evidence shows that Neanderthal mtDNA is significantly different not only from modern humans, but also from *Homo sapiens* dating back to 40,000 years ago. Neanderthals had a distinct pattern of resource exploitation, which includes scavenging and the hunting of large mammal herds and less likely the stalking of individual animals. They also appear to have relied on scavenging to a greater degree than do modern humans; indeed, a number of occurrences appear to be scavenging process sites, while others are associated with large mammal carcass processing. But like *Homo sapiens*, they buried their dead, and there is some slight evidence for the production of "art."

To a large degree, early modern *H. sapiens* and *H. neanderthalensis* had managed to avoid each other. Early modern humans depended on African fauna for food resources and followed its dispersals into the Levant and southern Europe, while Neanderthals were dependent on a European arctic fauna, whose migration pattern and territorial range expanded and shrank according to the fluctuations of Ice Age Europe. Eventually, early modern people moved into Europe, beginning sometime around 40,000 years ago. And when glacial conditions returned, this time, for whatever reason, they did not disperse south but remained in the region — perhaps coming into contact with the Neanderthals for the first time.

It is with the arrival of the people from the south that we see for the first time the full-blown expression of symbolism — mobile and fixed artistic expression, and a more refined template-based tool technology. They appear to have been more mobile, hunting and gathering over a broad area, while the Neanderthals tended to remain in their well-known and long-occupied valley systems. By 30,000 years ago, increased competition increased mortality rates, and a declining birthrate sent the Neanderthals in a downward spiral to extinction. There is no need to appeal to an argument of "prehistoric genocide" to explain their final disappearance from the earth some 27,000 years ago.

The final evidence supporting an "Out of Africa" origin for modern humans comes from Asia and Australia. The current fossil, archeological and molecular data overwhelmingly support the idea that Australia was populated by one population, which arrived some 60,000–40,000 years ago. This population shares no close biological affinities with either its European contemporary, *H. neanderthalensis*, or the Asian endemic *H. erectus*; they are very clearly *H. sapiens*. Is it just a coincidence that both the Neanderthal indigenous populations of Europe and the *erectus* populations of Asia became extinct at around the same time that populations of *H. sapiens* appear in these respective regions? We think not. We, like many others, believe that modern *H. sapiens* from Africa started their dispersal out of Africa around 120,000 years ago. Due to their advanced culture and a different way of "seeing things," they soon outcompeted the more archaic and endemic human populations — not through aggression but through a more efficient way of exploiting their environment for finite resources.

APPENDIX

Detailed Description of Characters (DWC)

APPENDIX TABLE 1 ▶ Faciodental Anatomical Characters for the Hominoi

Characters	Keny	Dryo	Graeco	Pongo	Gorilla	Pan	Sahel
1. Torus form	?	0	1	2	3	3	3
2. Supraorbital torus thickness	?	1	1	1	0	1	0
3. Glabella development	?	0	1	2	3	0	?
4. Supraorbital sulcus development	?	0	1	0	4	3	0
5. Temporal line orientation	?	0	0	0	0	1	1
6. Male sagittal crest development	?	?	?	0	0	2	2
7. Temporal fossae size	?	?	?	0	0	1	?
8. Postorbital constriction	?	1	?	1	0	1	1
9. Parietal overlap	?	?	?	0	0	0	?
10. Squamosal suture overlap development	?	?	?	0	0	0	?
11. Asterionic notch	?	?	?	0	0	0	?
12. Compound temporal/nuchal crest	?	0	?	3	3	3	3
13. Supraglenoid gutter development	?	?	?	1	0	1	?
14. Mastoid process inflation	?	0	?	0	0	0	1
15. Temporal squama pneumatization	?	0	?	0	0	0	0
16. External cranial base flexion	?	?	?	0	0	0	?
17. Nuchal plane orientation	?	?	?	0	0	0	2
18. Auditory meatus	?	?	?	0	1	0	0
19. External auditory meatus size	?	?	?	0	1	0	?
20. Articular tubercle height	?	?	?	0	0	0	?
21. Petrous orientation	?	?	?	0	0	1	1
22. Cranial base breadth	?	?	?	1	1	1	0
23. Basioccipital length	?	?	?	0	0	0	?
24. Glenoid fossa depth	?	0	?	2	2	3	3
25. Glenoid fossa area (size)	?	?	?	0	0	1	?
26. Postglenoid process development	?	1	?	0	0	0	0
27. Horizontal distance between TMJ and M2/M3	?	?	?	0	0	0	?
28. Eustachian process development	?	0	?	0	0	0	?
29. Tympanic shape	?	1	?	0	0	0	0
30. Vaginal process size	?	?	?	0	0	0	?
31. Digastric muscle insertion	?	?	?	0	0	0	?
32. Longus capitis insertion	?	?	?	0	0	0	?
33. Foramen magnum shape	?	?	?	0	0	0	0
34. Foramen magnum position	?	?	?	0	0	0	1
35. Foramen magnum inclination	?	?	?	0	0	0	?
36. Cranial capacity	?	?	?	0	0	0	0
37. Cerebellar morphology	?	?	?	0	0	0	?
38. Occipitomarginal sinus	?	?	?	0	0	0	?
39. Facial hafting	?	1	?	1	0	1	0
40. Interorbital breadth	?	2	2	0	1	1	?
41. Frontal sinus development	?	0	0	2	0	0	?
42. Lacrimal fossae location	?	0	1	0	0	0	0
43. Facial dishing	?	0	2	1	0	0	0
44. Upper facial breadth	?	1	3	0	1	1	1

	Praean	K. platy	K. rudol	Aust	Garhi grp	P. walk	P. boisei	P. rob	H. habilis	H. ergaster	H. sap
	3	0	0	0	0	3	3	3	0	0	4
	?	?	0	1	?	1	0	1	1	1	2
	4	?	3	4	4	3	3	3	4	4	4
	2	2	2	2	2	2	2	2	3	3	0
	1	2	2	1	1	0	0	0	2	2	2
	1	2	2	2	1	0	0	0	2	2	2
	?	?	?	1	?	0	0	1	1	1	2
	2	1	1	1	1	0	0	1	1	1	3
	0	?	0	0	?	2	1	0	0	0	0
	0	?	0	0	?	0	1	?	0	0	0
	0	?	2	2	?	0	2	2	1	2	2
	3	0	0	0	?	3	2	?	1	0	0
	?	?	2	2	?	0	1	1	2	2	2
	0	?	0	0	?	1	1	1	1	0	0
	0	?	1	0	?	0	1	2	2	2	2
	?	?	?	1	?	0	2	2	2	2	2
	1	?	2	2	?	2	2	2	2	2	2
	0	0	0	0	?	0	1	1	1	0	0
	0	0	0	0	?	0	1	1	1	0	0
	0	?	2	0	?	0	0	0	1	2	2
	1	?	2	1	?	2	2	2	2	2	2
	?	?	2	1	?	2	2	?	1	?	2
	1	?	?	1	?	1	2	2	2	2	2
	3	2	2	2	?	3	0	2	2	1	0
	1	?	1	1	?	0	0	0	1	1	2
	0	?	1	1	?	1	2	3	3	3	3
	0	?	0	0	?	0	0	0	1	1	1
	1	?	?	0	?	1	1	0	1	1	1
	0	0	0	1	?	1	2	1	1	1	1
	0	?	?	0	?	0	1	1	0	1	1
	0	1	?	0	?	?	0	1	1	1	1
	?	?	?	0	?	1	1	1	1	1	1
	0	0	?	0	?	1	1	0	0	0	0
	1	?	?	1	?	1	3	3	2	1	1
	?	?	?	0	?	?	1	1	1	2	1
	0	0	3	1	0	0	1	1	2	3	4
	0	?	1	0	?	0	1	1	1	1	1
	2	0	0	1	?	0	2	2	0	?	1
	1	1	1	1	1	0	0	0	1	1	2
	?	?	2	2	?	2	3	3	2	2	2
	?	?	0	1	?	?	0	0	1	0	0
	0	0	0	0	?	0	0	0	0	0	0
	0	0	0	0	0	3	3	3	0	0	0
	?	?	2	1	?	1	2	2	1	2	1

Characters	Keny	Dryo	Graeco	Pongo	Gorilla	Pan	Sahel
45. Mid-facial and upper facial breadth	?	?	0	0	0	0	?
46. Anterior zygomatic insertion	1	1	1	1	0	1	3
47. Height of masseter origin	?	?	?	1	0	0	?
48. Palate prognathism	?	?	0	0	0	0	?
49. Subnasal prognathism	?	1	0	0	0	0	2
50. Incisor alveoli prognathism	?	0	0	0	0	0	0
51. Anterior palate depth	0	1	2	2	0	0	?
52. Palate thickness	?	?	?	0	0	0	?
53. Palate breadth	?	?	0	1	1	1	?
54. Nasal bone projection	?	0	?	0	0	0	?
55. Nasal keel	?	1	0	1	0	1	1
56. Inferior orbital margin rounded	?	0	?	0	1	0	0
57. Orbital fissure configuration	?	?	?	0	0	0	?
58. Maxillary trigon	?	?	?	0	0	0	0
59. Malar/zygomatic orientation	1	0	?	1	0	2	?
60. Malar diagonal length	?	?	?	1	0	1	?
61. Infraorbital foramina location	0	?	?	0	0	0	?
62. Anterior nasal pillars	?	0	0	0	0	0	0
63. Nasal cavity entrance	?	0	0	2	0	0	?
64. Incisive canal development	?	0	0	2	0	0	?
65. Nasal clivus contour	?	0	0	0	0	0	1
66. Orbital shape	?	0	2	0	2	1	1
67. Inferior orbital margin	?	1	0	1	0	1	1
68. Subnasal length	?	?	1	1	1	1	0
69. Maxillary sinus size	0	0	?	1	1	1	?
70. Zygomatic insertion height	0	1	0	1	1	1	1
71. Canine fossa development	1	0	0	1	1	1	1
72. Mandibular symphysis orientation	0	0	0	0	0	0	?
73. Mandibular symphysis robusticity	1	?	?	0	1	1	?
74. Mandibular inferior transverse torus	0	1	1	0	2	2	?
75. Mandibular corpus robusticity	0	0	1	0	0	0	?
76. Mandibular premolar orientation	0	0	1	0	0	0	0
77. Mandibular mental foramen opening	0	1	2	1	1	0	?
78. Hollowing at mental foramen	0	0	0	0	0	0	?
79. Mandibular extramolar sulcus width	?	?	0	0	0	2	?
80. Upper incisal reduction	?	?	1	0	1	1	?
81. Upper incisor heteromorphy	?	?	0	0	1	1	?
82. Upper male canine size	0	0	0	0	0	0	1
83. Upper molar/premolar size	?	1	2	0	0	0	?
84. Upper M2 size	?	0	2	0	0	0	2
85. Molar cusp position	0	0	0	0	0	0	?
86. P3 metaconid development	0	0	1	0	0	0	?
87. P3 mesiobuccal enamel expansion	0	0	0	0	0	0	?
88. Molar enamel thickness	2	0	3	1	0	0	1
89. Upper molar lingual cingulum development	0	1	0	0	0	0	0
90. Upper M2 shape	1	0	0	0	1	0	1
91. Upper molar morphology	1	0	1	2	0	0	1
92. Deciduous M1 mesial crown profile	?	?	?	?	0	0	?

Praean	K. platy	K. rudol	Aust	Garhi grp	P. walk	P. boisei	P. rob	H. habilis	H. ergaster	H. sap
?	?	0	1	?	2	0	0	0	0	0
2	3	2	2	2	3	3	3	1	1	1
0	?	0	1	?	1	1	1	0	?	0
0	1	1	1	0	0	1	1	1	2	2
1	2	2	1	1	1	1	1	0	2	2
0	1	1	0	0	1	1	1	0	0	0
0	2	2	2	0	0	0	0	0	2	2
0	0	0	0	0	1	1	1	0	0	0
?	?	?	0	?	0	0	0	0	0	0
1	?	2	0	?	1	1	1	2	2	2
?	?	1	0	?	1	1	1	0	1	1
0	?	0	0	?	1	0	1	0	0	0
0	?	?	0	?	1	1	?	0	?	1
0	?	0	0	0	2	1	2	0	0	0
?	1	1	2	?	2	2	2	1	1	1
?	?	0	2	?	0	0	1	2	1	2
0	?	?	1	?	2	1	2	0	0	0
0	0	0	1	0	0	0	1	1	0	0
0	0	?	1	0	2	2	2	0	2	3
1	?	?	1	?	2	2	2	0	2	2
0	1	1	1	0	2	2	2	1	1	0
?	1	1	1	?	1	1	1	1	1	1
?	?	2	2	?	1	1	1	2	2	2
2	?	2	2	?	2	2	2	1	1	0
1	?	1	1	?	?	1	1	1	1	1
1	0	0	1	?	0	1	1	1	1	1
1	?	?	1	1	0	0	1	0	?	1
1	?	2	1	?	?	2	2	2	2.	3
2	?	3	2	?	?	3	?	3	3	1
2	?	3	2	?	?	0	0	3	3	3
1	?	1	2	?	?	2	2	1	1	0
0	?	2	2	?	?	2	2	2	2	2
1	?	3	1	?	?	3	3	3	3	2
0	?	1	1	?	?	2	2	2	2	2
2	?	2	1	?	?	0	0	2	2	2
1	?	?	1	1	1	2	1	1	?	1
1	2	?	1	?	?	1	1	1	?	1
1	1	1	1	1	1	2	2	1	2	2
1	?	?	1	?	?	1	0	1	?	2
2	0	2	2	2	2	2	2	1	0	0
1	?	1	3	?	?	4	4	1	1	1
1	?	2	2	?	?	2	2	2	2	2
2	?	3	3	3	?	3	3	3	3	3
2	2	2	2	2	?	3	3	2	2	2
0	0	?	0	1	?	0	0	0	0	0
0	0	?	0	0	?	?	0	0	0	0	0
1	0	?	1	1	?	?	3	3	1	1	1
1	1	?	?	1	?	?	2	2	?	1	1

APPENDIX TABLE 2 ▶ Hominid Character Definitions as Shown in Appendix Table 1

1. Supraorbital torus form	0 — Intermediate torus development, stronger laterally
	1 — Moderately developed tori — not bridge-like
	2 — Weak orbital rim-like structure
	3 — Strong bar-like torus with medial depression
	4 — Absent or reduced
2. Supraorbital torus thickness	0 — Broad (>0.29)
	1 — Intermediate (0.15–0.29)
	2 — Reduced (<0.15)
3. Glabella development	0 — Intermediate swelling
	1 — Depressed
	2 — Cannot be defined
	3 — Inflated
	4 — Broad in area but not inflated
4. Supraorbital sulcus development	0 — Not present
	1 — Mid-sulcus
	2 — Intermediate development
	3 — Moderately developed (between states 2 and 4)
	4 — Well developed/deep sulcus
5. Temporal line orientation	0 — Strong anteromedial incursion (frontal trigon)
	1 — Moderate anteromedial incursion
	2 — Weak
6. Male sagittal crest development	0 — Strongly developed
	1 — Present, but usually weakly developed
	2 — Absent
7. Temporal fossae size	0 — Large (>1.73)
	1 — Intermediate (1.73–0.85)
	2 — Small (<0.85)
8. Postorbital constriction	0 — Marked (<0.58)
	1 — Intermediate (0.58–0.74)
	2 — Reduced (>0.74 < 0.80)
	3 — Absolutely reduced (>0.80)
9. Parietal overlap	0 — No overlap of parietal at asterion
	1 — Variable
	2 — Overlap of parietal at asterion
10. Squamous suture overlap development	0 — Not extensive
	1 — Extensive
11. Asterionic notch	0 — Present
	1 — Variable
	2 — Absent

APPENDIX TABLE 2 ▶ Continued

12. Compound temporal/nuchal crest	0 — Absent 1 — Present laterally 2 — Variable 3 — Present
13. Supraglenoid gutter development	0 — Strongly developed (>26.0) 1 — Intermediate (18.0–26.0) 2 — Weakly developed (<18.0)
14. Mastoid process inflation	0 — Not inflated lateral to supramastoid crest 1 — Inflated to or beyond supramastoid crest
15. Temporal squama pneumatization	0 — Extensive 1 — Variable 2 — Reduced
16. External cranial base flexure	0 — Flat 1 — Moderate 2 — Flexed
17. Nuchal plane orientation	0 — Steep 1 — Intermediate 2 — Weak
18. Auditory meatus	0 — Anterior tympanic edge medial to porion 1 — Interior tympanic edge aligned with or lateral to porion
19. External auditory meatus size	0 — Small 1 — Large
20. Articular tubercle height	0 — Moderate to large 1 — Variable 2 — Small
21. Petrous orientation	0 — Sagittally oriented 1 — Intermediate 2 — Coronally orientated
22. Cranial based breadth	0 — Narrow (<3.00) 1 — Intermediate (3.00–3.54) 2 — Broad (>3.54)
23. Basioccipital length	0 — Long 1 — Intermediate 2 — Short
24. Glenoid fossa depth	0 — Deep 1 — Variable 2 — Intermediate (between states 0 and 3) 3 — Shallow
25. Glenoid fossa area (size)	0 — Large 1 — Intermediate 2 — Small
26. Postglenoid process development	0 — Large and unfused with tympanic 1 — Moderate and may be fused/unfused 2 — Small and may be fused/unfused 3 — Small and fused to tympanic
27. Horizontal distance between TMJ and M2/M3	0 — Long (58 mm or greater) 1 — Short (less than 58 mm)
28. Eustachian process development	0 — Prominent 1 — Weak/absent

APPENDIX TABLE 2 ▶ Continued

29. Tympanic shape	0 — Tubular
	1 — Crest with vertical plate
	2 — Crest with inclined plate
30. Vaginal process size	0 — Small
	1 — Moderate to large
31. Digastric muscle insertion	0 — Broad shallow fossa
	1 — Deep narrow notch
32. Longus capitis insertion	0 — Long oval
	1 — Small circle
33. Foramen magnum shape	0 — Heart shape
	1 — Oval/circular
34. Foramen magnum position	0 — Basion posterior to bi-tympanic
	1 — Basion close to bi-tympanic line
	2 — Basion variable at/or anterior to bi-tympanic
	3 — Basion anterior to bi-tympanic
35. Foramen magnum inclination	0 — Strongly inclined posteriorly
	1 — Horizontal
	2 — Strongly inclined anteriorly
36. Cranial capacity	0 — Small (<500 cc)
	1 — Intermediate (428–550 cc)
	2 — Increased (509–675 cc)
	3 — Large (750–1100 cc)
	4 — Absolutely large (>1100 cc)
37. Cerebellar morphology	0 — Lateral flare with posterior protrusion
	1 — Tucked
38. Occipitomarginal sinus	0 — Low frequency
	1 — Intermediate frequency
	2 — High frequency
39. Facial hafting	0 — Face set high relative to frontal
	1 — Intermediate condition
	2 — Face set low relative to frontal
40. Interorbital breadth	0 — Narrow (<0.34)
	1 — Intermediate (0.34–0.56)
	2 — Broad (>0.56 < 0.77)
	3 — Extremely broad (>0.77)
41. Frontal sinus development	0 — Large
	1 — Intermediate to small
	2 — Absent
42. Lacrimal fossae location	0 — Within orbit
	1 — Within infraorbital region
43. Facial dishing	0 — Prognathic mid face–premaxilla
	1 — Mid to lateral upper face vertical or slightly concave
	2 — Mid to lateral upper face concave (strong)
	3 — Facial dishing present
44. Upper facial breadth	0 — Narrow (<2.60)
	1 — Intermediate (2.60–3.28)
	2 — Broad (>3.28 < 3.81)
	3 — Extremely broad (>3.81)

APPENDIX TABLE 2 ▶ Continued

45. Upper and mid facial-breadth	0 — Intermediate (1.02–1.16) 1 — Broad mid-face (>1.16 < 1.25) 2 — Very broad mid-face (>1.25)
46. Anterior zygomatic insertion	0 — M1/M2 junction 1 — M1 2 — M1/P4 junction 3 — P4 or anterior
47. Height of masseter origin	0 — Zygomaticoalveolar height is low 1 — Zygomaticoalveolar height is high
48. Palate prognathism	0 — Strong (>0.57) 1 — Intermediate (0.57–0.30) 2 — Weak (<0.30)
49. Subnasal prognathism (angle)	0 — Low (<30°) 1 — Intermediate (30°–42°) 2 — High (>42°)
50. Incisor alveoli prognathism	0 — Beyond bi-canine line 1 — Within bi-canine line
51. Anterior palate depth	0 — Shallow 1 — Intermediate 2 — Deep
52. Palate thickness	0 — Thin 1 — Thick
53. Palate breadth	0 — Broad (>1.79) 1 — Intermediate (1.53–1.79) 2 — Narrow (<1.53)
54. Nasal bone projection	0 — Projected tapered 1 — Projected expanded 2 — Not projected
55. Nasal keel	0 — Present 1 — Absent
56. Inferolateral orbital margin rounded	0 — Yes 1 — No
57. Orbital fissure configuration	0 — Foramen 1 — Comma shaped
58. Maxillary trigon	0 — Absent 1 — Variable 2 — Present
59. Malar/zygomatic orientation	0 — Anterior–posterior bend 1 — Near vertical 2 — Anterior slope
60. Malar diagonal length	0 — Long (>1.20) 1 — Intermediate (0.90–1.20) 2 — Short (<0.90)
61. Infraorbital foramen location	0 — Within upper 50% of malar height 1 — Variable 2 — Within lower 50% of malar height

APPENDIX TABLE 2 ▶ Continued

62. Anterior nasal pillars	0 — Absent
	1 — Variable
63. Nasal cavity entrance	0 — Stepped
	1 — Intermediate (between states 0 and 2)
	2 — Smooth with overlap
	3 — Smooth with no overlap
64. Incisive canal development	0 — Slight canal
	1 — Intermediate
	2 — Extensive canal
65. Nasoalveolar clivus contour	0 — Convex
	1 — Flat
	2 — Concave (with gutter)
66. Orbital shape	0 — Oval/rhomboid (higher than broad)
	1 — Circular/rhomboid (similar dimensions)
	2 — Rectangular/rhomboid (broader than high)
67. Inferior orbital margin	0 — Well above superior nasal margin
	1 — Close to superior nasal margin
	2 — Well below superior nasal margin
68. Subnasal length	0 — Short (<0.59)
	1 — Intermediate ($0.59–0.85$)
	2 — Long (>0.85)
69. Maxillary sinus size	0 — Intermediate
	1 — Large
70. Zygomatic insertion height	0 — Low
	1 — High
71. Canine fossa development	0 — Shallow
	1 — Deep
72. Mandibular symphysis orientation	0 — Receding
	1 — Receding intermediate (between 0 and 2)
	2 — Vertical/near vertical
	3 — Vertical with chin
73. Mandibular symphysis robusticity	0 — Gracile (<0.33)
	1 — Intermediate ($0.33–0.43$)
	2 — Robust ($>0.43 < 0.51$)
	3 — Extremely robust (>0.51)
74. Mandibular inferior transverse torus	0 — Inferior torus stronger than superior torus
	1 — Both tori of similar development
	2 — Inferior torus weaker than superior torus
	3 — Both tori undeveloped
75. Mandibular corpus robusticity	0 — Intermediate ($0.48–0.64$)
	1 — Robust ($>0.64 < 0.75$)
	2 — Extremely robust (>0.75)
76. Mandibular premolar orientation	0 — U-shaped
	1 — V-shaped
	2 — Parabolic

APPENDIX TABLE 2 ► Continued

77. Mandibular mental foramen opening	0 — Opens anterosuperiorly
	1 — Variable
	2 — Opens posteriorly
	3 — Opens laterally
78. Mandibular mental foramen hollow	0 — Present
	1 — Variable
	2 — Absent
79. Mandibular extramolar sulcus width	0 — Broad
	1 — Variable
	2 — Narrow
80. Upper incisor reduction (area)	0 — Large (>0.17)
	1 — Intermediate (0.11–0.17)
	2 — Reduced (<0.11)
81. Upper incisor heteromorphy (area)	0 — Increased (>2.05)
	1 — Intermediate (1.22–2.05)
	2 — Reduced (<1.22)
82. Upper male canine size	0 — Robust and dagger-like
	1 — Intermediate
	2 — Reduced
83. Upper molar/premolar size (area)	0 — Intermediate (1.91–2.29)
	1 — Marked difference ($>2.29 - <2.62$)
	2 — Significant difference (>2.62)
84. Upper M2 size (area)	0 — Small (0.08–0.14)
	1 — Intermediate ($>0.14 < 0.18$)
	2 — Large (>0.18)
85. Molar cusp position	0 — Cusps close to buccal–lingual crown 'edge'
	1 — Lingual cusps near margin–buccal cusps slightly lingual
	2 — Lingual cusps at margin–buccal cusps moderately lingual
	3 — Lingual cusps laterally set buccal cusps moderately lingual
	4 — Lingual cusps laterally set buccal cusps strongly lingual
86. P_3 metaconid development	0 — Metaconid absent
	1 — Infrequent–variable
	2 — Metaconid present
87. P_3 mesiobuccal enamel expansion	0 — Strongly developed
	1 — Moderately developed
	2 — Variable
	3 — Weak or absent
88. Molar enamel thickness	0 — Thin
	1 — Intermediate thin to thick
	2 — Thick
	3 — Hyper-thick
89. Upper molar lingual cingulum development	0 — Weak to absent
	1 — Intermediate development
90. Upper M2 shape	0 — Broader than long (<0.95)
	1 — Square ($\pm95\%$)

APPENDIX TABLE 2 ▶   Continued

91. Upper molar morphology	0 — Well-developed cusps and cristae
	1 — Inflated cusps, limited cristae
	2 — Low cusps with enamel wrinkling
	3 — Flat/bunodont morphology
92. Deciduous M1 mesial crown profile	0 — MMR absent, protoconid anterior, fovea opening
	1 — MMR slight, protoconid anterior, fovea opening
	2 — MMR thick, protoconid even with metaconid, fovea closed

Note: Characters 1, 3, 29, 74, and 77 are always treated as unordered.

Appendix 3: Description of Characters

All characters are ordered unless stated otherwise. If a character is not discussed for a genus, then it is not preserved in any specimen and is coded with a question mark (?) within Appendix Table 1. A numeral within brackets; e.g., ($= 0$), indicates the coding structure used in Appendix Tables 1 and 2 to define the character state for the character in question.

In order to help determine likely duplication of associated characters, we were careful to note apparent similarities in the data matrix when a number of characters from one region were being examined; e.g., features associated with the temporomandibular joint (TMJ). We examined four characters associated with the TMJ (characters 24 to 27), and none of these features showed a consistent trend within taxa. If they had, we would have seriously considered deleting some of these characters so that there was no suggestion of bias. For example, a data matrix of one specific region (e.g. the TMJ) was presented like this:

Taxa:	A	B	C	D	E	F	G	H	I
Character I:	0	0	0	1	1	0	2	0	0
II:	0	0	0	1	1	0	2	0	0
III:	0	0	0	1	1	0	2	0	0
IV:	0	0	1	1	1	2	0	0	2

We would seriously have to consider combining Characters II and III because they duplicate the condition of the character polarity observed in

Character I, which may — but not necessarily — imply character duplication. This is emphasized because all four of these characters are part of one region — the TMJ. The distinct condition observed in character IV, however, suggests that this is a valid character.

When defining morphometric character states, an index was generated using orbital height as the constant. In order to come to some "objective" criterion, it was deemed desirable to construct a total extant great ape range for each character and to use the mean and one standard deviation to construct an average range (Cameron, 1997b). In the allocation of hominoids, mid-supraorbital thickness was divided by the orbital height. Combining all of the indices generated for *P. paniscus, P. troglodytes, G. gorilla,* and *P. pygmaeus* (n = 123), the mean and standard deviations were calculated; from this, an average range from the mean was generated (e.g., 0.22 [mean] ± 0.07 [S.D.], thus resulting in an average range of between 0.15 and 0.29). Any figure that is below the minimum (e.g., 0.15) is considered narrow, any character within the range is considered "average", and any figure greater than the maximum (e.g., 0.29) is considered broad. Mean scores (males and females combined) are used to allocate fossil and extant hominids to these conditions.

1. Supraorbital Torus Form

The condition in *Dryopithecus* (Begun, 1992, 1995; Moyà-Solà & Köhler, 1993), *Kenyanthropus* (M.G. Leakey *et al.,* 2001), *Australopithecus* (Rak, 1983), the "garhi group" (Asfaw *et al.,* 1999), *H. habilis* (Tobias, 1991), *K. rudolfensis,* and *H. ergaster* (Wood, 1991) is developed laterally, but is weaker medially (= 0). The supraorbital torus in *Graecopithecus* specimen XIR-1 is developed, but not bridge-like; thus, it has a developed torus above each orbit (= 1) (de Bonis *et al.,* 1990; de Bonis & Koufos, 1993; Dean & Delson, 1992; Cameron, 1997a). The torus region is rim-like in *Pongo* (= 2). Supraorbital torus form in *Sahelanthropus* is described as continuous and undivided (Brunet *et al.,* 2002), though it appears to be very similar to the condition in *Praeanthropus,* species of *Paranthropus, Gorilla,* and *Pan,* because there is evidence of a medial depression (= 3) (Tobias, 1967; Rak, 1983; Kimbel *et al.,* 1994). Thus, all of these taxa are allocated as having the same character state for this feature. The supraorbital in *H. sapiens* is undeveloped (= 4). Given the degree of variability of this feature, including within the outgroup, this character remains unordered in all analyses.

2. Supraorbital Torus Thickness

This feature is distinct from supraorbital torus form because it measures the thickness of the torus from its midpoint (inferosuperior chord distance). This value is divided by orbital height to give an index value. The extant hominoid mean of 0.22 and one standard deviation of 0.07 gives an intermediate range of 0.15–0.29. Data for the extant and fossil Miocene hominoids is unpublished data held by Cameron (unless stated otherwise), while all data of the fossil hominins has been calculated from the data provided in B.A. Wood (1991). It was also noted that the mean for *H. sapiens* was below the minimum value generated for the extant hominid (0.11). Thus, an additional character state of a very gracile torus ($<$ 0.11) is recognized; no hominids were beyond the maximum extant hominid value of 0.43.

A thick supraorbital torus ($= 0$) is observed in *P. boisei* with a mean index of 0.33 (n = 2), *K. rudolfensis* with a mean of 0.30 (n = 2), and *Gorilla* also with a mean index of 0.30 (n = 35). While no metric data are available for *Sahelanthropus*, the supraorbital is shown by Brunet *et al.* (2002) to be beyond the range observed in *Gorilla*. Thus, it is allocated to the same character state as this great ape. The intermediate state ($= 1$) defines the primitive hominid condition and is observed in *Dryopithecus* specimen RUD 77 (data from Kordos & Begun, 1997) with an index of 0.15, *Graecopithecus* specimen XIR-1 (original) with an index of 0.27, *Australopithecus* with a mean of 0.23 (n = 2), *P. walkeri* specimen KNM-WT 17000 with an index of 0.28, *P. robustus* with an index of 0.29 (n = 2), *H. habilis* with an index of 0.23 (n = 2), *H. ergaster* specimen KNM-ER 3733 with an index of 0.22, *Pongo* with a mean of 0.16 (n = 22), and *Pan* with an index of 0.20 (n = 66). A reduced torus ($= 2$) is present in *H. sapiens* with a mean of just 0.10 (n = 6).

3. Glabella Development

Dryopithecus (Begun, 1992, 1994; Kordos & Begun, 1997) and *Pan* both have limited or intermediate inflation of their glabella ($= 0$). While Brunet *et al.* (2002) state that glabella in *Sahelanthropus* is present, they give no description. Thus, this feature cannot be coded for this genus with any degree of certainty. *Graecopithecus* (de Bonis *et al.*, 1990; de Bonis & Koufos, 1993) has a sunken glabella ($= 1$), while glabella in *Pongo* cannot be defined ($= 2$). *Gorilla*, species within *Paranthropus*, and *K. rudolfensis* are defined by an inflated glabella ($= 3$) (Wood & Chamberlain, 1986; Chamberlain & Wood, 1987). *Praeanthropus* (Kimbel *et al.*, 1994), *Australopithecus* (Rak, 1983), the "garhi group" (Asfaw *et al.*, 1999),

H. habilis (Tobias, 1991), *H. ergaster*, and *H. sapiens* (Schwartz & Tattersall, 2002) have a broad but uninflated glabella region (= 4). Given the degree of variability of this feature, including within the outgroup, this character remains unordered in all analyses.

4. Supraorbital Sulcus Development

The frontal bone is defined by one of the following: sulcus is absent (0); mid sulcus (1); intermediate development of sulcus (2); or developed and deep sulcus (3). While it might be thought that the development of supra-orbital sulcus may be correlated with supraorbital development (reflecting the same feature), this need not always be the case. For example, while *Praeanthropus* specimen A.L. 444-2 has the primitive hominid develop-ment in its supraorbital development (e.g., *Gorilla* and *Pan*), its supra-orbital sulcus is described as being reduced, relative at least to *Pan* (Kimbel *et al.*, 1994; D.E. Lieberman *et al.*, 1996). Also, while *Sahelanthropus* has a strong supraorbital torus, it has a weak to non-existent frontal sulcus.

Dryopithecus (Begun, 1994; Kordos & Begun, 1997; Begun & Kordos, 1997) and *Pongo* (Andrews, 1992; Andrews *et al.*, 1996), *Sahelanthropus* (Brunet *et al.*, 2002), and *H. sapiens* do not have a supraorbital sulcus (= 0). *Graecopithecus* (de Bonis *et al.*, 1990; de Bonis & Koufos, 1993) is defined by a mid supraorbital sulcus, enclosed by the temporal lines (= 1). The hominins *Praeanthropus*, *Australopithecus*, the "garhi group," species within *Paranthropus* and *Kenyanthropus*, are defined by intermediate devel-opment of a supraorbital sulcus (= 2) (see Strait *et al.*, 1997; Asfaw *et al.*, 1999; M.G. Leakey *et al.*, 2002; Strait & Grine, 2001). *Pan*, *H. habilis*, and *H. ergaster* cannot easily be assigned to either condition observed in the other hominins as their sulcus is relatively deeper, though not to the degree observed in *Gorilla* (Strait *et al.*, 1997; Strait & Grine, 2001). Thus, they are assigned to their own character state (= 3), which is between the condition observed in the other hominids and *Gorilla*. Finally, *Gorilla* is defined by a strong supraorbital sulcus (= 4) (Cameron, 1997a).

5. Temporal Line Orientation

The development and orientation of the temporal lines is clearly associated with the development and orientation of the temporalis muscle, which itself will be influenced by sexual dimorphism (Kimbel, 1988; Rak, 1983). As such, only male specimens (where available) are considered. While some might argue against the inclusion of such "obviously functional" fea-tures within a hominin phylogenetic analysis (Skelton & McHenry, 1992), the concept of "phylogenetic niche conservatism" as discussed in the main text suggests that they are of phylogenetic interest. Other features, such as postorbital constriction, degrees of cranial base angulation, frontal lobe development, and neuro-orbital disjunction, will also influence the development and orientation of the temporal lines. As such, the development of this feature is probably quite complex, due to the numerous mosaic processes impacting the development of upper facial and frontal form.

Temporal lines converging anteriorly to help define an anterior frontal trigon ($= 0$) occur in *Dryopithecus* specimen RUD 44 (Begun, 1994, 2002) and species within *Paranthropus* (Rak, 1983), *Pongo*, and *Gorilla*. While this condition is not preserved entirely in either *Graecopithecus*, the surviving temporal lines, supraorbital tori, and frontal parts suggest that it was characterized by a marked frontal trigon (de Bonis & Koufos, 1993). Moderate anteromedial incursion of the temporal lines ($= 1$), but not resulting in a frontal trigon, is observed in *Sahelanthropus* (Brunet *et al.*, 2002), *Praeanthropus* (Kimbel, 1988; Kimbel *et al.*, 1994), *Australopithecus* (Rak, 1983; Kimbel *et al.*, 1994), the "garhi group" (Asfaw *et al.*, 1999), and most male specimens of *Pan*. The condition in *H. habilis* is undeveloped ($= 2$) (Strait *et al.*, 1997; Strait & Grine, 2001). *Kenyanthropus platyops* is described as having a similar pattern to that observed in the "1470 group" (*K. rudolfensis*), which is similar to *H. habilis* in this regard (M.G. Leakey *et al.*, 2001 — see Table 2). Thus, *Kenyanthropus* is coded the same condition as *H. habilis*, as is *H. ergaster* (B.A. Wood, 1991).

6. Male Sagittal Crest Development

The formation of a sagittal crest is clearly associated with masticatory considerations related to temporalis development relative to cranial size. In contrast with Skelton *et al.* (1986), Skelton and McHenry (1992) suggest phenotypic features associated with function should not automatically be dismissed as phylogenetically informative, given the concept of phylogenetic niche conservatism (see main text). While some may consider character 5 to be correlated with this character, an examination of Appendix Table 1 shows that this is not the case. For example, while there is a consistent correlation between strong anteromedial incursion of the temporal lines and sagittal crest formation, in some genera, moderate incursion may lead to a weak sagittal crest posteriorly (e.g., *Praeanthropus* [Kimbel *et al.*, 1994]) or to no sagittal crest at all (e.g., *Sahelanthropus* [Brunet *et al.*, 2002], *Australopithecus*, and *Pan* [D.E. Lieberman *et al.*, 1996; Strait & Grine, 2001]).

A well-developed sagittal crest (= 0) is observed in *Paranthropus* males (Tobias, 1967; Rak, 1983; Skelton *et al.*, 1986; Strait *et al.*, 1997; Strait & Grine, 2001), *Pongo*, and *Gorilla*. Males of *Praeanthropus* and the "garhi group" usually have a weak sagittal crest, which is more posteriorly located (= 1) (Kimbel *et al.*, 1994; Asfaw *et al.*, 1999; Strait & Grine, 2001). Where preserved, males of *Sahelanthropus* (Brunet *et al.*, 2002), *Pan*, *K. platyops* (M.G. Leakey *et al.*, 2001), *Australopithecus* (Rak, 1983; Skelton *et al.*, 1986; Skelton & McHenry, 1992), *H. habilis* (Tobias, 1991), *K. rudolfensis* (Strait *et al.*, 1997; Strait & Grine, 2001), *H. ergaster* (B.A. Wood, 1991), and *H. sapiens* usually do not have sagittal crests (= 2). Note that while the "australopithecine" material from Sterkfontein is supposed to display sagittal crests, it is not considered here due to the current taxonomic confusion of this material. See Strait *et al.* (1997) and Strait & Grine (2001) for a differing interpretation.

7. Temporal Fossa Size

This character is used to measure independently the relative degree of temporalis development, regardless of parietal/temporal bone size. It is distinct from sagittal crest development, which will be correlated directly with the size of these bones, relative to muscle development. For example, a small cranium with moderate to strong temporalis development (e.g., species of *P. robustus*) will require the formation of a temporal crest to help expand the attachment space requirements of the temporalis. A similar sized muscle in another hominid with increased cranial expansion (and thus increased parietal and temporal bones) will not need a sagittal crest. Because this character is likely to be of importance in terms of sexual dimorphism, only male specimens were examined where possible. In a few cases, however, only female fossil hominins could be used, due to problems of preservation (see later).

This character is defined using the same process as outlined earlier for supraorbital torus thickness (character 2). The index used here is temporal fossa size (area) divided by orbital size (area). An intermediate range is defined by 0.85–1.73 (mean of the four extant hominid species of 1.29 [n = 96] and standard deviation of 0.44). A large fossa is defined by an index greater than 1.73, and a small fossa is defined by an index less than 0.85. All data of extant specimens is taken from originals (unpublished data from DWC), while all data from fossil hominins has been taken from casts housed at the School of Archaeology and Anthropology, the Australian National University.

Males of *Gorilla* (mean = 1.90 [n = 13]), *Pongo* (mean = 1.75 [n = 7]), *P. walkeri* (KNM-WT 17000 = 1.77), and *P. boisei* (OH 5 = 1.81) are all defined by a large temporal fossa (= 0). Males of *Pan* (mean = 1.26 [n = 27]), *Australopithecus* (Sts 5 = 1.12 [female?]), *P. robustus* (SK 48 = 1.35), *H. habilis* (KNM-ER 1813 = 1.11 [female?]), and *H. ergaster* with a mean of 1.06 (n = 2) are all defined by the intermediate condition (= 1). While the sexing of some of the fossil hominins remains problematic, even accepting that some may represent females, they are still well within the intermediate range of the extant hominids. The fossa in *H. sapiens* has an average of just 0.76 (n = 8), which is small (= 2).

8. Postorbital Constriction

This character state is defined by an index of minimum frontal breadth (just posterior to the supraorbital torus) divided by maximum upper facial breadth (bi-frontomalare temporale). The extant hominids have a mean index value of 0.66, with one standard deviation of 0.08 (n = 122). This gives an intermediate range of 0.58–0.74. Thus, any value below 0.58 suggests increased postorbital constriction, while any value above 0.74 is reduced. The minimum extant hominid value is 0.46 and the maximum value is 0.80.

Increased postorbital constriction (= 0) is observed in *Gorilla*, with a mean value of 0.57 (n = 36), *P. walkeri* (KNM-WT 17000) with a value of 0.57, and *P. boisei* (KNM-ER 406), also with an index of 0.57. The intermediate condition (= 1) is observed in *Dryopithecus* specimen RUD 77 (original), with an estimated index of 0.73, *Sahelanthropus* specimen TM 266-01-060-1 with an index of 0.59 (data from Brunet *et al.*, 2002), *Australopithecus* with a mean index of 0.66 (Kimbel *et al.*, 1984), *P. robustus* with a mean index of 0.70 (data from B.A. Wood, 1991), *H. habilis* (OH 24 & KNM-ER 1813) with a mean index of 0.72, *K. rudolfensis* with a mean index of 0.70, *H. ergaster* with a mean index of 0.75 (data from B.A. Wood, 1991), *Pongo* with a mean index of 0.66 (n = 22), and *Pan* with a mean index of 0.70 (n = 64). While no data is available for the "garhi group," it is said to be similar to that observed in *Australopithecus* (Asfaw *et al.*, 1999; Strait & Grine, 2001). Also, while no raw data is available for *K. platyops*, postorbital constriction is said to be similar to that of the "1470 group" (*K. rudolfensis*); thus, it is considered to be within the range of *Pan* and *Pongo* (see Lieberman, 2001). Reduced postorbital constriction (= 2) is observed in *Praeanthropus* with an index of 0.80. Absolutely reduced postorbital constriction (= 3) is observed in *H. sapiens* with a mean index of 0.92 (n = 168 [data from Thorne, 1976]).

9. Parietal Overlap of Occipital at Asterion, at Least in Males

The feature refers to the expansion (increased osteoblast activity) of the bone table within the region of asterion (Kimbel & Rak, 1985). This results in a partial overlap of the temporal, parietal, and occipital bones (Strait *et al.*, 1997). This feature is included because of its association with nuchal and masticatory stress (Kimbel & Rak, 1985), which are of functional interest and also likely to be of phylogenetic significance, given the concept of phylogenetic niche conservatism (see text). Also, because of the complex interplay of sutural and bone growth patterns associated with the development of this feature, we suggest it is likely to have a strong heritability factor (see partly Enlow & Hans, 1996). Only in the East African species of *Paranthropus* has this phenotypic feature been expressed to varying degrees, and in some specimens of *P. boisei*, it is absent altogether; e.g. KNM-ER 23000 (B. Brown *et al.*, 1993).

10. Squamosal Suture Overlap Development

This character refers to the overall development of the posterosuperior and superior parts of the squamosal suture (Rak, 1978, 1983; Kimbel & Rak, 1985; Strait *et al.*, 1997). For the same reason just outlined for parietal overlap (character 9), this feature is retained as an informative character. While Rak and Kimbel (1991, 1993) consider *P. walkeri* (KNM-WT 17000) to have an extensive overlap, Walker *et al.* (1993) examined this region in detail and concluded that it was not extensive and was within most hominin ranges of variability. Unlike Strait and Grine (2001), *P. walkeri* is considered not be distinctive in this feature. Thus, the only species characterized by this feature is *P. boisei* because its sutures are extensively developed to the degree that it is tapered with a bevel edge (= 1) (see Kimbel & Rak, 1985). Whether this condition is also present in the South African species is currently unknown due to poor preservation.

11. Asterionic Notch

This character cannot be considered totally integrated with the two previous characters discussed above (9 and 10) because a number of taxa that do not express these morphologies are characterized by an asterionic notch. This feature is defined by the inferoposterior intrusion of the parietal between the occipital and the mastoid portion of the temporal bones (Kimbel & Rak, 1985; Aiello & Dean, 1990). This feature has been associated with the development and configuration of the temporalis muscle (Robinson, 1958; Tobias, 1967; Jolly, 1970; Kimbel *et al.*, 1984). *Pongo, Gorilla Pan, Praeanthropus*, and *P. walkeri* are all defined by the presence of an asterionic notch ($= 0$) (Kimbel & Rak, 1985; Walker *et al.*, 1986; Walker & Leakey, 1988). The condition observed in *H. habilis* is said to be variable ($= 1$), while all other taxa preserving this region lack an asterionic notch ($= 2$) (Kimble & Rak, 1985; Skelton & McHenry, 1992, Strait *et al.*, 1997; Strait & Grine, 2001).

12. Compound Temporal/Nuchal Crest in Males

Following Strait and Grine (2001), four character states are recognized. The development of this crest is related to overall cranial size and the requirements of the neck and masticatory musculature. As such, it is likely to be of interest from a phylogenetic perspective; that is, the phylogenetic niche conservatism. The developments of these muscles are of significance in relation to head support, and differential development may be associated with positional behavior. As such, they are of likely phylogenetic importance.

Dryopithecus, Kenyanthropus, Australopithecus, K. rudolfensis, H. ergaster, and *H. sapiens* are all defined by a complete absence of a crest ($= 0$). *Homo habilis* has a partial crest confined to the lateral third of the bi-asterionic breadth ($= 1$). *Paranthropus boisei* is defined by a variable condition in overall development ($= 2$). *Pongo, Gorilla, Pan, Sahelanthropus, Praeanthropus*, and *P. walkeri* have an extensively developed crest from inion to the lateral margin of the supramastoid crest ($= 3$) (Aiello & Dean, 1990; Kordos & Begun, 1997; Strait *et al.*, 1997; Strait & Grine, 2001; Brunet *et al.*, 2002; also see D.E. Lieberman *et al.*, 1996).

13. Supraglenoid Gutter Development

This feature is defined by an index of supraglenoid gutter width divided by bi-supramastoid breadth. *Pongo* has a mean index of 19.8 ($n = 3$), *Gorilla* has a mean of 28.3 ($n = 3$), and *Pan* has a mean of 19.4 ($n = 2$). Three states are recognized here. The first is defined as wide ($= 0$) in *Gorilla* and *P. walkeri* (with an index of 27.9); an intermediate stage ($= 1$) is defined by *Pongo, Pan, P. robustus* (SK 48 $= 24.2$), and *P. boisei* (KNM-ER 406 $= 22.9$, KNM-ER 732 $= 21.4$, and OH 5 $= 24.3$). A narrow gutter ($= 2$) defines *Australopithecus* (Sts 5 $= 14.4$), *H. habilis* (OH 24 $= 10.0$, KNM-ER 1813 $= 15.0$), *H. ergaster* (KNM-ER 3733 $= 13.4$), and *H. sapiens*. While no data were available for *K. rudolfensis*, it can best be considered very much reduced.

14. Mastoid Process Inflation

The overall inflation of the mastoid process will be associated with the development of the sternocleidomastoideus muscle, which is responsible for bending the neck laterally, head rotation as well as flexing the neck, and drawing the head ventrally. As such, this feature is likely to be influenced to some degree by differing patterns of locomotion, for it relates directly to the positioning of the head in relation to the clavicle and sternum (points of muscle origin). Two patterns are identified here. The lateral surface of the mastoid is not inflated beyond the supramastoid crest viewed frontally (= 0); the lateral surface of the mastoid is inflated to be level with, or beyond, the supramastoid crest (= 1).

Almost all species preserving this morphology are defined by a reduction in this feature (see Olsen, 1981; B. Brown et al., 1993; Strait et al., 1997). While not totally preserved in *Dryopithecus*, specimen RUD 77 suggests that the mastoid was probably not strongly developed relative to the supraglenoid gutter (see Kordos & Begun, 1997). Only species within *Paranthropus* as well as *H. habilis* display lateral inflation of the mastoid, though this may also be the case for *Sahelanthropus*, which is described as being large and pneumatized (Brunet et al., 2002). While Strait and Grine (2001) recognize *H. habilis* as being variable in this feature (i.e., KNM-ER 1805 has lateral inflation, while the inflation observed in KNM-ER 1813 is aligned to the crest), we considered these character states to be similar and thus place them within state 1 (not recognizing a third "variable" character state).

15. Temporal Squama Pneumatization

The primitive character state for this feature is for an extensively pneumatized temporal (Sherwood, 1999). As defined by both Strait et al. (1997) and Sherwood (1999), pneumatization of the temporal squama is associated with the development of pneumatic tracts extending to the squamosal suture, thickening the squamous temporal (squamous antrum). When squamous antrum is absent, the temporal bone is not considered pneumatized.

The temporal squama is inflated (= 0) in *Dryopithecus* (Kordos & Begun, 1997), *Pongo, Gorilla, Pan, Sahelanthropus* (Brunet et al., 2002), *Ardipithecus*, the "anamensis group," *Praeanthropus*, and *P. walkeri* (Strait & Grine, 2001). While Skelton and McHenry (1992) considered *Australopithecus* (*A. africanus*) as being weakly pneumatized, Strait et al. (1997) and Strait and Grine (2001) consider the squama to be extensively inflated, which is the condition adopted here. The condition is variable within *P. boisei* and *K. rudolfensis* (see B.A. Wood, 1991; M.G. Leakey et al., 2001;

partly Strait & Grine, 2001). *Paranthropus robustus, H. habilis, H. ergaster* and *H. sapiens* are all defined by reduced pneumatization (B.A. Wood 1991; Strait *et al.*, 1997; Strait & Grine, 2001).

16. External Cranial Base Flexion

This feature is developmentally significant because it relates to a number of mosaic influences that will affect a number of morphological patterns including facial orientation, neuro-orbital disjunction, postorbital con striction (and thus indirectly influencing the development of the tempo ralis through the size of the temporal fossae), and development of the supraorbital torus, to mention just a few (see main text for further details) Given the usually poor preservation of the phenotypic landmarks in fossi specimens that are usually used to define cranial base flexure/extension we can only partly define this feature by observing the degree of angula tion of the external cranial base. The feature is defined here by the angle between the Frankfurt horizontal and the basion-hormion chord length that is, the inclination of the basioccipital and the basisphenoid measured externally (see Strait *et al.*, 1997; Strait, 2001). As in the measure of nuchal plane inclination, the character states and their allocation to taxa have been taken directly from Strait and Grine (2001).

  Pongo, Gorilla, Pan, and *P. walkeri* are defined by the primitive condi tion of a flat external cranial base ($= 0$). *Australopithecus* is defined by a moderately flexed cranial base ($= 1$), while *Paranthropus boisei, P. robus tus, H. habilis, H. ergaster*, and *H. sapiens* are defined by a strongly flexed anterior cranial base ($= 2$) (see Strait & Grine, 2001). For the same reason outlined by Strait *et al.* (1997), *Praeanthropus* was not scored because the specimen used by other researchers to define this species as displaying a weak inclination (e.g., Skelton & McHenry, 1992, and presumably D.E. Lieberman *et al.*, 1996) is based on specimen AL 333–105, which is an infant.

17. Nuchal Plane Orientation

This feature will obviously be affected by the development of the nuchal muscles, including the rectus capitis posterior (major and minor) and the obliquus capitis superior. These muscles are involved in head rotation and elevation. As in the inflation of the mastoid, this character may also be influenced by differential patterns of locomotion (see earlier). Of second ary importance in the development of this feature will be the degree/pattern of cranial base angulation. Following Kimbel *et al.* (1984) and Strait *et al.*

(1997), this character is defined by the angle between inion-opisthion chord length and the Frankfurt horizon. All character states have been taken from Strait *et al.* (1997) and Strait and Grine (2001).

Three states are recognized: steeply inclined, angle >60° (= 0) as observed in *Pongo, Gorilla,* and *Pan*; intermediately inclined, angle between 60° and 45° (= 1), as observed in *Praeanthropus*; and weakly inclined, angle <45° (= 2), as observed in *Australopithecus,* all species of *Paranthropus,* and *Homo* as well as *K. rudolfensis.* Unfortunately, post-depositional distortion of *Kenyanthropus* specimen KNM-WT 400000 does not enable this character to be defined for this genus. While no data were directly available for *Sahelanthropus,* Brunet *et al.* (2002) state that the nuchal plane is near horizontal; thus, it is coded as the same condition observed in most hominins.

18. Auditory Meatus

This character defines the relative position of the inferior tympanic edge relative to porion. This feature can be partially related to the length of the auditory tube, lateral expansion of the cranial base, lateral inflation of the mastoid process, and the development of the supramastoid crest. It is difficult to disentangle this feature from a number of functional and developmental processes, and it is for this reason that it is maintained as a separate feature.

The condition observed in *Pongo, Pan, Sahelanthropus,* the "anamensis group", *Praeanthropus, Australopithecus, P. walkeri,* both species of *Kenyanthropus, H. ergaster,* and *H. sapiens* is defined by a tympanic edge that is medially positioned, relative to porion (= 0) (see Kimbel *et al.,* 1994; M.G. Leakey *et al.,* 1995; Strait *et al.,* 1997; Strait & Grine, 1998, 2001; M.G. Leakey *et al.,* 2001; Brunet *et al.,* 2002). The condition observed in *Gorilla, Ardipithecus, P. boisei, P. robustus,* and *H. habilis* is for the tympanic to be laterally placed relative to porion (= 1) (see B. Brown *et al.,* 1993; T.D. White *et al.,* 1994; Strait & Grine, 2001; Strait *et al.,* 1997). Because the South African specimens of *H. habilis* are not considered here, and because the inferior tympanic edge is laterally placed in the east African *H. habilis* specimen KNM-ER 1805 (Strait & Grine, 1998; Strait *et al.,* 1997), *H. habilis* is allocated as displaying the derived condition.

19. External Auditory Meatus Size

The size of the external auditory meatus is defined as being either small (= 0) or large (= 1). This is a relative measure, and the character states and taxa allocations follow Strait and Grine (2001).

20. Articular Tubercle Height

This is defined by the perpendicular distance between the tip of the articular tubercle and the plane of the zygomatic arch, divided by the cranial base geometric mean, which is calculated as a geometric mean of four chords (basion–opisthion, basion–sphenobasion, bi-entoglenoid, and bi-cartoid canal [see Strait, 2001]). This index defines the relative development of the articular tubercle, which is related to the requirements of the outer oblique bands of the temporomandibular ligament, which converge onto an area at the back of the neck of the mandibular condyle. This ligament helps keep the condyle, disc, and temporal bone firmly opposed (Aiello & Dean, 1990). As such, it is an important functional component of the TMJ. The character definitions and allocations to hominin taxa have been taken directly from Strait (2001).

The primitive character state is for a moderate to large tubercle, as seen in *Pongo, Gorilla, Pan, Praeanthropus, Australopithecus, P. walkeri, P. robustus*, and *P. boisei* (= 0). The condition in *H. habilis* is said to be variable (= 1), while in *K. rudolfensis, H. ergaster*, and *H. sapiens*, it is undeveloped (= 2) (see Strait, 2001). While preservation means that a value cannot be generated for *Ardipithecus* or members of the "anamensis group", it is described as undeveloped in both taxa and is coded as small (see T.D. White *et al.*, 1994; M.G. Leakey *et al.*, 1995).

21. Petrous Orientation

A more coronal orientation of the petrous usually implies that the cranial base is relatively broad and short, as opposed to a longer and narrower base, which is emphasized by more sagittal orientation of the petrous. Numerous processes may be involved in the overall orientation of the petrous and the shape of the cranial base, including head-balancing requirements and differential patterns of encephalization and brain morphology. Functional requirements associated with the digastric muscle (digastric fossa) may also affect the overall orientation of the petrous, though this is likely to be of secondary importance.

Three character states are recognized: the primitive condition with a sagittally oriented petrous as observed in *Pongo*, and *Gorilla*, which is oriented to almost 90° (= 0); an intermediate condition, as observed in specimens of *Pan, Sahelanthropus* (Brunet *et al.*, 2002), *Praeanthropus*, and *Australopithecus* (= 1) (M.C. Dean & Wood, 1982; Strait *et al.*, 1997); coronally oriented petrous as observed in all other hominins preserving this feature (= 2) (M.C. Dean & Wood, 1981, 1982; M.C. Dean, 1986; Walker

et al., 1986; B. Brown *et al.*, 1993; Strait *et al.*, 1997; Keyser, 2000).
Brunet *et al.* (2002) describe petrous orientation in *Sahelanthropus* as
being within the *Pan, Australopithecus* (*Praeanthropus*), and *Ardipithecus*
range. Strait and Grine, however, suggest that *Gorilla* and *Pan* share a more
sagittal orientation as compared to *Praeanthropus*. On examining material
of *Pan* and *Gorilla*, it is suggested here that the condition in *Pan* is intermediate between the conditions observed in *Pongo* and *Gorilla* at one
extreme and the later hominins at the other. Thus, *Gorilla* and *Pongo* alone
are considered to have a sagittal orientation of the petrous bone, while *Pan*
shares a similar intermediate condition to that observed in the early
hominins, including *Praeanthropus* and *Australopithecus*.

22. Cranial Base Breadth

This character is an index of bi-porion divided by orbital height (chords).
The method used for defining the three character states used here is
described earlier (see character 2). The overall breadth of the cranial base
is significant in terms of head support and will have an impact on the orientation of numerous other basicranial morphologies (e.g., tympanic plate
and the petrous bone). All data for the fossil hominins has been taken from
B.A. Wood (1991); data for the extant hominids is unpublished and collected by Cameron. The index for *P. walkeri* (KNM-WT 17000) was generated from a cast. The mean index for the extant hominids ($n = 122$) is
3.27, with one standard deviation $= 0.27$. Thus, a narrow cranial base is
defined by an index of less than 3.00, intermediate is between 3.00 and
3.54, and a broad cranial base is defined by an index greater than 3.54.

Only *Sahelanthropus*, with an estimated index of 2.83, has a narrow
cranial base ($= 0$) (data from Brunet *et al.*, 2002). *Pongo* with a mean
index of 3.05 ($n = 22$), *Gorilla* with a mean index of 3.30 ($n = 35$), *Pan*
with a mean index of 3.33 ($n = 65$), *Australopithecus* with a mean index of
3.26 ($n = 2$), and *H. habilis* specimen KNM-ER 1813 with an index of
3.33 are all characterized by an intermediate breadth of the cranial base
($= 1$). Finally, a broad cranial base ($= 2$) is seen in *P. walkeri* specimen
KNM-WT 17000 with an estimated index of 3.72, *P. boisei* with a mean
index of 3.59 ($n = 2$), *K. rudolfensis* specimen KNM-ER 1470 with an
index of 3.63, and *H. sapiens* with a mean index of 3.65 ($n = 8$).

23. Basioccipital Length

This is an index of basion–sphenobasion length divided by cranial base
geometric mean (see Strait, 2001). This index gives a relative length of

the basioccipital. This feature relates directly to facial prognathism relative to the cranial base. The authors did not calculate any of these indices, except for *Pongo* (n = 2) but did use the coding structure and taxa allocation of Strait (2001). The primitive condition is defined by a long basioccipital, as seen in *Gorilla* and *Pan* (= 0), an intermediate condition as calculated for *Praeanthropus, Australopithecus*, and *P. walkeri* (= 1), and a short length for all other taxa preserving this feature (= 2) (Strait, 2001). *Pongo*, with an index of 0.20, was most similar to *Gorilla* with an index of 0.18, thus sharing the same condition as *Gorilla* and *Pan*.

24. Glenoid Fossa Depth

Four character states are defined: deep, variable, intermediate (between deep and shallow), and a shallow fossa. The depth of the glenoid process is clearly associated with the requirements of the masticatory apparatus (temporomandibular joint) and is of functional significance. This is also aligned to developmental processes associated with osteoclast/osteoblast activity and bone drift, which will affect mandibular and lower facial form (Enlow & Hans, 1996). This character is measured as an index of depth perpendicular to the Frankfurt horizontal, that is from the base of the articular eminence to the apex of the fossa/breadth of the eminence from the articular tubercle to the entoglenoid process.

The allocation of hominids to character states follows Strait *et al.* (1997) and Strait and Grine (2001). The Miocene hominids help define the primitive condition as deep (= 0). In the case of the Miocene hominids, no such measurements were generated (data not available). In all cases, the Miocene hominids are described as having a deep fossa, when preserved; e.g., *Dryopithecus* (Begun, 1994, 2002; Kordos & Begun, 1997). A deep fossa is also present in *H. sapiens*. Glenoid fossa depth is variable (= 1) in *H. ergaster* (Strait *et al.*, 1997; Strait & Grine, 2001). The glenoid fossa is of intermediate depth (= 2) in *Pongo, Gorilla, K. platyops, K. rudolfensis, Australopithecus, P. robustus*, and *H. habilis* (B.A. Wood, 1991; Strait & Grine, 2001; M.G. Leakey *et al.*, 2001). *Sahelanthropus* is described as having a shallow glenoid fossa, similar to the condition observed in *Pan* (= 3) (T.D. White *et al.*, 1994; Strait *et al.*, 1997; Brunet *et al.*, 2002). The same applies to *Ardipithecus*, members of the "anamensis group," *Praeanthropus*, and *P. walkeri* (Strait & Grine, 2001).

25. Glenoid Fossa Area (Size)

The size of the fossa is defined as the square root of the area of the triangular plane between: (a) tips of the postglenoid process; (b) entoglenoid process; and (c) the articular tubercle (see Strait, 2001). While the size (area) of the glenoid fossa may be considered by some to be similar to or strongly correlated with glenoid fossa depth (character 24), we argue that this feature, as defined by Strait (2001) is distinct, for size (area) need not correlate with depth. This is reinforced by the character state allocations shown in Appendix Table 1. For example, while *Gorilla* and *P. walkeri* both have a large glenoid fossa, in terms of depth they are distinct, because *Gorilla* has the intermediate condition, while in *P. walkeri* it is shallow. The same character states and allocations suggested by Strait (2001) are used here.

Pongo (n = 2) falls between the values generated for *Pan* (n = 2) and *Gorilla* (n = 3); however, it was close to the value generated for a cast of KNM-WT 17000. Thus, *Pongo* is allocated here as having a large glenoid fossa. The primitive condition is represented by a large fossa as seen in *Pongo, Gorilla, P. walkeri, P. robustus*, and *P. boisei* (= 0). An intermediate state is defined for *Pan, Praeanthropus, Australopithecus, K. rudolfensis, H. habilis*, and *H. ergaster* (= 1). A small fossa is defined for *H. sapiens* only (= 2).

26. Postglenoid Process Development

This feature is associated with the requirements of the masticatory apparatus. While it might be thought that this feature should be integrated with glenoid fossa depth, this is not the case, for the depth of the fossa and the development of the process are clearly not coupled. For example, both *Gorilla* and *Pan* have a similar development of the postglenoid process (large and anteriorly set), yet while *Gorilla* has intermediate depth of the glenoid fossa, it is shallow in *Pan*. Differential patterns between these two character states may suggest distinct excursive movements of the mandible or differential adaptive responses in TMJ morphology (dependent on two different ancestral conditions) to help stabilize the joint that is a functional convergence as opposed to a morphological convergence.

Four conditions are recognized here: a large process that is anteriorly set and not fused to the tympanic (= 0), as seen in *Pongo, Gorilla, Pan*, and *Praeanthropus* (Strait et al., 1997); a medium-sized process that may or may not be fused to the tympanic (= 1), as observed in *Dryopithecus, Australopithecus, P. walkeri*, and *K. rudolfensis* (Kordos & Begun, 1997; Strait et al., 1997; Begun, 2002); a small process that may be fused or unfused (= 2), as observed in *P. boisei* (Strait & Grine, 2001); and a small process that is fused to the tympanic (= 3), as observed in *P. robustus, H. habilis*, and *H. sapiens*. A small and fused process defines all East

African specimens of *H. habilis* (Strait *et al.*, 1997). While *Sahelanthropu* is described as having a large postglenoid process, no information is pro vided regarding whether it is fused to the tympanic (Brunet *et al.*, 2002) It is allocated here as displaying the primitive hominid condition.

27. Horizontal Distance Between TMJ and M2/M3

As discussed by Strait *et al.* (1997), this feature is of functional signifi cance given that it approximates the load arm of molar bite force durin mastication (Grine *et al.*, 1993; also see M.A. Spencer, 1998, 1999). Strai *et al.* (1997) recognize two character states: long distance from TMJ t molar complex (= 0) and short distance (= 1). As such, all taxa have relatively long distance between the TMJ and molar complex except fo *H. habilis, H. ergaster*, and *H. sapiens*, in which it is considered short.

28. Eustachian Process Development

The Eustachian process is a major attachment site for the tensor and leva tor palati muscles in the great apes. These muscles are responsible fo pulling the soft palate upward and thus aid in closing the nose off from th mouth during swallowing and for changes in vocalization patterns (Aiello & Dean, 1990). This process is present (= 0) in *Dryopithecus* (Kordos & Begun, 1997), *Pongo, Gorilla, Pan, Australopithecus*, and *P. robustus*, anc is absent or only weakly developed (= 1) in all other taxa preserving thi region.

29. Tympanic Shape

As discussed by Strait *et al.* (1997), the shape of the tympanic canal and th orientation of the anterior tympanic plate should be considered a single fea ture because a tubular tympanic will always be associated with a horizonta plate (= 0), while a vertical plate cannot exist without a crest (= 1) (Strai *et al.*, 1997; Strait & Grine, 2001; see also Keyser, 2000). The primitive con dition of a tubular tympanic (= 0) is observed in *Pongo, Gorilla*, and *Pan Sahelanthropus, Ardipithecus*, the "anamensis group," *Praeanthropus*, anc *K. platyops*, are described as tubular lacking a crest, similar to that observec in *Ardipithecus*; thus, it has the primitive hominid condition (T.D. White *et al.*, 1994; Strait *et al.*, 1997; M.G. Leakey *et al.*, 1995, 2001; Strait & Grine, 2001; Ward *et al.*, 2001; Brunet *et al.*, 2002). They also describe th "1470 group" (*K. rudolfensis*) as having the same condition in Table 2 though they give no additional information. All other hominids preserv ing this morphology have a crest with a vertical plate (= 1), although i *boisei*, the crest and plate are inclined ($= 2$) (see Kordos & Begun, 1997; trait *et al.*, 1997; Strait & Grine, 2001).

0. Vaginal Process Size

he vaginal process forms part of the tympanic plate, basal to the styloid rocess. The position and orientation of the styloid–vaginal process are of unctional significance in vocalization patterns, because it will influence the osition of the hyoid bone and associated soft tissues (Aiello & Dean, 990). The process is absent to small ($= 0$) in *Pongo, Gorilla, Pan, raeanthropus, P. walkeri, Australopithecus*, and *H. habilis*, while being noderate to large ($= 1$) in *P. robustus, P. boisei, H. ergaster*, and *H. sapiens* Strait & Grine, 2001).

1. Digastric Muscle Insertion

The digastric fossa is the site of origin for the digastric muscle. This muscle is responsible for raising the hyoid bone as well as the opening action of the mandible (Aiello & Dean, 1990). As such, it is of functional significance, especially in terms of the masticatory apparatus. A broad and shallow fossa is observed in *Pongo* (which is very broad), *Gorilla, Pan, raeanthropus, Australopithecus*, and *P. boisei* ($= 0$). A deep and narrow notch is observed in *Ardipithecus, K. platyops, P. robustus, H. habilis*, and *H. sapiens* ($= 1$). The condition for *K. platyops* is taken from M.G. Leakey *et al.* (2001), who describe the digastric fossa in the temporal bone specimen KNM-WT 40001 as being well developed, deep, and narrow.

2. Longus Capitis Insertion

This muscle is associated with flexing and rotation of the head (Aiello & Dean, 1990). The muscle insertion is located just anterior to the foramen magnum, with its origin at the thoracic and cervical vertebrae. The primitive condition is for a large and oval insertion ($= 0$), which is present in *Pongo, Gorilla, Pan*, and *Australopithecus*, while in species of *Paranthropus* and *Homo*, it is small and circular ($= 1$) (see Strait & Grine, 2001; Strait, 2001).

3. Foramen Magnum Shape

Unlike Strait *et al.* (1997), only two character states are recognized: oval/round ($= 0$) and heart shaped ($= 1$). Strait *et al.* (1997) and Strait and Grine (2001) recognize a "variable" state, as described for *H. ergaster*, because the sub-adult specimen KNM-WT 15000 is said to have a heart-shaped foramen (Walker *et al.*, 1986), while adult specimens of this same species, KNM-ER 3733 and KNM-ER 3883, have an oval foramen.

Because of the specimens sub-adult status, the condition observed i KNM-WT 15000 is not considered.

An oval-round foramen is observed in *Pongo, Gorilla, Pan Sahelanthropus, Praeanthropus, K. platyops, Australopithecus, P. robustu* and all species of *Homo* (see Tobias, 1991; McHenry & Skelton, 1992 Strait *et al.*, 1997; Strait & Grine, 2001; M.G. Leakey *et al.*, 2001; Strai 2001). Only *P. walkeri* and *P. boisei* are recognized as having a hear shaped foramen magnum (see Tobias, 1967; Walker *et al.*, 1986; Brow *et al.*, 1993).

34. Foramen Magnum Position

This character is defined by the position of basion, relative to th bi-tympanic line. This feature relates to cranial base length as well as th configuration of the tympanic. Following Strait *et al.* (1997), four charac ter states are recognized. *Pongo, Gorilla*, and *Pan* define the primitiv condition with basion positioned posterior to the bi-tympanic line (= 0) *Ardipithecus, Praeanthropus, Australopithecus, P. walkeri, H. ergaster* and *H. sapiens* are all defined by "anterior migration" of basion to b aligned with the bi-tympanic (= 1) (T.D. White *et al.*, 1995; Strait *et al.* 1997; Strait & Grine, 2001). Brunet *et al.* (2002) consider the position of th foramen magnum in *Sahelanthropus* to be similar to that observed i *Ardipithecus*, though they define the relative position from the bi-caroti chord. As such, it is allocated to the same condition as *Ardipithecus*. Th next character state is defined by a variable state, as defined by *H. habili* alone, because basion either is aligned with the bi-tympanic (KNM-EF 1813) or is placed anterior to it (= 2) (KNM-ER 1805). The last characte state is defined by *P. boisei* and *P. robustus*, with basion placed anterior t the bi-tympanic (= 3).

35. Foramen Magnum Inclination

The primitive condition is maintained by *Pongo, Gorilla*, and *Pan* with th foramen magnum strongly inclined posteriorly (= 0). This is also the cas for *Australopithecus* (Aiello & Dean, 1990). The intermediate condition i defined by *P. robustus, P. boisei, H. habilis*, and *H. sapiens* (= 1). *Homo ergaster* is unique, in that the foramen magnum is strongly inclined ante riorly (= 2) (Strait *et al.*, 1997; Strait & Grine, 2001; Strait, 2001).

36. Cranial Capacity

While there is some overlap between the ranges used here, allocation i based on recognizing the range within and between differing taxonomi nits; and if the upper range is moderately greater (even though there may be overlap at the lower range), we define a separate character state. For example, while both *Gorilla* and *Pan* are defined by an upper range of 500 cc ($= 0$), a separate character state is recognized for a range between 428 and 550 cc ($= 1$) because the upper range is moderately larger then the upper range observed in the extant hominids (and partially define distinct taxonomic units). Cranial capacities published by Aiello and Dean 1990), Strait *et al.* (1997), Asfaw *et al.* (1999), M.G. Leakey *et al.* (2001), and Brunet *et al.* (2002) were used to allocate taxa to character states.

Pongo, Gorilla, Pan, Sahelanthropus, Praeanthropus, the "garhi group," *K. playops*, and *P. walkeri* are all defined by relatively small cranial capacities, less than 500 cc ($= 0$). Next are *Australopithecus*, *P. boisei*, and *P. robustus*, with a capacity between 428 and 550 cc ($= 1$). This is followed by *H. habilis*, with a capacity between 509 and 675 cc ($= 2$). *Kenyanthropus rudolfensis* and *H. ergaster* are defined by a capacity between 750 and 1,100 cc ($= 3$), while *H. sapiens* is defined by a capacity larger than 1100 cc ($= 4$).

37. Cerebellar Morphology

The cerebellum is not tucked beneath the cerebrum; rather, it flares laterally with posterior protrusion ($= 0$) in *Pongo, Gorilla, Pan, Praeanthropus, Australopithecus*, and *P. walkeri* (Strait *et al.*, 1997; Strait & Grine, 2001). The cerebellum is tucked ($= 1$) in *P. boisei, P. robustus, K. rudolfensis*, and species of *Homo* (Holloway, 1972, 1988; Begun & Walker, 1993; Strait *et al.*, 1997; Strait & Grine, 2001).

38. Occipitomarginal Sinus

The development of this derived vascular pattern has been equated with the changing gravitational pressures associated with bipedalism (Falk & Conroy, 1984; Falk, 1986, 1988). It is suggested that this early pattern would enable increased flow of blood to the vital organs, given the changed posture. Tobias (1967), however, suggests that the patterns identified in these hominins may be associated with the early growth during ontogeny of the cerebellum, which may have forced blood into the marginal sinus system, which then became the established path of blood supply during adulthood. In either case it is of developmental and functional significance because the likelihood of heritability is strong.

An occipitomarginal sinus is absent in *Pongo, Gorilla, Pan, P. walkeri* (Walker *et al.*, 1986; Skelton & McHenry, 1992), *K. platyops* (M.G. Leakey *et al.*, 2001), *K. rudolfensis* (Kimbel, 1984; Tobias, 1991), and *H. habili* (Aiello & Dean, 1990; Tobias, 1991; Strait *et al.*, 1997). The condition i *Australopithecus* and *H. sapiens* is that it is intermediately expressed (= 1 (Kimbel, 1984; Aiello & Dean, 1990). An occipitomarginal sinus is, how ever, most frequently present in *Praeanthropus*, *P. boisei*, and *P. robustu* (= 2) (Falk, 1986, 1988; Skelton & McHenry, 1992; Strait *et al.*, 1997).

39. Facial Hafting (Upper Facial Position Relative to Bregma)

Originally described by Tobias (1967) and Howell (1978) as the positioi of the face relative to the neurocranium, that is, the position of the uppe face (supraorbital) relative to the highest point of the cranium (usuall} bregma), this feature is of developmental interest because it is to som degree the result of differential patterns of anterior cranial base angula tion, neuro-orbital disjunction, and frontal lobe development/morphology

A low, flat squama, with the upper face being set high relative t the cranium (= 0), defines *Gorilla*, *Sahelanthropus*, and species withii *Paranthropus* (Cameron, 1997a). A relatively rounded squama se above the face (= 1) defines *Dryopithecus, Pongo, Pan, Praeanthropus Australopithecus*, the "garhi group," both species of *Kenyanthropus H. habilis*, and *H. ergaster*. Finally, *H. sapiens* is defined by the uniqu condition of a face set low relative to the high, rounded cranium (= 2).

40. Interorbital Breadth

The character is generated from metric data and is defined by an index o mid-interorbital breadth divided by orbital height (see earlier). A narrow interorbital has often been argued to by a synapomorphy of the Ponginae and is associated with patterns of airorynchy (see Shea, 1985, 1988; Andrews, 1992; Cameron, 2002). As such, this character is of developmen tal and phylogenetic interest. The extant hominid mean is 0.45 (n = 123), with one standard deviation of 0.11. Thus, the intermediate range is 0.34–0.56. A narrow interorbital is an index less than 0.34, while a broad interorbital is an index beyond 0.56. It was noted that an additiona character state was required because a few hominids were beyond the extant hominid maximum value for this feature; that is, an index greater than 0.77. Thus, a fourth character state is recognized as an "extremely broad" interorbital. All indices for the fossil hominins have been generated from the data provided in B.A. Wood (1991), unless stated otherwise.

A narrow interorbital (= 0) is observed in *Pongo* with a mean index of 0.32 (n = 22). The intermediate condition (= 1) is observed in *Gorilla* with a mean index of 0.50 (n = 35), and *Pan* with a mean index of 0.46 (n = 66). A broad interorbital (= 2) is observed in *Dryopithecus* specimen RUD 77 (data from Kordos & Begun, 1997) with an index of 0.58; *Graecopithecus* specimen XIR-1 (original) has an index of 0.60; *Australopithecus* with a mean of 0.63 (n = 2); *P. walkeri* specimen KNM-WT 17000 (cast) with an index of 0.60; *K. rudolfensis* with a mean index of 0.73 (n = 2); *H. habilis* with a mean index of 0.71 (n = 2); *H. ergaster* with a mean index of 0.63 (n = 2); and *H. sapiens* with a mean index of 0.65 (n = 6). An extremely broad interorbital (= 3) is observed in *P. boisei*, with a mean index of 0.72 (n = 2), and *P. robustus*, with a mean index of 0.83 (n = 2); that is, they are beyond the maximum values observed in the extant hominids.

41. Frontal Sinus Development

Three character states are recognized: large (0), intermediate to small (1), and absent (2). The frontal sinus in *Dryopithecus* is developed because it extends below nasion (Begun, 2002). A well-developed frontal sinus is observed in *P. boisei* (Tobias, 1967), *P. robustus* (Tobias, 1991), *K. rudolfensis* as defined by KNM-ER 1470 (B.A. Wood, 1991), and *H. ergaster* specimen KNM-WT15000 (Walker & Leakey, 1993b). *Gorilla, Pan,* and *H. sapiens* share a similar development (Spoor & Zonneveld, 1999), which Begun (1992) and Rae (1999) consider similar to *Dryopithecus.* Thus, all of the foregoing genera are allocated to the primitive hominoid condition of a developed frontal sinus system. *Graecopithecus* also appears to have a moderately developed frontal sinus (see Begun, 2002). In the "garhi group", the frontal sinus is restricted to the mid-frontal just above glabella; thus, it is intermediate to small (= 1) (Asfaw *et al.*, 1999). This also applies to *Australopithecus* and *H. habilis*, in which it is described as small (Tobias, 1991). A frontal sinus is absent in *Pongo* (= 3), and while some pneumatization of the frontal may occur, it is regarded as an extension of the maxillary sinus (Cave, 1940; Koppe & Ohkawa, 1999).

42. Lacrimal Fossa Location

The primitive condition is for the lacrimal fossa to be located within the orbit. The morphology of the lacrimal fossa is of functional, developmental, and phylogenetic interest, for the "derived" pattern requires a major reconfiguration of the mid-facial region. Only in a *Graecopithecus* is the derived pattern present, defined by the lacrimal fossa located anterior to the inferomedial orbital margin (see Cameron, 2002, in press, Begun, 2002).

43. Facial Dishing

This character is of functional and developmental interest because it relates to masseter muscle development, mid-facial prognathism, as well as patterns of bone growth within this area. This character is also defined by the morphological pattern originally noted by Robinson (1962). Robinson defined facial dishing as a result of the expanded and anteriorly positioned zygomatic prominence, relative to the more posterior "sunken" nasal aperture and infraorbital plane (see also Tobias, 1967; Rak, 1983). Using this morphological pattern, four conditions are recognized: absent (a developed premaxilla; i.e, prognathic to varying degrees, which negates the formation of mid-facial depression) ($= 0$); lateral and upper face near vertical and mid-face concave ($= 1$); the whole facial region is strongly convex ($= 2$); and facial dishing as defined by Robinson is present ($= 3$).

A developed mid-face–premaxilla ($= 0$) is observed in *Dryopithecus, Gorilla, Pan, Sahelanthropus*, both species of *Kenyanthropus, Praeanthropus, Australopithecus*, the "garhi group," and species of *Homo*. Shallow mid and lateral upper facial dishing defines *Pongo* ($= 1$). The condition in *Graecopithecus* is unique because it is marked by strong convexity and a mid-face that is more prognathic relative to *Pongo*, but it also lacks the "snout-like" appearance seen in the African hominids ($= 2$) (Cameron 1997a; partly de Bonis & Koufos, 1993). Facial dishing as described by Robinson (1962) is observed only within species of *Paranthropus* ($= 3$) (Asfaw *et al.*, 1999; Skelton *et al.*, 1986; Skelton & McHenry, 1992).

44. Upper Facial Breadth

The method used for defining the indices and ranges of variability and resulting character states are described earlier (see character 2). The overall breadth of the upper face is important because it gives a relative statement concerning the lateral expansion of the upper face. All data for the fossil hominins have been taken from B.A. Wood (1991); data for the Miocene and extant hominids are unpublished data collected by Cameron, though partly provided in Cameron (1997a, 1997b, 1998). The index is bi-frontomalare temporale/orbital height. The extant hominid mean is 2.94 (n = 122), with one standard deviation of 0.34. Thus, the intermediate condition is between 2.60 and 3.28, while broad is larger than 3.28, and narrow is less than 2.60. It was also noted that *Graecopithecus*, with an index of 4.44, lies well beyond the extant hominid maximum value of 3.81, so it is allocated to an additional character state, "extremely broad" (see next paragraph).

A narrow upper face ($= 0$) defines *Pongo* with a mean index of 2.47 ($n = 20$). Intermediate upper facial breadth ($= 1$) is observed in *Dryopithecus* specimen RUD 77, with an estimated index of 2.81 (data supplied in Kordos & Begun, 1997), *Sahelanthropus* specimen TM 266-01-060-1 has an index of 2.83 (data from Brunet *et al.*, 2002), *Australopithecus* has a mean index of 3.11 ($n = 2$), *P. Walkeri* specimen KNM-WT 17000 (cast) has an index of 2.86, *H. habilis* has a mean index of 3.23 ($n = 2$), *Gorilla* has a mean of 3.11 ($n = 36$), *Pan* has a mean index of 2.99 ($n = 65$), and *H. sapiens* has a mean index of 3.13 ($n = 176$ [data from Thorne, 1976]). A broad face ($= 2$) is observed in *P. boisei*, with a mean index of 3.29 ($n = 2$), *P. robustus* specimen SK 49, with a mean index of 3.57, *K. rudolfensis*, with a mean index of 3.57 ($n = 2$), and *H. ergaster*, with a mean index of 3.32 ($n = 2$). While there is evidence of some post depositional distortion in *Graecopithecus* specimen XIR-1, it clearly has a broad face. An estimate of facial breadth and orbital height taken from the original specimen gives a significant value of 4.44, which is well beyond the maximum value generated for the extant hominid, at 3.81 ($n = 122$). Thus, it is allocated to its own character state, an "extremely broad" upper face ($= 3$).

45. Mid-Facial–Upper Facial Breadth

This is defined by an index of bijugal/bi-frontomalare temporale, using the same method described for character 2, to construct the character states. This feature is distinct from character 44, because it describes the breadth of the upper face relative to the mid-face, while the previous character measures the relative breadth of the upper face between hominid taxa. This feature is likely to be of developmental and functional interest because it relates to musculature development (e.g., temporalis and masseter). All data for the fossil hominins have been taken from B.A. Wood (1991); data for the extant hominids are unpublished data collected by Cameron. The mean index for the extant hominids ($n = 121$) is 1.09, with one standard deviation of 0.07. Thus, any value less than 1.02 indicates that the upper face and lower face have a similar breadth; any value between 1.02 and 1.16 indicates an intermediate condition; while any value greater than 1.16 indicates that mid-facial breadth is significantly greater than that of the upper face. The extant hominid minimum value is 0.94, and the maximum value is 1.25.

No specimens were defined here by having a similar breadth as defined by the indices generated here. Upper and mid-facial breadth are defined by the intermediate condition ($= 0$) in *Graecopithecus* specimen

XIR-1 (original) with an index of 1.02, *P. boisei* with a mean index of 1.15 (n = 2), *P. robustus* with a mean of 1.09 (n = 2), *K. rudolfensis* with a mean of 1.07 (n = 2), *H. habilis* specimen KNM-ER 1813 with an index of 1.02, *H. ergaster* with a mean index of 1.05 (n = 2), *Pongo* has a mean index of 1.15 (n = 21), *Gorilla* with a mean index of 1.14 (n = 36), *Pan* with a mean of 1.04 (n = 64), and *H. sapiens* with a mean index of 1.10 (n = 8). A broad mid-face, relative to upper facial breadth (= 1), is observed in *Australopithecus* specimen Sts 5 with an index of 1.18. A very broad mid-face (= 2) is observed in *P. walkeri* specimen KNM-WT 17000 with an index of 1.28, which is beyond the extant hominid range.

46. Position of Anterior Zygomatic Relative to Upper Dentition

The first is defined by the zygomatic insertion being at the M^1/M^2 junction (= 0) and is usually observed in specimens of *Gorilla*. Insertion at the M^1 (= 1) is observed in *Kenyapithecus* (L.S.B. Leakey, 1962; McCrossin & Benefit, 1997; Cameron, 1998), *Dryopithecus* (Cameron, 1995), *Graecopithecus* (de Bonis & Koufos, 1993), the "anamensis group" (C.V. Ward *et al.*, 2001), *H. habilis*, (Tobias, 1991), *H. ergaster* (B.A. Wood, 1991), *Pongo, Pan*, and *H. sapiens*. Insertion at the P^4/M^1 junction (= 2) is observed in *Praeanthropus* (Kimbel *et al.*, 1994), *Australopithecus*, the "garhi group" (Asfaw *et al.*, 1999), and *K. rudolfensis* defined by KNM-ER 1470 (B.A. Wood, 1991). Insertion at the P^4 or anterior to it (= 3) is observed in *Kenyanthropus* (M.G. Leakey *et al.*, 2001) and species of *Paranthropus* (Tobias, 1967; B.A. Wood, 1991). While no information is supplied by Brunet *et al.* (2002) for this feature in *Sahelanthropus*, their Figure 1 indicates that it is anterior to the molar complex; thus, it is considered as being at P^4.

47. Height of Masseter Origin

This is used here as an important feature relating to masseter muscle leverage; that is, it is implied that the shorter the index, the greater the leverage. This character is defined using the index zygomaticoalveolar height relative to orbitoalveolar height (Kimbel *et al.*, 1984). The results and allocations for the fossil hominins and extant African apes have been taken directly from Strait *et al.* (1997) and Strait and Grine (2001), while the author generated the indices for *Pongo* from unpublished data. Most are considered as having a low origin for the master (an index of over 56), which is the primitive condition. Only *Pongo* with a mean of 57.2 (n = 19), *Australopithecus*, and species within *Paranthropus*, are considered to have a high origin (see Strait *et al.*, 1997; Kimbel *et al.*, 1984).

48. Palate Prognathism Relative to Sellion

This is calculated as an index of palate length divided by plate protrusion (relative to sellion) as defined by Rak (1983; see his Plate 34 and Table 3). A prognathic premaxilla is defined by an index greater than 0.57; intermediate prognathism is between 0.30 and 0.57, and weak prognathism is an index less than 0.30. The mean indices for hominin taxa have been taken from Rak (1983). All hominin and extant African ape allocations follow Strait et al. (1997) and Strait & Grine (2001), except for *Australopithecus*, which has the intermediate condition. While no metric data are available for *Kenyanthropus*, M.G. Leakey et al. (2001) state that the overall degree of facial prognathism within this species (taking into account post-depositional distortion) is within the range observed in *K. rudolfensis*. Increased prognathism as observed in *Pongo* is also defined by increased prognathism with a mean index of 0.71 (n = 21 [see Cameron, 1997b]). While *Graecopithecus* specimen XIR-1 has been affected by some post-depositional distortion, an index of 0.61 (original specimen) is estimated as a minimum, so it is defined by increased prognathism.

49. Subnasal Prognathism (Nasal Clivus)

This character has been taken from M.G. Leakey et al. (2001). Their Figure 3 indicates three patterns of subnasal prognathism as measured by the angulation of the nasal clivus (prosthion–nasospinale). This character is of interest not only because it relates to subnasal prognathism, but this itself is associated with the requirements of the anterior dentition as well as palate depth, which may be associated with differential patterns of cranial base flexure (see D.E. Lieberman et al., 2001).

Increased prognathism (= 0) is observed in *Graecopithecus, Pongo, Gorilla*, and *Pan*, which fall below 30° (= 0). The intermediate condition (= 1) is observed in *Dryopithecus, Praeanthropus, Australopithecus*, species within *Paranthropus*, and *H. habilis* between 30° and 42° (= 1). This also appears to be the case for BOU-VP-12/130 ("garhi group"). Reduced prognathism (= 2) is observed in both species of *Kenyanthropus, H. ergaster*, and *H. sapiens*, with reduced prognathism at above 42° (= 0). *Sahelanthropus* is characterized by weak subnasal prognathism (Brunet et al., 2002), so it is allocated to this character state.

50. Incisor Alveoli Prognathism

This character is defined by the prognathism of the incisor complex relative to the bi-canine line; this is distinct from the previous character (subnasal prognathism–angle) because this character measures the degree of anterior migration of the incisor complex, while the previous character measures degree of subnasal prognathism, relative to height of the nasal clivus. As such, this feature helps measure degrees of incisor crowding as well as the anterior or posterior migration of the incisor complex relative to the bi-canine line.

Only two character states are recognized here. The primitive condition of the incisor alveoli anterior to the bi-canine line (= 0) is observed in *Dryopithecus, Graecopithecus, Pongo, Gorilla, Pan, Sahelanthropus,* the "anamensis group" (C.V. Ward *et al.*, 2001), *Praeanthropus, Australopithecus,* the "garhi group," *H. habilis, H. ergaster,* and *H. sapiens.* The derived condition of the incisor alveolar being close to or aligned with the bi-canine line (= 1) is observed in both species of *Kenyanthropus* and the species within *Paranthropus* (see Strait *et al.*, 1997; Strait & Grine, 2001; M.G. Leakey *et al.*, 2001; Brunet *et al.*, 2002).

51. Anterior Palate Depth

This character is of functional and developmental significance because it will be affected by (1) masticatory demands; (2) cranial base angulation; and (3) requirements of maintaining an appropriate orbital horizon (which will have an impact on cranial base and palate floor angulation) (see D.E. Lieberman *et al.*, 2001). As such, the origins of this feature are quite complex. A shallow anterior palate, defined by the palate grading posteriorly from the base of the incisors as a flat surface or in which there is a slight depression anterior to the incisive foramen (= 0), is observed in *Kenyapithecus, Gorilla, Pan,* the "anamensis group," *Praeanthropus,* the "garhi group," species of *Paranthropus,* and *H. habilis* (Leakey, 1962; Cameron, 1997a, 1998; C.V. Ward *et al.*, 2001; see partly Strait & Grine, 2001). Intermediate anterior palate depth (= 1) defines *Dryopithecus* (partly Cameron, 1997a, 1998). A deep anterior palate is defined by a sharp vertical rise before turning posteriorly at the incisive foramen (= 2) and is observed in *Graecopithecus, Pongo, Australopithecus,* both species of *Kenyanthropus, H. ergaster,* and *H. sapiens* (see Cameron, 1997a; Strait & Grine, 2001).

52. Palate Thickness

This character measures the absolute thickness of the palate posterior to the incisive fossa. Character states and taxa allocations follow Strait *et al.* (1997) and Strait (2001). A thin palate represents the primitive condition that is under 7 mm (= 0); a thick palate is observed only in *Paranthropus* species that is over 7 mm ($= 1$) (see also McCollum, 1997). Even though no data are available for *Kenyanthropus*, Leakey *et al.* (2001) describe the palate roof as being thin, so it is allocated as such.

53. Palate Breadth

Palate breadth is defined by an index of external palate breadth at M^2 divided by orbital height. The method used for defining the three character states used here is described earlier (see character 2). All data for the fossil hominins have been taken from B.A. Wood (1991), and data for the extant and Miocene hominids are unpublished data collected by Cameron. The mean index for the extant hominids (n $= 124$) is 1.66, with one standard deviation equal to 0.13. Thus, a narrow palate is defined by an index of less 1.53, intermediate breadth is between 1.53 and 1.79, while a broad palate is defined by an index greater than 1.79.

A broad palate ($= 0$) is observed in *Graecopithecus* specimen XIR-1 (original) with an index of 2.25, *Australopithecus* (specimen Sts 5) with an index of 2.17, *P. walkeri* specimen KNM-WT 17000 (cast) with an index of 2.15, *P. boisei* with a mean index of 2.20, *P. robustus* (specimen SK 48) with an index of 2.20, *H. habilis* with a mean index of 2.13 (n $= 2$), *H. ergaster* specimen KNM-ER 3733 with an index of 1.81, and *H. sapiens* with a mean index of 1.82 (n $= 8$). Intermediate palate breadth ($= 1$) is defined in *Pongo* with a mean index of 1.67 (n $= 22$), *Gorilla* with a mean index of 1.66 (n $= 36$), and *Pan* with a mean index of 1.67 (n $= 66$).

54. Projection of Nasal Bones above Frontomaxillary Suture

The morphology of the nasal bones has often been considered of phylogenetic significance (Robinson, 1954; Tobias, 1967; Olsen, 1978, 1985; Rak, 1983). This is emphasized because the nasal region and its relationship to nasion and glabella are formed early in ontogeny, within late fetal growth (Olson, 1985); thus, it is less likely influenced by functional considerations. First, the nasal bones extend superior to the frontomaxillary suture, resulting in a tapering of the nasal bones above the suture ($= 0$), as observed in *Dryopithecus* (Kordos & Begun, 1997), *Pongo*, *Gorilla*, *Pan*, and *Australopithecus*. Next, they are defined by the projection of the nasal bones above the frontomaxillary suture, widening laterally as they pass the suture line ($= 1$), as observed in *Praeanthropus* and species within *Paranthropus*. The last condition is defined by the nasal bone not projecting above the frontomaxillary suture ($= 2$), observed in *K. rudolfensis*, *H. ergaster*, *H. habilis*, and *H. sapiens* (Strait & Grine, 2001).

55. Nasal Keel

A nasal keel is defined by the presence of a distinct vertical "pinching" of the nasal bones along its midline, defining a nasal keel (= 0). When present, a nasal keel is usually pronounced in the region between the orbits, flattening out inferiorly before the superior nasal margin. As discussed earlier (character 54), the morphology of the nasal bones appears early in ontogeny and is thus considered of developmental and phylogenetic interest. A nasal keel is observed only in *Graecopithecus* (de Bonis & Koufos, 1993), *Gorilla, Australopithecus* (Rak, 1983), and *H. habilis* (Tobias, 1967).

56. Inferior Orbital Margin Is Rounded Laterally

The inferior orbital margin is considered either rounded laterally or not rounded; that is, the inferolateral orbital margin is blunted, with a shallow bend posteriorly into the orbit. The overall significance of this feature in terms of function, development, and phylogeny remains obscure. But given that Strait *et al.* (1997) and Strait and Grine (2001) have included it in their detailed studies, this feature is retained here. Rounded corners in the taxa preserving this region are present in *Gorilla, P. walkeri,* and *P. robustus* only (= 1).

57. Superior Orbital Fissure Configuration

The superior orbital fissures separate the greater wings of the sphenoid bone, which are located just inferior to them (Aiello & Dean, 1990). This feature is of likely developmental importance because it is associated with a complex interplay of numerous bones within the region of the orbit–sphenoid. For example, many other separate bones help define the orbit, including the maxilla, ethmoid, lacrimal, frontal, and the zygomatic, which will all to varying degrees be affected by differing amounts of bone remodelling and displacement. Thus, their development adds substantially to the development of this region (Enlow & Hans, 1996). The character states and hominin allocations follows Strait and Grine (2001). The primitive condition is defined by the fissure representing a fossa, as observed in *Pongo, Gorilla, Pan, Praeanthropus, Australopithecus,* and *H. habilis* (= 0), while the same condition is "comma" shaped in *P. walkeri, P. boisei,* and *H. sapiens* (= 1). This feature could not be observed in other hominins due to problems of preservation.

58. Maxillary Trigon

This feature has been discussed at length in Rak (1983). It is described as a furrow-like feature that occupies much of the infraorbital region and inferior malar. It is consistently present in *P. walkeri* and *P. robustus,* though it is variable in *P. boisei* (see Suwa *et al.*, 1997). This feature is distinct from our other character, "facial dishing" (character 43), because it defines the depression within the infraorbital region only and not the mid-face as a whole. Indeed, while *P. boisei* is characterized by strong facial dishing, it is not always marked by an infraorbital furrow.

59. Malar–Zygomatic Orientation

Emphasis is placed on the orientation of the zygomatic viewed laterally, whereas previously it was defined by the position of the inferior zygomatic (zygomaticoalveolar crest) relative to the inferior orbital margin (viewed laterally). Three conditions are defined. First, an anterior/posterior bend (= 0) defines *Dryopithecus* (Cameron, 1997a) and *Gorilla*. A near-vertical slope (= 1), is observed in *Pongo, Kenyanthropus* (M.G. Leakey *et al.*, 2001), *H. habilis* (Tobias, 1991; B.A. Wood, 1991), the "1470 group," *H. ergaster* (B.A. Wood, 1991), and *H. sapiens*. The preserved inferior parts of the zygomatic in *Kenyapithecus* specimen KNM-FT 46 indicate that it too was most likely near vertical. Finally, there is an anterior slope from the region of the inferior orbital margin (= 2). This is observed in *Pan, Australopithecus* (Rak, 1983), and species of *Paranthropus* (Rak, 1983).

60. Diagonal Length of the Malar

Malar length is defined as an index of orbitale–zygomaxillare divided by orbital height. The mean for the extant hominids is 1.05, with one standard deviation of 0.15 (n = 122 specimens). Thus, the intermediate range falls between 0.90 and 1.20. All data for the fossil hominins was taken from B.A. Wood (1991). The minimum extant hominid value is 0.75 and the maximum value is 1.56. A long malar (= 0) is observed in *P. walkeri* specimen KNM-WT 17000 with an index of 1.23, *P. boisei* with a mean index of 1.22 (n = 3), *K. rudolfensis* defined by KNM-ER 1470 with an index of 1.23, and *Gorilla* with a mean of 1.21 (n = 35). The intermediate condition (= 1) is observed in *P. robustus* defined by specimen SK 48 with an index of 1.00, *Pongo* with a mean index of 1.01 (n = 21), *H. ergaster* with a mean index of 1.12 (n = 2), and *Pan* with a mean index of 0.97 (n = 66). A short malar (= 2) is observed in *Australopithecus* with a mean index of 0.88 (n = 2), *H. habilis* specimen OH 24 with an index of 0.84, and *H. sapiens* with a mean index of 0.88 (n = 8).

61. Infraorbital Foramen(ina) Location

This character is likely of developmental interest in terms of its inferior migration and implications for differential patterns in the hominin capsular system. Three states are recognized. The infraorbital foramen is located within the top 50% of the malar ($= 0$), observed in *Pongo, Gorilla, Pan, Praeanthropus, H. habilis, H. ergaster*, and *H. sapiens*. Also, Andrews & Walker (1976) suggest that the infraorbital foramen in *Kenyapithecus* specimen KNM-FT 46 must have been high set, given that it, and its resulting inferior furrow, are not present in its preserved parts. Next is a variable state; that is, the foramen may be located within the top or bottom 50% of the malar ($= 1$), as observed in specimens of *Australopithecus* and *P. boisei* (Strait & Grine, 2001). The third condition is where the foramen is located inferiorly within the bottom half of the malar ($= 2$), as observed in *P. walkeri* and *P. robustus*.

62. Anterior Nasal Pillars
Rak (1983) has discussed this feature extensively. He associates this feature with *P. robustus*, arguing that it evolved as a functional response to help absorb stresses associated with the anterior dental complex. This feature, however, should not be confused with canine jugum, which is merely the result of a large robust canine. A true "nasal pillar" is composed of a column-like buttress that extends well beyond the canine root, helping to define the lateral nasal aperture borders (Rak, 1983). Only two character states are recognized: absent ($= 0$) and variable ($= 1$). All taxa preserving this region lack anterior pillars, except for *Australopithecus, P. robustus*, and *H. habilis*, in which they may, or may not, be present (see also Tobias, 1991; Asfaw *et al.*, 1999; M.G. Leakey *et al.*, 2001; C.V. Ward *et al.*, 2001).

63. Nasal Cavity Entrance
This feature is of developmental interest because it relates to patterns of cranial base angulation (Shea, 1988, 1993) as well as differential patterns of bone growth and associated drift (Enlow & Hans, 1996; D.E. Lieberman, 2000). The primitive condition is defined by a distinct step into the nasal floor from the posterior nasal clivus ($= 0$) as observed in *Dryopithecus* (Begun, 1994; Cameron, 1997a), *Graecopithecus* (de Bonis & Melentis, 1987; de Bonis *et al.*, 1990; de Bonis & Koufos, 1993; Cameron, 1997a), the "anamensis group" (C.V. Ward *et al.*, 2001), *Praeanthropus* (Strait *et al.*, 1997; Strait & Grine, 2001), *K. platyops* (M.G. Leakey *et al.*, 2001), the "garhi group" (Asfaw *et al.*, 1999; Strait & Grine, 2001), *H. habilis* (Tobias, 1991; McCollum *et al.*, 1993), *Gorilla*, and *Pan*. The intermediate condition ($= 1$) between states 0 and 2 is present in *Australopithecus* (McCollum *et al.*, 1993). A smooth transition between the posterior clivus and the nasal floor with considerable overlap ($= 2$) is observed in *Pongo* (McCollum & Ward, 1997),

species of *Paranthropus* (McCollum *et al.*, 1993), and *H. ergaster* (Strait & Grine, 2001). Finally, *H. sapiens* alone is defined by a smooth entrance but with no overlap between the clivus and the palate (= 3) (see McCollum & Ward, 1997; Strait *et al.*, 1997; Strait & Grine, 2001).

64. Incisive Canal Development

This character is similar to that discussed in Cameron (1997a). This feature is related to the development in length and caliber of the incisive canal. This feature is also of interest in terms of the conditions associated with airorynchy and klinorynchy (Shea, 1985, 1988). It is also of interest in the impacts that its reconfiguration has on the capsular system. It is distinct from the previous character in that a species with a distinct step to the nasal floor can have a moderately developed canal or it may have no canal at all, but, rather, a large incisive foramen. This character does, however, combine incisive foramen size and canal development, which have been treated as distinct characters in the past (e.g., Begun, 1992).

The primitive condition is for a broad "canal," as observed in *Dryopithecus* (Cameron, 1997a), *Graecopithecus* (de Bonis & Melentis, 1987), *Gorilla*, and *Pan* (McCollum & Ward, 1997). Also, *H. habilis* (defined by specimens OH24 and OH 62) is said to be similar to the chimpanzee in this feature (Tobias, 1991; see also Kimbel *et al.*, 1997). An intermediate condition (= 1) defines members of the "anamensis group" (C.V. Ward *et al.*, 2001), *Praeanthropus* specimen A.L. 200-1a (Kimbel *et al.*, 1997), and *Australopithecus*, which has a developed canal (Rak, 1983) but not to the extant observed in *H. sapiens*. An extensive canal (= 2) defines *Pongo*, species of *Paranthropus* (Rak, 1983; McCollum *et al.*, 1993), *H. ergaster* (Walker & Leakey, 1993b), and *H. sapiens* (McCollum & Ward, 1997).

65. Nasal Clivus Contour in Coronal Plane

The clivus is convex (= 0) in *Dryopithecus* (Begun, 1992; Cameron, 1995), *Graecopithecus, Pongo, Gorilla, Pan*, members of the "anamensis group," *Praeanthropus*, the "garhi group," and *H. sapiens* (see Strait *et al.*, 1997; Strait & Grine, 2001; C.V. Ward *et al.*, 2001). The clivus is flat (= 1) in *Sahelanthropus*, both species of *Kenyanthropus, Australopithecus, H. habilis*, and *H. ergaster* (Strait *et al.*, 1997; M.G. Leakey *et al.*, 2001; Strait & Grine, 2001; Brunet *et al.*, 2002). The clivus is concave with a developed gutter (= 2) in species of *Paranthropus* (Strait *et al.*, 1997; Strait & Grine, 2001).

66. Orbital Shape

Orbital shape will be affected by numerous processes, including cranial base angulation (Shea, 1985, 1988), facial height and breadth considerations (Rak, 1983), and capsular requirements (Enlow & Hans, 1996). This feature is thus rather complex and considered to be of developmental, functional, and phylogenetic interest. Three character states are recognized. Orbits, higher than broad (= 0) are observed in *Dryopithecus* (Begun, 1994a; Kordos & Begun, 1997) and *Pongo*. Circular–rhomboid shaped orbits (= 1) are observed in *Sahelanthropus* (as observed from Brunet *et al.*, 2002, Figure 1, even though distorted), *Pan*, and all hominins preserving this feature (see Tobias, 1991; Wood, 1991; M.G. Leakey *et al.*, 2001). Finally, rectangular orbits, broader than high, are observed in *Graecopithecus*, taking into account damage and some post-depositional distortion (de Bonis *et al.*, 1990; de Bonis & Koufos, 1993), and *Gorilla*.

67. Inferior Orbital Margin

This feature helps define the depth of the face (Cameron, 1997a). It is defined by the horizontal position of the inferior orbital margins against the superior nasal aperture margin (viewed anteriorly). Three states are recognized. The first is defined by inferior orbital margins that are positioned well above the superior nasal aperture margin, indicating a tall midface (= 0); next is the intermediate condition (= 1); and finally, the inferior orbital margins are aligned to or below the superior nasal aperture margin, indicating a short mid-face (= 2). A deep mid-face is observed in *Graecopithecus* and *Gorilla*. The intermediate condition is observed in *Dryopithecus, Pongo, Pan, Sahelanthropus*, and all three species of *Paranthropus*. A short mid-face (2) is observed in *Australopithecus, H. habilis, K. rudolfensis, H. ergaster*, and *H. sapiens*.

68. Subnasal Length

This is a chord distance from nasospinale to prosthion, so it is not a measure of subnasal prognathism. Subnasal prognathism is measured in character 49. Subnasal length is defined as an index against orbital height. The mean extant hominid value for this index is 0.72, with 1 standard deviation of 0.13. Thus, an intermediate range is defined by values falling between 0.59 and 0.85. Data for fossil hominins has been taken from B.A. Wood (1991), while all other values have been constructed from unpublished data collective by DWC, unless otherwise specified.

A short clivus (= 0) is observed in *H. sapiens* with a mean index of 0.53 (n = 8). Intermediate clivus length (= 1) is observed in *Graecopithecus* specimen XIR-1 (original) with an index of 0.87, *H. habilis* specimen OH 24 with an index of 0.81, *Pongo* with a mean index of 0.68 (n = 22), *Gorilla* with a mean index of 0.73 (n = 35), and *Pan* also with a mean index of 0.73 (n = 66). While no data are available for *Sahelanthropus*, the subnasal is described as short (Brunet *et al.*, 2002). A relatively long clivus (= 2) is observed in *Australopithecus* with a mean index of 0.93 (n = 2), *P. walkeri* specimen (cast) KNM-WT 17000 with an index of 0.95, *P. boisei* with a mean index of 1.02 (n = 2), *P. robustus* specimen SK 48, also with an index of 0.93, *K. rudolefensis* as represented by specimen KNM-ER 1470 with an index of 1.03, and *H. ergaster* specimen KNM-ER 3733 with an index of 1.00. While no data is available for either *Praeanthropus* or the "garhi group," the sub-nasal region is said to be even longer than that observed in *Australopithecus* (Rak, 1983; Asfaw *et al.*, 1999).

69. Maxillary Sinus

The same character states and allocations to taxa for the extant and Miocene hominids have also been taken straight from Begun *et al.* (1997). Intermediate development of the maxillary sinus (= 0) is observed in *Kenyapithecus* (Leakey, 1962; Andrews & Walker, 1976) and *Dryopithecus* (Begun, 1994a/b; Begun & Gülec, 1998). A large maxillary sinus (= 2) defines *Pongo, Gorilla, Pan, Praeanthropus, Australopithecus P. robustus, P. boisei, K. rudolfensis, H. habilis,* and *H. ergaster* (Tobias, 1967, 1991; Walker & Leakey, 1993a; Kimbel *et al.*, 1997). Finally, *H. sapiens* is shown in the studies of Spoor & Zonneveld (1999) and Rae & Koppe (2000) as having a relatively similar maxillary sinus size compared to *Gorilla* and *Pan*, which also appears to be the case for the "anamensis group," which has a sinus system defined by a large anterior cavity and smaller middle cavity superior to M^3 (C.V. Ward *et al.*, 2001).

70. Zygomatic Insertion Height

This feature relates to the insertion of the zygomaticoalveolar crest onto the alveolar border. This feature is of functional interest, because it will bear directly on the leverage of the masseter muscles as well as partly defining facial height. Two conditions are recognized, a low insertion (= 0) and a high insertion (= 1). A low insertion is observed in *Kenyapithecus* (McCrossin & Benefit, 1997 [*contra* Begun & Gülec, 1997; Begun *et al.*, 1997]), *Graecopithecus* (Begun *et al.*, 1997), both species of *Kenyanthropus*, and *P. walkeri* (M.G. Leakey *et al.*, 2001). A high insertion is observed in *Dryopithecus* (Begun *et al.*, 1997), *Sahelanthropus* (Brunet *et al.*, 2002), *Australopithecus* (Rak, 1983), *P. boisei* (Tobias, 1967), *P. robustus*

(M.G. Leakey *et al.*, 2002), *H. ergaster* (Walker & Leakey, 1993), *Pongo, Gorilla, Pan*, and *H. sapiens*. While Kimbel *et al.* (1994) do not discuss the height of the insertion in *Praeanthropus*, specimen A.L. 444-2 is observed to have a relatively high insertion (their Figure 2). Also, while B.A. Wood (1991) considers *H. habilis* specimen KNM-ER 1813 to have a low insertion relative to specimen KNM-ER 1470, it is still relatively high; that is, within the range of the extant hominids.

71. Canine Fossa Development

A canine fossa is defined by a distinct depression or hollow of the bone table posterior to the canine within the lateral maxillary wall. In the extant hominids it is usually associated with a developed transverse buttress, located immediately above the infraorbital foramen (see Rak, 1983). While this feature in some taxa may be associated with sexual dimorphism (i.e., greater development in males), there is evidence that it can also be well defined in extant hominid females (personal observation). A distinct canine fossa can also be identified in Miocene female hominids; e.g., *Kenyapithecus* female specimen KNM-FT 46 (see Leakey, 1962). Wherever possible, however, only male specimens were examined. A shallow canine fossa defines *Dryopithecus* (Begun, 1994a/b), *Graecopithecus* (Koufos, 1995; Begun & Gülec, 1997), *P. walkeri, P. boisei* (Rak, 1983), and *H. habilis* (Tobias, 1991). A deep canine fossa is observed in *Kenyapithecus* (Leakey, 1962), *Sahelanthropus* (Brunet *et al.*, 2002), the "anamensis group" (C.V. Ward *et al.*, 2001), *Praeanthropus* (Rak, 1983), *Australopithecus* (Rak, 1983), the "garhi group" (Asfaw *et al.*, 1999), *Pongo, Gorilla*, and *Pan* (contra Begun *et al.*, 1997). While variable in specimens of *H. sapiens*, it is usually developed. The maxillary fossula as defined by Rak (1983) for *P. robustus* is considered here as a canine fossa. The reason Rak defines these characters as separate features appears to be the association of a transverse buttress in *Australopithecus* and a maxillary trigon in *P. robustus*. These are, however, separate feature, as such, the canine fossa in *Australopithecus* and the maxillary fossula of *P. robustus* are homologous features.

72. Mandibular Symphysis Orientation

This feature is of developmental interest because it relates to patterns of bone growth and drift within the mandible (see Enlow & Hans, 1996). Symphyseal development probably is also influenced by twisting requirements of this region during mastication (see partly Hylander, 1988). The character states and allocations for the fossil hominins are taken directly from Strait *et al.* (1997), with one exception, *H. sapiens*, which is defined here as having a unique pattern, that is, a chin. A receding external symphyseal region (= 0) is observed in *Kenyapithecus, Dryopithecus, Graecopithecus, Pongo, Gorilla, Pan,* and members of the "anamensis group" (Andrews, 1971; Walker & Andrews, 1973; McCrossin & Benefit, 1997; B. Brown, 1997; C.V. Ward *et al.*, 2001; Begun, 2002). The intermediate condition (= 1) between a receding symphysis and a near-identical symphysis can be observed in *Praeanthropus* and *Australopithecus* (Strait *et al.*, 1997). The external mandibular symphyseal orientation is near vertical (= 2) in *P. boisei, P. robustus, K. rudolfensis, H. habilis,* and *H. ergaster* (Strait *et al.*, 1997). *Homo sapiens* is unique in its near vertical orientation with a distinct chin (= 3).

73. Mandibular Symphysis Robusticity (Breadth/Height)

This feature measures the general robusticity of the mandibular symphysis. That is, a high index means the symphysis is robust, while a lower index means it is more gracile. As suggested by Hylander (1984, 1988), Hylander and Johnson (1994), and Hylander *et al.* (2000), the thickness of the symphysis has important implications associated with wishboning and its impact on the symphyseal region. All data for the fossil hominins have been taken from B.A. Wood (1991), and data for the extant hominids are either unpublished data collected by DWC, unless stated otherwise. The mean index for the extant hominids (n = 99) is 0.38, with one standard deviation of 0.05. A gracile symphysis is defined by an index of less than 0.33, an intermediate breadth is between 0.33 and 0.43, while a robust symphysis is defined by an index greater than 0.43. In addition to this, the minimum value for the extant hominids is 0.26, while the maximum value is 0.51.

A relatively gracile symphysis (= 0) is observed in *Pongo* with a mean index of 0.32 (n = 22). The intermediate condition (= 1) is observed in *Kenyapithecus* specimen KNM-FT 45 (reconstruction) with an index of 0.40 (data from Andrews, 1971), *Gorilla* with a mean index of 0.42 (n = 23), *Pan* with a mean index of 0.38 (n = 54), and *H. sapiens* with a mean index of 0.43 (n = 128 [data from Thorne, 1976]). A robust mandible (= 2) is observed in the "anamensis group" with a mean index of 0.50 (n = 3 [data from C.V. Ward *et al.*, 2001]), *Praeanthropus* with a mean index of 0.50 (n = 4), and *Australopithecus* specimen Sts 52 with an index of 0.51. Finally, an extremely robust symphysis (= 3) is seen in *P. boisei* with a mean index of 0.56 (n = 7), *K. rudolfensis* with a mean value of 0.63 (n = 4), *H. habilis* with a mean index of 0.69 (n = 2), and *H. ergaster* with a mean index of 0.57 (n = 4).

74. Inferior Mandibular Transverse Torus

A developed inferior transverse torus helps withstand stresses associated with mastication, related to medial and lateral wishboning of the mandible, as well as increasing resistance to dorsoventral shear (Hylander, 1984, 1988; Hylander & Johnson, 1994; Hylander *et al.*, 2000). This character is distinct from the previous character, because a deep symphysis may or may not have a developed inferior transverse torus. Thus, there is no clear association between these features, even though they both help withstand masticatory processes.

The less-developed superior transverse torus relative to the developed inferior torus (= 0) is observed in *Kenyapithecus* (Pickford, 1986; McCrossin & Benefit, 1997; Ward & Duren, 2002), *Pongo*, members of the "anamensis group" (M.G. Leakey *et al.*, 1995; C.V. Ward *et al.*, 2001), *P. robustus*, and *P. boisei* (*P. boisei* [B.A. Wood, 1991]). Mandibular transverse tori of similar development (= 1) are observed in *Dryopithecus* (Smith-Woodward, 1914) and *Graecopithecus* (Martin & Andrews, 1984; de Bonis & Koufos, 1994; Begun, 2002). A more developed superior transverse torus (= 2) is observed in *Praeanthropus* (Johanson *et al.*, 1978), *Australopithecus* (Tobias, 1980), *Gorilla*, and *Pan*. The absence of either torus (= 3) is observed in *K. rudolfensis* (B.A. Wood, 1991), *H. habilis* (B.A. Wood, 1991), *H. ergaster* specimen KNM-ER 992 (B.A. Wood, 1991), and *H. sapiens*. While tori are observed in *H. ergaster* specimen KNM-WT 15000, they are only weakly developed (Walker & Leakey, 1993a). This feature remains unordered, as its morphocline cannot be ascertained with any degree of certainty.

75. Mandibular Corpus Robusticity (Breadth/Height)

This character measures the robusticity of the corpus at M_2. This feature is of functional and developmental interests because it relates to bone deposition associated with tensile and compressive stress during the masticatory cycle (see Hylander, 1984, 1988; Hylander & Johnson, 1994; Hylander *et al.*, 2000; B. Brown, 1997). All data for the hominins has been taken from B.A. Wood (1991), while all extant hominid data is from DWC (unpublished). Most of the values for the Miocene hominids have been taken from Begun (1994b). The mean hominid index for this character is 0.56 (n = 103), with one standard deviation equal to 0.08. Thus, the intermediate condition is defined by values between 0.48 and 0.64. The minimum extant hominid value is 0.38, and the maximum extant value is 0.75. Note, while Begun (1994b) gives values for *Kenyapithecus*, the specimens used to define this genus (from Maboko Island) have since been reallocated to *Equatorius* (see Ward *et al.*, 1999).

Intermediate corpus robusticity ($= 0$) is seen in *Kenyapithecus* with a mean index of 0.55 (taken from M_1 [Andrews, 1971]), *Dryopithecus* with a mean index of 0.64 ($n = 7$), *Pongo* with a mean index of 0.57 ($n = 18$), *Gorilla* with a mean index of 0.56 ($n = 22$), *Pan* with a mean index of 0.55 ($n = 63$), and *H. sapiens* with a mean index of 0.56 ($n = 8$). While no data were available for the corpus at M_2 for members of the "anamensis group," specimen KNM-KP 29281 has a robusticity of 0.58 from the P_4/M_1 junction (data taken from C.V. Ward *et al.*, 2001). Also, Strait and Grine (2001) allocate the "anamensis group" to the same condition observed in *H. sapiens* and *Pan*. Thus, this group is defined here by having the intermediate condition. A relatively robust corpus ($= 1$) is observed in *Graecopithecus* with a mean index of 0.66 ($n = 3$), *Praeanthropus* with a mean index of 0.71 ($n = 7$), *H. habilis* with a mean index of 0.71 ($n = 3$), and *K. rudolfensis* with a mean index of 0.70 ($n = 3$). An extremely robust corpus ($= 2$) is observed in *Australopithecus* with a mean index of 0.76 ($n = 5$), *P. boisei* with a mean index of 0.78 ($n = 22$), and *P. robustus* with a mean index of 0.79 ($n = 4$).

76. Mandibular Premolar Orientation

This character helps define the general shape of the mandible by the orientation of the premolar tooth row. The primitive condition is defined by a U-shaped post canine tooth row ($= 0$), as a result of a near-parallel orientation of each premolar tooth row that is observed only in *Kenyapithecus, Dryopithecus, Pongo, Gorilla, Pan, Sahelanthropus,* and *Praeanthropus* (Begun, 2002; Brunet *et al.*, 2002). Next is the dental arcade being V-shaped as the post canine tooth rows diverge laterally ($= 1$), as observed in *Graecopithecus* (see Begun, 2002). The mandibular dental arcade is more parabolic ($= 2$) due to a more laterally diverging tooth row, which characterizes all other hominins examined here, preserving this morphology (Strait & Grine, 2001).

77. Mandibular Mental Foramen Opening

This character is retained in this analysis because it relates to the vascular system and is thus likely to be of developmental and functional interest. The same character states and species allocations provided by Strait *et al.*

(1997) have been retained. Four states are recognized. The primitive condition is defined by the foramen opening anterosuperiorly ($= 0$) as seen in *Kenyapithecus* (Andrews & Walker, 1976), *Pan*, and *Ardipithecus* (Strait & Grine, 2001). This is followed by some intra-species variability ($= 1$) in *Dryopithecus* (Begun, 2002), *Pongo, Gorilla, Praeanthropus*, and *Australopithecus* (Strait & Grine, 2001). In *Graecopithecus* and in most specimens of *H. sapiens* the foramen opens posteriorly ($= 2$). The foramen opens laterally ($= 3$) in the "anamensis group," *P. boisei, P. robustus, K. rudolfensis, H. habilis*, and *H. ergaster* (Strait & Grine, 2001). This feature remains unordered, as its morphocline cannot be ascertained with any degree of certainty.

78. Hollowing at the Mental Foramen

This feature is clearly the result of differential patterns of bone growth (e.g., increased osteoblast activity in the region of the foramen and/or increased osteoblast activity anterior to it) and is likely associated with mandibular robusticity requirements. The same character definitions and character state allocations provided by Strait *et al.* (1997) and Strait and Grine (2001) are followed for the fossil hominins. The primitive character state is seen in *Kenyapithecus, Dryopithecus, Graecopithecus, Pongo, Gorilla, Pan, Ardipithecus*, the "anamensis group," and *Praeanthropus*, in that there is distinct hollowing within the external bone table, above and behind the mental foramen ($= 0$). *Australopithecus* and *K. rudolfensis* show variability in the development of this feature; intra-specifically ($= 1$), *P. boisei, P. robustus, H. habilis, H. ergaster*, and *H. sapiens* are defined by the absence of such a hollow.

79. Mandibular Extramolar Sulcus Width

This feature has been related to a more parabolic shape in the dental arcade, resulting in an M_3 placed more medially then the M_2, and will result in a large space between the M_3 and the ramus, which is said to be essential in accommodating transverse movements of the M^3, both of which lie posterior to the ramus (see Aiello & Dean, 1990). The character state definitions and most species allocations as provided by Strait *et al.* (1997) and Strait and Grine (2001) are followed here, with only one exception (see later).

The primitive state is represented by a broad sulcus ($>6.5\,mm$) ($= 0$), which is observed in *Pongo, Gorilla, P. robustus*, and *P. boisei*. While no metrical data are available for *Graecopithecus*, they are clearly very broad (de Bonis & Koufos, 1993). Intraspecies variability ($= 1$) is observed only in *Australopithecus*. The sulcus is narrow (= 2) in *Pan, Praeanthropus, K. rudolfensis, H. habilis, H. ergaster*, and *H. sapiens. Homo habilis* is allocated as having a narrow sulcus, for all East African specimens are narrow. Strait *et al.* (1997) include the South African specimen SK 15 in their definition of this species, which has a wide sulcus. Thus, unlike DWC, they define *H. habilis* as variable.

80. Upper Incisor Reduction (I^1–I^2 Area/Orbital Area)

Incisor size is usually related to dietary considerations and is thus of functional and phylogenetic interest (see Ungar & Kay, 1995; Kay & Ungar, 1997; Ungar, 2002). The index used here is the sum of I^1 and I^2 mesial-distal length divided by orbital area. All data for fossil hominins has been taken from Wood (1991). The mean for the extant hominids is 0.14 (n = 92) with one standard deviation of 0.03. This gives an intermediate range of 0.11–0.17. The minimum extant hominid value is 0.08 and maximum is 0.23.

Increased incisor area (= 0) is observed in *Pongo* with a mean index of 0.18 (n = 18). The intermediate condition (= 1) is observed in *Graecopithecus* specimen XIR-1 (original) with an index of 0.15, *Australopithecus* with a mean index of 0.13, *P. robustus* with a mean index of 0.12, *H. habilis* with a mean index of 0.13, *Gorilla* with a mean index of 0.12 (n = 23), *Pan* with a mean index of 0.13 (n = 51), and *H. sapiens* with a mean index of 0.12 (data from Thorne, 1976). The broad incisor alveolar border in *P. walkeri* specimen KNM-WT 17000 suggests that the incisors were probably within the range of *Gorilla*. While no data is available for *Ardipithecus*, the "anamensis group," *Praeanthropus*, and members of the "garhi group," they are said to be similar to either *Australopithecus* or early *Homo*, which are allocated here to the intermediate condition (Asfaw *et al.*, 1999; Strait & Grine, 2001). Reduced incisor size (= 2) is observed in *P. boisei* specimen OH 5 with an index of 0.10.

81. Upper Incisor Heteromorphy (I1 Area/I2 Area)

The intermediate condition is defined by index values of 1.27–2.05 (n = 92). Reduced heteromorphy is defined by a value less than 1.27; increased heteromorphy has a value greater than 2.05 (Cameron, 1997a). The minimum value is 1.09 and the maximum value is 2.84. Increased heteromorphy (= 0) is observed in *Graecopithecus* with a mean index of 2.62 (Cameron, 1998) and *Pongo* with a mean of 2.31 (n = 18). The intermediate condition (= 1) is observed in *Praeanthropus* with a mean index of 1.62 (n = 5/8 [data from B.A. Wood, 1991; see also Tobias, 1980]), *Australopithecus* with a mean index of 1.93 (n = 3 to 5 [B.A. Wood, 1991; see also Tobias, 1980]), *P. boisei* specimen OH 5 with an index of 1.54, *P. robustus* with a mean index of 1.48, *H. habilis* with a mean index of 1.73, *Gorilla* with a mean index of 1.57 (n = 23), *Pan* with a mean index of 1.45 (n = 51), and *H. sapiens* with a mean index of 1.40 (n = 56/126 [data from Thorne, 1976]). While no data are directly available for *K. platyops*, M.G. Leakey *et al.* (2002) describe the preserved central and lateral incisor roots with an index of just 90%, whereas they are typically 50–70% in other known hominin taxa. Thus *K. platyops* is allocated as having reduced heteromorphy (= 2).

82. Male Canine Size

A robust dagger-like canine (= 0) is observed in *Kenyapithecus* (Pickford, 1985), *Dryopithecus* (Begun, 1994a; Cameron, 1997a), *Graecopithecus* (Kelley, 2002), *Pongo, Gorilla*, and *Pan*. The intermediate condition (= 1), between states 0 and 2, is observed in *Sahelanthropus* (Brunet *et al.*, 2002), *Ardipithecus*, members of the "anamensis group," *Praeanthropus, K. platyops, Australopithecus*, the "garhi group," *K. rudolfensis*, and *H. habilis*. The preserved canine roots in *P. walkeri* specimen KNM-WT 17000 suggest that it was within the intermediate range. An absolutely reduced male canine (= 3) is observed in *P. boisei, P. robustus, H. ergaster*, and *H. sapiens*.

83. Molar/Premolar Size

This feature is defined using an index of molar area (M^1–M^3) divided by premolar area (P^3–P^4). The relationship between premolar and molar size is likely to be associated with masticatory demands. All data for the fossil hominins has been taken from B.A. Wood (1991); data for the extant hominids are unpublished data collected by DWC. The mean index for the extant hominids (n = 122) is 2.10, with one standard deviation of 0.19. Thus, the intermediate condition is between 1.91 and 2.29. The minimum extant hominid value is 1.64, and the maximum value is 2.62 (see Cameron, 1998).

The intermediate condition (= 0) is observed in members of the "anamensis group" with a mean index of 2.16 (Ward *et al.*, 2001), *P. robustus* with a mean index of 2.24, *Pongo* with a mean index of 2.03 (n = 21), *Gorilla* with a mean index of 2.10 (n = 35), and *Pan* with a mean index of 2.14 (n = 66). They are characterized by a reduction in differential size between the molar and premolar complex (whether this means the molars have decreased or the premolars have increased will be examined in the next character). A marked difference (= 1) is observed in *Dryopithecus* with a mean index of 2.35 (Cameron, 1998), *Praeanthropus* with a mean index of 2.44, *Australopithecus* with a mean index of 2.41, *P. boisei* specimen OH 5 with an index of 2.55, and *H. habilis* with a mean index of 2.52. Significant difference between the premolar and molar complexes (= 2) is observed in *Graecopithecus* with a mean value of 2.75 (Cameron, 1998) and *H. sapiens* with a mean index of 2.80 (n = 8), which are both beyond the extant great ape maximum value.

84. Upper Second Molar Area/Orbital Area

This character gives a relative statement of the molar complex as defined by the M^2 area (divided by orbital area). Combined with the previous character, we can determine whether any identified changes in the premolar/molar complex are the result of increased molar size or reduced premolar size. That is, the previous character on its own does not enable us to determine whether the premolar complex has increased or reduced in size. All data to generate the fossil hominin has been taken from B.A. Wood (1991) unless stated otherwise. The extant hominid mean value is 0.11 (n = 124), with one standard deviation of 0.03. The intermediate range is between 0.08 and 0.14. The minimum extant hominid value was 0.06 and the maximum value was 0.18.

A reduced molar complex (= 0) is observed in *Dryopithecus* specimen RUD 77 with an estimated index of 0.11 (data from Kordos, 1988), *Sahelanthropus* specimen TM266-01-060-1 with an index of 0.13 (data from Brunet *et al.*, 2002), *H. ergaster* specimen KNM-ER 3733 with an index of 0.12, *Pongo* with a mean index of 0.12 (n = 22), *Gorilla* with a mean index of 0.13 (n = 36), *Pan* with a mean index of 0.09 (n = 66), and *H. sapiens* with a mean index of 0.13 (values generated from Thorne, 1976). While no data are available for *Kenyanthropus*, M.G. Leakey *et al.* (2001; see also their Figure 3a) state that it has among the smallest M^2 relative to all other Pliocene hominins. A large molar complex (= 1) is observed only in *H. habilis* with a mean index of 0.18. An extremely large molar complex (= 2) is observed in *Graecopithecus* specimen XIR-1 (original) with an index of 0.23, *Australopithecus* with a mean index of 0.21, *P. boisei* specimen OH 5 with an index of 0.27, *P. robustus* with a mean value of 0.23, and the "1470 group" with a mean index of 0.20. While no values could be generated using this index for either *Praeanthropus* or the "garhi group," Asfaw *et al.* (1999) demonstrate that *Praeanthropus* and specimen BOU-VP-12/130 (the "garhi group") are within the size range of both *Australopithecus* and *K. rudolfensis*. This also applies to *Ardipithecus* and members of the "anamensis group" because M.G. Leakey *et al.* (2001) show that both groups in overall area are within the range observed in *Praeanthropus, Australopithecus*, and *K. rudolfensis*. Finally, while no upper molars are available for *P. walkeri*, the massive tooth roots observed in KNM-WT 17000 clearly indicate that the molars were extremely large.

85. Molar Cusp Position

The position of the lingual and buccal cusps is of dietary-functional importance and likely to be of phylogenetic significance, taking into account phylogenetic niche conservatism (see text). Strait and Grine (2001) are used in the definition of the character states and allocations for the hominins. In the case of the Miocene hominoids, because some species have extensive lingual cingulum (character 89), the location of the cusps ignores this feature, because the enamel extension gives a "false" reading of the position of the cusps relative to the occlusal surface.

The primitive condition is observed in *Kenyapithecus* (Andrews & Walker, 1976), *Dryopithecus* (Begun, 1994a), *Graecopithecus* (de Bonis & Koufos, 1993), *Pongo, Gorilla*, and *Pan*. The cusps are located close to their respective crown edges; that is, the cusps are not close together (= 0). The next condition observed in *Praeanthropus, K. rudolfensis, H. habilis, H. ergaster*, and *H. sapiens* is for the lingual cusps to be located close to the margin of the crown but the buccal cusps have moved medially (= 1). *Ardipithecus* and members of the "anamensis group" have lingual cusps that are close to the margin and buccal cusps moderately lingual to margin (= 2). *Australopithecus* has lateral movement of its lingual cusps and increased medial movement of its buccal cusps (= 3). Finally, in *P. boisei* and *P. robustus* is found the continued crowding of the occlusal surface with increased medial and lateral movement of the buccal and lingual cusps (= 4).

86. P₃ Metaconid Development

Three states are recognized. This feature is likely of phylogenetic significance given its apparent lack in the Miocene hominoids, which helps differentiate the earlier hominoids from later hominids/hominins. The primitive condition is defined by an undeveloped metaconid or the total absence of a metaconid (= 0) as observed in *Kenyapithecus* (Andrews &

Walker, 1976). *Dryopithecus* is said to display a metaconid (Begun, 1992, 1994a), though it is undeveloped and within the range observed in the extant African hominids that are also defined by this condition. Next is an infrequent-to-variable presence of a metaconid (= 1) observed in *Graecopithecus* (Koufos, 1993; de Bonis & Koufos, 2001) and *Praeanthropus* (= 1), with all hominins preserving this morphology frequently having a metaconid (= 2).

87. P3 Mesiobuccal Enamel Extension

This feature relates to the mesiobuccal enamel expansion. The character states and hominin allocations have been taken directly from Strait and Grine (2001). In all Miocene hominoids preserving this morphology, it is strongly developed. The condition in *H. ergaster* specimen KNM-ER 992 is clearly not expanded (see B.A. Wood, 1991) and thus reflects the condition of the later hominins as defined by Strait and Grine (2001).

88. Molar Enamel Thickness

This character is of developmental and functional (dietary) interest. The character state definitions and taxa allocation are taken directly from Strait *et al.* (1997) and Strait and Grine (2001), with some modifications for the Miocene hominoids as defined by Andrews and Martin (1991). Thin to intermediate thin molar enamel (= 0) is observed in *Dryopithecus, Ardipithecus, Gorilla*, and *Pan* (see T.D. White *et al.*, 1994; Strait & Grine, 2001). Intermediate-thick to thick enamel (= 1) is observed in *Pongo*. The condition in *Sahelanthropus* is said to lie between the thinner condition observed in *Pan* and the thicker condition observed in *Australopithecus* (Brunet *et al.*, 2002). Thus, it is allocated here as having intermediate to thick molar enamel. Thick molar enamel (= 2) is observed in members of the "anamensis group," *Praeanthropus, Australopithecus*, both species of *Kenyanthropus*, and all species of *Homo*. While no data are available for *Kenyapithecus*, Pickford (1985, 1986) states that the enamel is thick. Hyper-thick molar enamel (= 3) is observed in *Graecopithecus, P. boisei*, and *P. robustus*.

89. Upper Molar Lingual Cingulum Development

This feature, like the metaconid, is of phylogenetic interest as its presence/absence appears to be of some phylogenetic importance, given its common presence in the earliest hominoids and its later reduction/absence in the later hominids/hominins. It is also of functional interest given that the cingulum may to some degree be associated with trying to increase the occlusal surface. Lingual cingulum is only weakly developed or absent (= 0) in *Kenyapithecus* (Leakey, 1962; Andrews & Walker, 1976; Cameron, 1998; Ward & Durren, 2002), *Ankarapithecus* (Andrews & Tekkaya, 1980), *Graecopithecus* (de Bonis *et al.*, 1990; de Bonis & Koufos, 1993, 2001), *Sahelanthropus* (Brunet *et al.*, 2002), *Praeanthropus* (T.D. White *et al.*, 1983), *P. robustus* (Tobias, 1967), *K. rudolfensis* (B. Brown *et al.*, 1993), *H. habilis* (Tobias, 1991), *Pongo, Gorilla, Pan,* and *H. sapiens*. While no direct mention is made of cingulum in the discussion of *Ardipithecus*, White *et al.* (1994) state that the overall morphology of the teeth is very similar to *Praeanthropus*; thus, like this genus, it is considered to have extremely reduced or no lingual cingulum. Lingual cingulum is present (= 1) in *Dryopithecus* (Begun, 1994a) and *Australopithecus* (Tobias, 1967; T.D. White *et al.*, 1983).

90. M^2 Shape

This feature is of functional interest because it relates to tooth crowding of the palate. For instance, an increase in molar breadth may be a requirement to increase its occlusal surface while maintaining reduced palate prognathism. As such, it is also of likely phylogenetic interest. A square tooth is defined by a length/breadth value that is within 5% of both dimensions, with a broad tooth less than 0.95 and a long tooth beyond 1.05 (see Cameron, 1997a). The values for the Miocene hominoids and fossil and extant hominids have been taken from Cameron (1998) unless stated otherwise, while all of the fossil hominin values have been generated from B.A. Wood (1991).

A relatively broad molar (= 0) is observed in *Dryopithecus* with a mean of 0.93, *Graecopithecus* with a mean of 0.91 (n = 6 [Koufos, 1995]), *Ardipithecus* with a mean index of 0.81 (n = 2), and members of the "anamensis group" with a mean index of 0.94 (n = 3) (data from Ward *et al.*, 2001), *Praeanthropus* with a mean value of 0.88, *Australopithecus* with a mean of 0.90, *P. boisei* specimen OH 5 with an index of 0.82, *P. robustus* with an index of 0.90, *H. habilis* with a mean index of 0.88, *K. rudolfensis* with a value of 0.83, *H. ergaster* specimen KNM-ER 3733 with an index of 0.92, *Pongo* with a mean index of 0.90 (n = 72), *Pan* with a mean index of 0.88 (n = 89), and *H. sapiens* with a mean index of 0.83 (n = 168 to 241, values generated from Thorne, 1976). Square molars (= 1) are observed in *Kenyapithecus* specimen KNM-FT 46 with an index of 1.02 (Leakey, 1962), *Sahelanthropus* with an index of 1.02 (Brunet *et al.*, 2002), and *Gorilla* with a mean index of 0.97 (n = 74).

91. Upper Molar Morphology

This character is related to masticatory demands and is of phylogenetic significance, not only given the concept of phylogenetic niche conservatism, but also because of the apparent high heritability of dental development processes and occlusal morphology (see Jernvall, 1995; Teaford, 2000). This character, while undoubtedly influenced by molar enamel thickness, is not strongly correlated, and each represents a distinct feature. For example, *Kenyapithecus* has relatively thick enamel with inflated cusps and cristae, while *Ankarapithecus* also has relatively thick enamel. It, however, has undeveloped cusps and cristae (see Andrews & Alpagut, 2002).

The primitive condition is observed in *Dryopithecus*, which has high and developed cusps–cristae formation (= 0) (Begun, 1994a/b; Andrews *et al.*, 1996), *Gorilla, Pan*, and *Praeanthropus* (T.D. White *et al.*, 1983). *Ardipithecus* is described as having a similar overall morphology to that of *Praeanthropus* (T.D. White *et al.*, 1994). Low inflated cusps with limited or no cristae development (= 1) are observed in *Kenyapithecus* (Leakey, 1962; Andrews & Walker, 1976; Ward & Durren, 2002), *Graecopithecus* (de Bonis *et al.*, 1990; de Bonis & Koufos, 1993, 2001), *Sahelanthropus* (Brunet *et al.*, 2002), the "anamensis group" (M.G. Leakey *et al.*, 1995; C.V. Ward *et al.*, 2001), *Australopithecus* (T.D. White *et al.*, 1983), *H. habilis* (Tobias, 1991), *K. rudolfensis* (B.A. Wood, 1991), *H. ergaster* (B.A. Wood, 1991), and *H. sapiens*. Low cusps with extensive enamel wrinkling (= 2) are observed in *Pongo* (S. Ward, 1997); while a low, flat "grinding stone-like" occlusal surface (= 3) is observed in *P. boisei* and *P. robustus* (Tobias, 1967; T.D. White *et al.*, 1983).

92. Deciduous M1 Mesial Crown Profile

This character is of developmental interest, especially given the ontological significance, relative to the adult condition. The example provided by Raff (1996) will suffice to establish to some degree the likely significance of these two deciduous dental characters. He reminds us that early in the 19th century, ascidians (marine invertebrates that as adults are saclike filter feeders) were classified as molluscs, but they were later shown to have larvae unlike any mollusc. Their larvae would resemble tiny tadpoles, complete with dorsal chord and notochord. These larvae are built along the same overall body plan as vertebrates and as such were reclassified. While perhaps an extreme example relative to the deciduous dentition, it does, however, highlight the need to examine processes that can be partly identified by examining stages of ontogeny.

The character state definitions and taxa allocation used here have been taken directly from Strait *et al.* (1997) and Strait and Grine (2001). Three states are recognized. The primitive condition is defined by the absence of a mesial marginal ridge (MMR), the protoconid is situated well mesial, and the anterior fovea has a wide lingual opening ($= 0$), as observed in *Ardipithecus, Gorilla,* and *Pan.* The next condition is defined by the primitive condition, but with the MMR being slightly developed ($= 1$), as seen in the "anamensis group," *Praeanthropus, Australopithecus, H. ergaster,* and *H. sapiens.* Finally, the MMR is strongly developed, the protoconid is aligned with the metaconid, and the fovea is closed ($= 2$), as observed in *P. robustus* and *P. boisei.*

REFERENCES

Abbate, E., Albianelli, A., Azzaroli, A., Benventuti, M., Tesfamariam, B., Bruni, P., Cipriana, N., Clarke, R.J., Ficcarelli, G., Macchiarelli, R., Napoleone, G., Paini, M., Rook, L., Sagri, M., Tecle, T.M., Torre, D. & Villa, I. 1998. A one-million-year-old *Homo* cranium from the Danakil (Afar) Depression of Eritrea. *Nature.* 393: 458–460.

Abbie, A.A. 1968. The homogeneity of Australian aborigines. *Archeology & Physical Anthropology in Oceania.* 3: 223–231.

Abbie, A.A. 1976. Morphological variation in the adult Australian Aboriginal. In R. Kirk & A.G. Thorne (eds.). *The Origin of the Australians*, pp. 211–214. Australian Institute of Aboriginal Studies, Canberra.

Adcock, G.J., Dennis, E.S., Easteal, S., Huttley, G.A., Jermiin, L.S., Peacock, W. James & Thorne, A.G. 2001. Mitochondrial DNA sequences in ancient Australians: Implications for modern human origins. *Proceedings of the National Academy of Science.* 98: 537–542.

Agusti, J. & Antón, M. 2002. *Mammoths, Sabertooths, and Hominids: 65 Million Years of Mammalian Evolution in Europe.* Columbia University Press, New York.

Agusti, J., Köhler, M., Moyà-Solà, S., Cabrera, L., Garces, M., & Pares, J.M. 1996. Can Llobateres: the pattern and timing of the Vallesian hominoid radiation reconsidered. *Journal of Human Evolution.* 31: 143–155.

Agusti, J., Cabrera, L. & Garces, M. 2001. Chronology and zoogeography of the Miocene hominoid record in Europe. In L. de Bonis, G.D. Koufos & P.J. Andrews (eds.). *Hominoid Evolution and Climate Change in Europe*, Vol. 2: *Phylogeny of the Neogene Hominoid Primates of Eurasia*, pp. 1–18. Cambridge University Press, Cambridge, UK.

Aiello, L. & Dean, C. 1990. *An Introduction to Human Evolutionary Anatomy.* Academic Press, New York.

Aiello, L. & Wheeler, P. 1995. The expensive-tissue hypothesis: The brain and the digestive system in human and primate evolution. *Current Anthropology*. 36: 199–221.

Alexeev, V.P. 1986. *The Origin of the Human Race*. Progress Publishers, Moscow.

Alpagut, B., Andrews, P.J., Fortelius, M., Kappelman, J., Temizsoy, I., Celebi, H. & Lindsay, W. 1996. A new specimen of *Ankarapithecus meteai* from the Sinap Formation of Central Anatolia. *Nature*. 382: 349–351.

Andersson, M. 1982. Female choice selects for extreme tail length in a widowbird. *Nature*. 299: 818–820.

Andrews, P.J. 1984. On the characters that define *Homo erectus*. *Courier Forschungsinstitut Senckenberg*. 69: 167–175.

Andrews, P.J. 1985. Family group systematics and evolution among catarrhine primates. In E. Delson (ed.). *Ancestors: The Hard Evidence*, pp. 14–22. Alan R. Liss, New York.

Andrews, P.J. 1992. Evolution and environment in the Hominoidea. *Nature*. 360: 641–646.

Andrews, P.J. 1995. Ecological apes and ancestors. *Nature*. 376: 555–556.

Andrews, P. & Cronin, J. 1982. The relationship of *Sirapithecus* and *Ramapithecus* and the evolution of the orangutan. *Nature*. 297: 541–546.

Andrews, P.J. & Bernor R.L. 1999. Vicariance biogeography and paleo-ecology of Eurasian Miocene hominoid primates. In J. Agusti, L. Rook & P.J. Andrews (eds.). *Hominoid Evolution and Climatic Change in Europe: The Evolution of Neogene Terrestrial Ecosystems in Europe*, Vol. 1, pp. 454–487. Cambridge University Press. Cambridge, UK.

Andrews, P.J. & Humphrey, L. 1999. African Miocene environments and the transition to early hominins. In T.G. Bromage & F. Schrenk (eds.). *African Biogeography, Climate Change & Human Evolution*, pp. 282–300. Oxford University Press, Oxford, UK.

Andrews, P.J. & Martin, L.B. 1991. Hominoid dietary evolution. *Philosophical Transactions of the Royal Society of London, Series B*. 334: 199–209.

Andrews, P.J. & Tekkaya, I. 1980. A revision of the Turkish Miocene hominoid *Sivapithecus meteai*. *Paleontology*. 23: 85–95.

Andrews, P.J. & Van Couvering, J.A.H. 1975. Palaeoenvironments in the East African Miocene. In F. Szlay (ed.). *Approaches to Primate Palaeobiology*, Vol. 5, pp. 62–103. Karger, Basel.

Andrews, P.J. & Walker, A.C. 1976. The Primate and Other Fauna from Fort Ternan, Kenya. In G.L. Isaac & E.R. McCown (eds.). *Human*

Origins: Louis Leakey and the East African Evidence, pp. 279–306. W.A. Benjamin, Inc. Menlo Park, California.

Andrews, P.J., Harrison, T., Delson, E., Bernor, R.L. & Martin, L.B. 1996. Distribution and biochronology of European and Southwest Asian Miocene Catarrhines. In R.L. Bernor, V. Fahlbusch & H.W. Mittmann (eds.). *The Evolution of Western Eurasian Neogene Mammal Faunas*, pp. 168–295. Columbia University Press, New York.

Andrews, P.J., Begun, D.R. & Zylstra, M. 1997. Interrelationships between functional morphology and palaeoenvironments in Miocene Hominoids. In D.R. Begun, C.V. Ward & M.D. Rose (eds.). *Function, Phylogeny, and Fossils: Miocene Hominoid Evolution and Adaptations*, pp. 29–58. Plenum Press, New York.

Antón, S.C. & Weinstein, K.J. 1999. Artificial cranial deformation and fossil Australians revisited. *Journal of Human Evolution*. 36: 195–209.

Arambourg, B. 1989. New skeletal evidence concerning the anatomy of Middle Palaeolithic populations in the Middle East: The Kebara skeleton. In P. Mellars & C. Stringer (eds.). *The Human Revolution: Behavioral and Biological Perspectives on the Origins of Modern humans*, pp. 165–171. Princeton University Press, Princeton, NJ.

Arambourg, B. & Coppens, Y. 1968. Découverte d'un australopithécine nouveau dans les gisements de l'Omo (Ethiopie). *South African Journal of Science*. 64: 58–59.

Arsuaga, J.L. 2002. *The Neanderthal's Necklace: In Search of the First Thinkers*. Four Walls, Eight Windows, New York.

Arsuaga, J.L., Martinez, I., Gracia, A. & Lorenzo, C. 1997. The Sima de los Huesos crania (Sierra de Atauerca, Spain). A comparative study. *Journal of Human Evolution*. 33: 219–281.

Ascenzi, A., Mallegni, F., Manzi, G., Serge, A. & Serge-Naldini, E. 2000. A reappraisal of Ceprano calvarial affinities with *Homo erectus* after the new reconstruction. *Journal of Human Evolution*. 39: 443–450.

Asfaw, B., White, T., Lovejoy, O., Latimer, B., Simpson, S. & Suwa, G. 1999. *Australopithecus garhi*: A new species of early hominid from Ethiopia. *Science*. 284: 629–635.

Asfaw, B., Gilbert, W.H., Beyene, Y., Hart, W.K., Renne, P.R., Gabriel, G.W., Vrba, E.S. & White, T.D. 2002. Remains of *Homo erectus* from Bouri, middle Awash, Ethiopia. *Nature*. 416: 317–320.

Balme, J. 1983. Prehistoric fishing in the lower Darling, western New South Wales. In G. Grigson & J. Clutton-Brock (eds.). *Animals and Archaeology*, Vol. 2, pp. 19–32. Oxford.

Barry, J.C., Johnson, N.M., Raza, S.M. & Jacobs, L.L. 1985. Neogen mammalian faunal change in southern Asia: Correlations with climatic tectonic, and eustatic events. *Geology*. 13: 637–640.

Bar-Yosef, O. 1995. The role of climate in the interpretation of human movements and cultural transformation in western Asia. In E.S. Vrba G.H. Denton, T.C. Partridge, & L.H. Burckle (eds.). *Paleoclimate and Evolution, with Emphasis on Human Origins*, pp. 507–523. Yale University Press, New Haven, CT.

Bar-Yosef, O. 1998. The chronology of the Middle Paleolithic of the Levant. In T. Akazawa, K. Aoki, & O. Bar-Yosef (eds.). *Neandertals and Modern Humans in Western Asia*, pp. 39–56. Plenum Press, New York.

Bar-Yosef, O. 2000. The Middle and Early Upper Paleolithic in Southwest Asia and Neighboring Regions. In O. Bar Yosef & D. Pilbeam (eds.) *The Geography of Neanderthals and Modern Humans in Europe and the Greater Mediterranean*, pp. 107–130. Peabody Museum Bulletin No. 8. Harvard University, Cambridge, MA.

Begun, D.R. 1992a. Miocene fossil hominids and the chimp–human clade. *Science*. 247: 1929–1933.

Begun, D.R. 1992b. European Miocene catarrhine diversity. *Journal of Human Evolution*. 20: 521–526.

Begun, D.R. 1994a. Relations among the great apes and humans: New interpretations based on the fossil great ape *Dryopithecus*. *Yearbook of Physical Anthropology*. 37: 11–63.

Begun, D.R. 1994b. The significance of *Otavipithecus nambiensis* to interpretations of hominoid evolution. *Journal of Human Evolution*. 27: 385–394.

Begun, D.R. 2001. African and Eurasian Miocene hominoids and the origins of the Hominidae. In L. de Bonis, G.D. Koufos & P.J. Andrews (eds.). *Hominoid Evolution and Climate Change in Europe*, Vol. 2. *Phylogeny of the Neogene Hominoid Primates of Eurasia*, pp. 231–253. Cambridge University Press, Cambridge, UK.

Begun, D.R. 2002. European Hominids. In W.C. Hartwig (ed.). *The Primate Fossil Record*, pp. 339–368. Cambridge University Press, Cambridge, UK.

Begun, D.R. & Gülec, E. 1998. Restoration of the type and palate of *Ankarapithecus metei*: taxonomic and phylogenetic implications. *American Journal of Physical Anthropology*. 105: 279–314.

Begun, D.R & Kordos, L. 1997. Phyletic affinities and functional convergence in *Dryopithecus* and other Miocene and living hominids. In

D.R. Begun, C.V. Ward & M.D. Rose (eds.). *Function, Phylogeny, and Fossils: Miocene Hominoid Evolution and Adaptations*, pp. 291–316. Plenum Press, New York.

Begun, D.R. & Walker, A.C. 1993. The Endocast. In A.C. Walker & R.E.F. Leakey (eds.). *The Nariokotome* Homo erectus *Skeleton*, pp. 326–358. Harvard University Press, Cambridge, MA.

Begun, D.R., Ward, C.V. & Rose, M.D. 1997. Events in hominoid evolution. In D.R. Begun, C.V. Ward & M.D. Rose (eds.). *Function, Phylogeny, and Fossils: Miocene Hominoid Evolution and Adaptations*, pp. 389–415. Plenum Press, New York.

Bellwood, P. 1997. *Prehistory and the Indo-Malaysian Archipelago.* 2nd Edition. University of Hawaii Press, Honolulu.

Bermúdez de Castro, J., Arsuaga, J.L., Carbonell, E., Rosas, A., Martínez, I. & Mosquera, M. 1997. A hominid from the Lower Pleistocene of Atapuerca, Spain: A possible ancestor to Neanderthals and modern humans. *Science.* 276: 1392–1395.

Bernor, R.L., Brunet, M., Ginsburg, L., Mein, P., Pickford, M., Rögl, F., Sen, S., Steininger, F. & Thomas, H. 1987. A consideration of some major topics concerning Old World Miocene mammalian chronology, migrations and paleogeography. *Geobios.* 20: 431–439.

Bernor, R.L., Fahlbusch, V., Andrews, P.J., de Bruijn, H., Fortelius, M., Rögl, F., Steininger, F.F. & Werdelin, L. 1996. The evolution of Western Eurasian neogene mammal faunas: A chronologic, systematic, biogeographic, and palaeoenvironmental synthesis. In R.L. Bernor, V. Fahlbusch & H.W. Mittmann (eds.). *The Evolution of Western Eurasian Neogene Mammal Faunas*, pp. 449–470. Columbia University Press, New York.

Biegert, J. 1963. The evaluation of characteristics of the skull, hands and feet for primate taxonomy. In S.L. Washburn (ed.). *Classification and Human Evolution*, pp. 116–145. Aldine Publishing Co., Chicago.

Binford, L. 1981. *Bones: Ancient Men and Modern Myths*. Academic Press, New York.

Binford, L. 1983. *In Pursuit of the Past: Decoding the Archaeological Record*. Thames & Hudson, London.

Binford, L. 1985. Human ancestors: Changing views of their behaviour. *Journal of Anthropological Archaeology.* 4: 292–327.

Binford, L. 1988. Fact and fiction about *Zinjanthropus* floor: Data, arguments, and interpretations. *Current Anthropology.* 29: 123–135.

Binford, L. & Binford, S.R. 1966. A preliminary analysis of functional variability in the Mousterian of Levallois facies. *American Anthropologist.* 68: 238–239.

Birdsell, J.H. 1949. The racial origins of the extinct Tasmanians. *Records of the Queen Victorian Musem.* 2: 105–122.

Birdsell, J.H. 1967. Preliminary data on the trihybrid origin of the Australian aborigines. *Archaeology and Physical Anthropology in Oceania.* 2: 100–155.

Bocquet-Appel, J.P. & Masset, C. 1982. Farewell to paleodemography. *Journal of Human Evaluation.* 11: 321–333.

Bonis, L. de & Koufos, G. 1993. The face and the mandible of *Ouranopithecus macedoniensis*: Description of new specimens and comparisons. *Journal of Human Evolution.* 24: 469–491.

Bonis, L. de & Koufos, G. 1994. Our ancestors: *Ouranopithecus* is a Greek link in human ancestry. *Evolutionary Anthropology.* 3: 75–83.

Bonis, L. de & Koufos, G. 1999. The Miocene large mammal succession in Greece. In J. Agusti, L. Rook & P.J. Andrews (eds.). *Hominoid Evolution and Climatic Change in Europe*, Vol. 1: *The Evolution of Neogene Terrestrial Ecosystems in Europe*, pp. 205–237. Cambridge University Press, Cambridge, UK.

Bonis, L. de & Koufos, G. 2001. Phylogenetic relationships of *Ouranopithecus macedoniensis* (Mammalia, Primates, Hominoidea, Hominidae) of the late Miocene deposits of Central Macedonia (Greece). In L. de Bonis, G.D. Koufos & P.J. Andrews (eds.). *Hominoid Evolution and Climate Change in Europe,* Vol. 2: *Phylogeny of the Neogene Hominoid Primates of Eurasia*, pp. 254–268. Cambridge University Press, Cambridge, UK.

Bonis, L. de & Melentis, J. (1987). Intérêt de l'anatomie nasomaxillaire pour la phylogénie de Hominoidea. *Comptes rendus de l' Académie de Sciences de paris.* 304: 767–769.

Bonis, L. de, Bouvrain, G. & Koufos, G. 1988. Late Miocene mammal localities of the Lower Axios Valley (Macedonia, Greece) and their stratigraphic significance. *Modern Geology.* 13: 141–147.

Bonis, L. de, Bouvrain, G., Geraads, D. & Koufos, G. 1990. New hominid skull material from the late Miocene of Macedonia in Northern Greece. *Nature.* 345: 712–714.

Bonis, L. de, Bouvrain, G., Geraads, D. & Koufos, G. 1992. Diversity and paleoecology of Greek late Miocene mammalian faunas. *Palaeogeography, Palaeoclimatology, Palaeoecology.* 91: 99–121.

Bordes, F. 1950. Principles d'une méthode d'étude des techniques de débitage et de la typologie du Paléolithique ancien et moyen. *L'Anthropologie*. 54: 19–34.

Bordes, F. 1961. Typologie du Paléolithique ancien et moyen. *Mémoires de l'Institut Préhistoriques de l'Université de Bordeaux*, Vol. 1. Delmas, Bordeaux.

Bordes, F. 1969. Reflections on typology and techniques in the Paleolithic. *Arctic Anthropology*. 6: 1–29.

Bordes, F. & Sonneville-Borders, D. de. 1970. The significance of variability in Paleolithic assemblages. *World Archaeology*. 2: 61–73.

Bowdler, S. 1997. The pleistocene Pacific. In D. Denoun (ed.). *The Cambridge History of the Pacific Islandes*, pp. 41–50. Cambridge: Cambridge University Press.

Bowler, J.M., Johnston, H., Olley, J.M., Prescott, J.R., Roberts, R.G., Shwacross, W. & Spooner, N.A. 2003. New ages for human occupation and climatic change at Lake Mungo, Australia. *Nature*. 421: 837–840.

Brain, C.K. 1970. New finds at the Swartkrans australopithecine site. *Nature*. 225: 1112–1119.

Brain, C.K. 1981. *The hunters or the hunted? An introduction to African cave taphonomy*. University of Chicago Press, Chicago.

Bräuer, G. 1984. A craniological approach to the origin of anatomically modern *Homo sapiens* in Africa and implications for the appearance of modern humans. In F.H. Smith and F. Spencer (eds.). *The Origins of Modern Humans*, pp. 327–410. Alan R. Liss, New York.

Bräuer, G. 1989. The evolution of modern humans: A comparison of the African and non-African evidence. In P. Mellars and C. Stringer (eds.). *The Human Revolution: Behavioral and Biological Perspectives on the Origins of Modern Humans*, pp. 123–154. Princeton University Press, Princeton, NJ.

Bräuer, G. & Schultz, M. 1996. The morphological affinities of the Plio-Pleistocene mandible from Dmanisi, Georgia. *Journal of Human Evolution*. 30: 445–481.

Broadfield, D.S., Holloway, R.L., Mowbray, K., Sivers, A., Yuan, M.S. & Marquez, S. 2001. Endocast of Sambungmacan 3 (Sm 3): A new *Homo erectus* from Indonesia. *Anatomical Record*. 262: 369–379.

Brooks, D.R. & McLennan, D.A. 1991. *Phylogeny, Ecology, and Behavior: A Research Program in Comparative Biology*. University of Chicago Press, Chicago.

Brose, D.S. & Wolpoff, W.H. 1971. Early Upper Palaeolithic man and Late Middle Palaeolithic tools. *American Anthropologist*. 73: 1156–1194.

Brown, B. 1997. Miocene hominoid mandibles: Functional and phylogenetic perspectives. In D.R. Begun, C.V. Ward & M.D. Rose (eds.). *Function, Phylogeny, and Fossils: Miocene Hominoid Evolution and Adaptations*, pp. 153–172. Plenum Press, New York.

Brown, B. & Walker, A.C. 1993. The Dentition. In A.C. Walker & R.E.F. Leakey (eds.). *The Nariokotome* Homo erectus *Skeleton*, pp. 161–193. Harvard University Press, Cambridge, MA.

Brown, B., Walker, A.C. & Leakey, R.E.F. 1993. New *Australopithecus boisei* calvaria from East Lake Turkana, Kenya. *American Journal of Physical Anthropology*. 91: 137–159.

Brown, F. & Mcdougall, I. 1993. Geologic setting and age. In A.C. Walker & R.E.F. Leakey (eds.). *The Nariokotome* Homo erectus *skeleton*, pp. 9–20. Harvard University Press, Cambridge, MA.

Brown, P. 1981. Artificial cranial deformation: A component in the variability in Pleistocene Australian aboriginal crania. *Archaeology in Oceania*. 16: 156–167.

Brown, P. 1987. Pleistocene homogeneity and Holocene size reduction: The Australian human skeletal evidence. *Archaeology in Oceania*. 22: 41–67.

Brown, P. 1989. *Coobool Creek: A Morphological and Metrical Analysis of the Crania, Mandibles and Dentitions of a Prehistoric Australian Human Population: Terra Australis*. Department of Prehistory, Australian National University, Canberra.

Brown, P. 1992. Recent human evolution in East Asia and Australasia. In M.J. Aitken., C.B. Stringer & P. Mellars (eds.). *The Origin of Modern Humans and the Impact of Chronometric Dating*, pp. 217–233. Princeton University Press, Princeton, NJ.

Bruijn, de, H. 1986. Is the presence of the African family Thryonomyidae in the Miocene deposits of Pakistan evidence of fauna exchange? *Palaeontological Proceedings B*. 89: 125–134.

Brunet, M., Beauvilain, A., Coppens, Y., Heintz, E., Aladji, H.E & Pilbeam, D.R. 1995. The first australopithecine 2,500 kilometers west of the Rift Valley (Chad). *Nature*. 378: 273–275.

Brunet, M., Beauvilain, A., Coppens, Y., Heintz, E., Aladji, H.E. & Pilbeam, D.R. 1996. *Australopithecus bahrelghazali*, une nouvelle espèce d'Hominidé ancien de la region de Koro Toro (Tchad). *C.R. Acad. Sci. Paris, t.* 322, série IIa, 907–913.

Brunet, M., Guy, F., Pilbeam, D.R., Mackaye, H.T., Likius, A., Ahounta, D., Beauvilains, A., Blondel, C., Bocherens, H., Boisserie, JR., de Bonis, L., Coppens, Y., Dejax, J., Denys, C., Duringer, P., Eisenmann, V., Fanone, G., Fronty, P., Geraads, D., Lehmann, T., Lihoreau, F., Louchart, A., Mahamat, A., Merceron, G., Mouchelin, G., Otero, O., Champomanes, P.P., Leon, M.P.D., Rage, J.C., Sapanet, M., Schuster, M., Sudre, J., Tassy, P., Valentin, X., Vignaud, P., Viriot, L., Zazzo, A. & Zollikofer, C. 2002. A new hominid from the Upper Miocene of Chad, Central Africa. *Nature*. 418: 145–151.

Bunn, H.T. 1981. Archaeological evidence for meat-eating by Plio-Pleistocene hominids from Koobi Fora and Olduvai George. *Nature*. 291: 574–577.

Bunn, H.T. 2002. Hunting, power scavenging, and butchering by Hadza Forages and by Plio-Pleistocene *Homo*. In C.B. Stanford & H.T. Bunn (eds.). *Meat-Eating and Human Evolution*, pp. 199–218. Cambridge University Press, Cambridge, UK.

Bunn, H.T. & Kroll, E.M. 1986. Systematic butchery by Plio-Pleistocene hominids at Olduvai George, Tanzania. *Current Anthropology*. 27: 431–452.

Cameron, D.W. 1993a. The Pliocene hominid and proto-chimpanzee behavioral morphotypes. *Journal of Anthropological Archaelogy*. 12: 386–414.

Cameron, D.W. 1993b. Uniformitarianism and prehistoric archaeology. *Australian Archaeology*. 36: 42–49.

Cameron, D.W. 1995. *The Systematics of the European Miocene Faciodental Fossils Ascribed to the Family Hominidae: Aspects of Anatomical Variability, Taxonomy and Phylogeny*, pp. 1–700. Ph.D. Dissertation, School of Archaeology and Anthropology, Australian National University.

Cameron, D.W. 1997a. A revised systematic scheme for the Eurasian Miocene fossil Hominidae. *Journal of Human Evolution*. 33: 449–477.

Cameron, D.W. 1997b. Sexual dimorphic features within extant great-ape faciodental skeletal anatomy and testing the single species hypothesis. *Z. Morph. Anthrop.* 81: 253–288.

Cameron, D.W. 1998. Anatomical variability and systematic status of the hominoids currently allocated to the African Dryopithecinae. *Journal of Comparative Human Biology*. 49: 101–137.

Cameron, D.W. 2001. The taxonomic status of the Siwalik late Miocene hominid *Indopithecus* (= "*Gigantopithecus*"). *Himalayan Geology*. 22: 29–34.

Cameron, D.W. 2002. The morphology of the European Miocene hominid frontal bone: functional, developmental and phylogenetic considerations. *Anthropologischer Anzeiger*. 60: 137–159.

Cameron, D.W. In press a. *Hominid Adaptations and Extinctions*. University of New South Wales Press.

Cameron, D.W. In press b. Early Hominin speciation at the Plio/Pleistocene transition. *Journal of Comparative Human Biology*.

Cameron, D.W. & Groves, C.P. 1993. Pliocene hominins: Hunters and/or scavengers? *Perspectives of Human Biology*. 3: 23–31.

Cameron, D.W., Patnaik, R. & Sahni, A. In press. The phylogenetic significance of the Middle Pleistocene Narmada hominin cranium from Central India. *Intern. J. Osteoarch*.

Campbell, B.G. 1972. Sexual Selection and the Descent of Man 1871–1971. Chicago: Chicago University Press.

Cann, R.L., Stoneking, M. & Wilson, A.C. 1987. Mitochrondrial DNA and human evolution. *Nature*. 325: 31–36.

Carbonell, E., Bermudez de Castro, J.M., Arsuga, J.L., Diez, J.C., Cuenca-Bescos, G., Sala, R., Mosquera, M. & Rodriguez, X.P. 1995. Lower Pleistocene hominids and artifacts from Atapuerca-TD 6 (Spain). *Science*. 269: 826–830.

Carrier, D.R. 1984. The energetic paradox of human running and hominid evolution. *Current Anthropology*. 25: 483–495.

Cartmill, M. 1992. New views on primate origins. *Evolutionary Anthropology*. 1: 105–111.

Cave, A.J.E. & Straus, W.L. 1957. Pathology and posture of Neanderthal man. *Quarterly Review of Biology*. 32: 348–363.

Cerling, T.E. 1992. Development of grasslands and savannas in East Africa during the Neogene. *Palaeogeography, Palaeoclimatology, Palaeoecology*. 97: 241–247.

Cerling, T.E., Harris, J.M., MacFadden, B.J., Leakey, M.G., Quade, J., Eisenmann, V., & Ehleringer, J.R. 1997. Global vegetation change through the Miocene/Pliocene boundary. *Nature*. 389: 153–158.

Cerling, T.E., Harris, J.M., MacFadden, B.J., Leakey, M.G., Quade, J., Eisenmann, V. & Ehleringer, J.R. 1998. Reply to Köhler, Moyá-Solá & Agusti. *Nature*. 393: 127.

Chamberlain, A. & Wood, B.A. 1987. Early hominid phylogeny. *Journal of Human Evolution*. 16: 119–133.

Chang, S., Gu, Y., Bao, Y., Shen. W., Wang, Z., Wang, C. & Du, Z. 1980. *Atlas of Primitive Man in China*. Science Press, Beijing.

Chappell, J. 1976. Aspects of late Quaternary palaeogeography of the Australian–East Indonesian region. In R.L. Kirk & A.G. Thorne (eds.). *The origin of the Australians*, pp. 11–22. Australian Institute of Aboriginal Studies, Canberra.

Chase, P.G. & Dibble, H.L. 1987. Middle Palaeolithic symbolism: A review of current evidence and interpretations. *Journal of Anthropological Archaeology*. 6: 263–296.

Chen, F.-C. & Li, W.-H. 2001. Genomic divergences between humans and other hominoids and the effective population size of the common ancestor of humans and chimpanzees. *American Journal of Human Genetics*. 68: 444–456.

Chivers, D.J. & Langer, P. (eds.). 1994. *The Digestive System in Mammals: Food, Form and Function*. Cambridge University Press, Cambridge, UK.

Ciochon, R.L. & Etler, D.A. 1994. Reinterpreting past primate diversity. In R.S. Corruccini & R.L. Ciochon (eds.). *Integrative Paths to the Past: Palaeoanthropological Advances in Honor of F. Clark Howell*, pp. 37–68. Prentice Hall, Englewood Cliffs, NJ.

Ciochon, R.L. & Olson, J.W. 1986. Paleoanthropological and archaeological research in the Socialist Republic of Vietnam. *Journal of Human Evolution*. 15: 623–633.

Clark, J.D. 1977. *Prehistory*. Cambridge University Press, Cambridge, UK.

Clark, J.D., de Heinzelin, J., Schick, K., Hart, W., White, T.D., WoldeGabriel, G., Walter, R., Suwa, G., Vrba, B. & Hail-Selassie, Y. 1994. African *Homo erectus*: Old radiometric ages and young Oldowan assemblages in the Middle Awash Valley, Ethiopia. *Nature*. 264: 1907–1910.

Clark, J.D., Beyene, Y., WoldeGabriel, G., Hart., W.K., Renne, P.R., Gilbert, H., Defleur, A., Suwa, G., Katoh, S., Ludwig, K.R., Boisseries, J.R., Asfaw, B. & White, T.D. 2003. Stratigraphic, chronological and behavioral contexts of Pleistocene *Homo sapiens* from Middle Awash, Ethiopia. *Nature*. 423: 747–752.

Clarke, R.J. 1988. A New *Australopithecus* cranium from Sterkfontein and its bearing on the ancestry of *Paranthropus*. In F.E. Grine (ed.). *Evolutionary History of the "Robust" Australopithecines*, pp. 285–292. Aldine de Gruyter, New York.

Clarke, R.J. 2000. A corrected reconstruction and interpretation of the *Homo erectus* calvaria from Ceprano, Italy. *Journal of Human Evolution*. 39: 433–442.

Colgan, D.J. 2001. Commentary on D.J. Adcock *et al.*, 2001. Mitochondrial DNA sequences in ancient Australians: Implications for modern human origins. *Archaeology in Oceania*. 36: 168–169.

Collard, M. & Wood, B.A. 1999. Grades among the African early hominids. In T.G. Bromage & F. Schrenk (eds.). *African Biogeography, Climate Change, and Human Evolution*, pp. 316–327. Oxford University Press, New York.

Collard, M. & Wood, B.A. 2000. How reliable are human phylogenetic hypotheses? *Proceedings of the National Academy of Science USA*. 97: 5003–5006.

Collard, M. & Wood, B.A. 2001a. Homoplasy and the early hominid masticatory system: Inferences from analyses of extant hominoids and papionins. *Journal of Human Evolution*. 41: 167–194.

Collard, M. & Wood, B.A. 2001b. How reliable are current estimates of fossil catarrhine phylogeny? An assessment using great apes and Old World monkeys. In L. de Bonis, G. Koufos & P.J. Andrews (eds.). *Hominoid Evolution and Climatic Change in Europe*, Vol. 2: *Phylogeny of the Neogene Hominoid Primates of Eurasia*, pp. 118–150. Cambridge University Press, Cambridge, UK.

Conroy, G.C., Pickford, M.H., Senut, B., Van Couvering, J. & Mein, P. 1992. *Otavipithecus namibiensis*, first Miocene hominoid from Southern Africa. *Nature*. 356: 144–147.

Conroy, G.C., Pickford, M.H., Senut, B. & Mein, P. 1993. Diamonds in the desert: The discovery of *Otavipithecus namibiensis*. *Evolutionary Anthropology*. 2: 46–52.

Coon, C.S. 1962. *The Origin of Races*. Knopf, New York.

Cope, D.A. 1988. Dental variation in three sympatric species of *Cercopithecus*. (Abstract). *American Journal of Physical Anthropology*. 75: 198–199.

Cope, D.A. 1993. Measure of dental variation as indicators of multiple taxa in samples of sympatric *Cercopithecus* species. In W.H. Kimbel & L.B. Martin (eds.). *Species, Species Concepts, and Primate Evolution*, pp. 211–237. Plenum Press, New York.

Cope, D.A. & Lacy, M.G. 1992. Falsification of a single-species hypothesis using the coefficient of variation. A simulation approach. *American Journal of Physical Anthropology*. 89: 359–378.

Coppens, Y. 1994. East side story: The origin of humankind. *Scientific American*. 270: 62–69.

Coppens, Y. 1999. Theory — Introduction. In T.G. Bromage & F. Schrenk (eds.). *African Biogeography, Climate Change and Human Evolution*, pp. 13–18. Oxford University Press, Oxford, UK.

Crompton, R.H. 1995. Visual predation, habitat structure, and the ancestral primate niche. In L. Alterman, G.A. Doyle & M.K. Izard, (eds.).

Creatures of the Dark: The Nocturnal Prosimians, pp. 11–30. Plenum Press, New York.

Curnoe, D. & Thorne, A. 2003. Number of ancestral human species: A molecular perspective. *Journal of Comparative Human Biology.* 53: 201–224.

Currocini, R.S. 1992. Metrical reconstruction of Skhul IV and IX and Border Cave I crania in the context of modern human origins. *American Journal of Physical Anthropology.* 97: 433–445.

Dart, R.A. 1925. *Australopithecus africanus*: The man-ape of South Africa. *Nature.* 115: 195–199.

Dart, R.A. 1957. The Makapansgat australopithecine osteodontokeratic culture. *Proceedings of the Third Pan-African Congress on Prehistory.* Chatto & Windus, London.

Dart, R.A. 1959. Further light on australopithecine humeral and femoral weapons. *American Journal of Physical Anthropology.* 17: 87–94.

Dart, R.A. 1960. The bone tool–manufacturing ability of *Australopithecus prometheus*. *American Anthropologist.* 62: 134–143.

Darwin, C.R. 1859. *On the Origin of Species by Means of Natural Selection.* London: John Murray.

Darwin, C.R. 1872. *The Descent of Man, and Selection of Means of Sex.* London: John Murray.

Davidson, I. & Noble, W. 1989. The archaeology of perception. *Current Anthropology.* 30: 125–155.

Davidson, I. & Noble, W. 1993. Tools and language in human evolution. In K.R. Gibson & T. Ingold (eds.). *Tools, Language and Cognition in Human Evolution*, pp. 363–388. Cambridge University Press, Cambridge, UK.

Day, M.H. & Stringer, C.B. 1982. A reconsideration of the Omo Kibish remains and the *erectus–sapiens* transition. In H. de Lumley (ed.). *L'Homo erectus et la Place de l'Homme de Tautavel parmi les Hominides Fossiles*, Vol. 2, pp. 814–816. Louis-Jean, Nice.

Day, M.H & Wickens, E.H. 1980. Laetoli Pliocene hominid footprints and bipedalism. *Nature.* 286: 385–387.

Day, M.H., Leakey, R.E.F., Walker, A.C & Wood, B.A. 1975. New hominids from East Rudolf, Kenya I. *American Journal of Physical Anthropology.* 42: 461–476.

Day, M.H., Leakey, M.D. & Olson, T.R. 1980. On the status of *Australopithecus afarensis*. *Science.* 207: 1102–1103.

Deacon, H.J. & Deacon, J. 1999. *Human Beginnings in South Africa: Uncovering the Secrets of the Stone Age.* Altamira Press, CA.

Dean, D. & Delson, E. 1992. Second gorilla or third chimp? *Nature.* 359: 676–677.

Dean, M.C. 1986. *Homo* and *Paranthropus*: Similarities in the cranial base and developing dentition. In B.A. Wood, L.B. Martin & P.J. Andrews (eds.). *Major Topics in Primate and Human Evolution*, pp. 249–265. Cambridge University Press, Cambridge, UK.

Dean, M.C. 2000. Incremental markings in enamel and dentine: What they can tell us about the way teeth grow. In M.F. Teaford, M. Smith & M.W.J. Ferguson (eds.). *Development, Function and Evolution of Teeth*, pp. 119–130. Cambridge University Press, Cambridge, UK.

Dean, M.C. & Wood, B.A. 1981. Metrical analysis of the basicranium of extant hominids and *Australopithecus*. *American Journal of Physical Anthropology.* 54: 63–71.

Dean, M.C. & Wood, B.A. 1982. Basicranial anatomy of Plio-Pleistocene hominids from East and South Africa. *American Journal of Physical Anthropology.* 59: 157–174.

Defleur, A., White, T.D., Valensi, P., Slimak, L. & Crégut-Bonnoure, E. 1999. Neanderthal cannibalism at Moula-Guercy, Ardéche, France. *Science.* 286: 128–131.

Degusta, D., Gilbert, W.H. & Turner, S.P. 1999. Hypoglossal canal size and hominid speech. *Proceedings of the National Academy of Sciences, USA.* 86: 1800–1804.

Delson, E. 1975. Paleoecology and zoogeography of the Old World Monkeys. In R.H. Tuttle (ed.). *Primate Functional Morphology and Evolution*, pp. 37–64. Mouton, Paris.

Delson, E. 1988. Chronology of South African Australopith site units. In F.E. Grine (ed.). *Evolutionary History of the "Robust" Australopithecines*, pp. 317–324. Aldine de Gruyter, New York.

Delson, E. 1994. Evolutionary history of the colobine monkeys in palaeoenvironmental perspective. In A.G. Davies & J.F. Oates (eds.). *Colobine Monkeys: Their Ecology, Behavior and Evolution*, pp. 11–43. Cambridge University Press, Cambridge, UK.

Delson, E., Harvati, K., Reddy, D., Marcus, L., Mowbray, K., Sawyer, G.J., Jacob, T. & Marquez, S. 2001. The Sambungmacan 3 *Homo erectus*: Morphometric and morphological analysis. *Anatomical Record.* 262: 380–397.

Dennell, R.W., Rendell, H.M. Hurcombe, L. & Hailwood, E.A. 1994. Archaeological evidence for hominids in Pakistan before one million years ago. *Courier Forschungs-Institut Senckenberg.* 171: 151–155.

Denton, G.H. 1999. Cenozoic climate change. In T.G. Bromage & F. Schrenk (eds.). *African Biogeography, Climate Change, & Human Evolution*, pp. 94–114. Oxford University Press, Oxford, UK.

Denys, C. 1999. Of mice and men: Evolution in East and South Africa during Plio-Pleistocene times. In T.G. Bromage & F. Schrenk (eds.). *African Biogeography, Climate Change, & Human Evolution*, pp. 226–252. Oxford University Press, Oxford, UK.

Donoghue, M.J., Doyle, J.A. & Gauthier, J. 1989. The importance of fossils in phylogeny reconstruction. *Annual Review of Ecology and Systematics*. 20: 431–460.

Dover, G. 1999. Human evolution: Our turbulent genes and why we are not chimps. In B. Sykes (ed.). *The human Inheritance: Genes, Language, and Evolution*, pp. 75–92. Oxford university Press, Oxford, UK.

Duarte, C., Maurício, J., Pettitt, P.B., Souto, P., Trinkaus, E., van der Plicht, H. & Zilhao, J. 1999. The early Upper Paleolithic human skeleton from the Abrigo do Lagar Velho (Portugal) and modern human emergence in Iberia. *PNAS*. 96: 7604–7609.

Dunsworth, H. & Walker A.C. 2002. Early genus *Homo*. In W.C. Hartwig (ed.). *The Primate Fossil Record*, pp. 419–436. Cambridge University Press, Cambridge, UK.

Eckhardt, R.B. 2000. *Human Paleobiology*. Cambridge University Press, Cambridge, UK.

Eldredge, N. 1989. *Macroevolutionary Dynamics*. McGraw-Hill Publishing Company, New York.

Eldredge, N. & Gould, S.J. 1972. Punctuated equilibria: An alternative to phyletic gradualism. In T.J.M. Schopf (ed.). *Models in Paleobiology*, pp. 82–115. Freeman, Cooper & Company, California.

Enlow, D.H. 1982. *Handbook of Facial Growth*. W.B. Saunders Company, Philadelphia.

Enlow, D.H. & Hans, M.G. 1996. *Essentials of Facial Growth*. W.B. Saunders Company, Philadelphia.

Etler, D.A. & Li, T. 1994. New archaic human fossil discoveries in China and their bearing on hominid species definition during the Middle Pleistocene. In R.S. Corrucini & R.L. Ciochon (eds.). *Integrative Paths to the Past: Palaeoanthropological Advances in Honor of F. W. Clark Howell*, pp. 639–676. Prentice Hall, Englewood Cliffs, NJ.

Falguéres, C. 1999. Earliest humans from Europe: The age of TD6 Gran Dolina, Atapuerca, Spain. *Journal of Human Evolution*. 37: 343–352.

Falk, D. 1986. Evolution of cranial blood drainage in hominids: Enlarged occipital/marginal sinuses and emissary foramina. *American Journal of Physical Anthropology.* 70: 311–324.

Falk, D. 1988. Enlarged occipital/marginal sinuses and emissary foramina: Their significance in hominid evolution. In F. Grine (ed.). *Evolutionary History of the "Robust" Australopithecines,* pp. 85–96. Aldine de Gruyter, New York.

Falk, D. & Conroy, G. 1984. The cranial venous sinus system in *Australopithecus afarensis. Nature.* 306: 779–781.

Farris, J. S. 1989. The retention index and rescaled consistency index. *Cladistics.* 5: 417–419.

Feibel, C.S. 1999. Basin evolution, sedimentary dynamics, and hominid habitats in East Africa: An ecosystem aproach. In T.G. Bromage & F. Schrenk (eds.). *African Biogeography, Climate Change, & Human Evolution,* pp. 276–281. Oxford University Press, Oxford, UK.

Ferguson, W.W. 1989. A new species of the genus *Australopithecus* (Primates: Hominidae) from Plio/Pleistocene depsoits west of Lake Turkana in Kenya. *Primates.* 30: 223–232.

Fisher, R.A. 1930. *The Genetical Theory of Natural Selection.* Oxford: Clarendon Press.

Fleagle, G.J. 1999. *Primate Adaptation and Evolution.* 2nd ed. Academic Press, New York.

Flood, J. 1999. *Archaeology of the Dreamtime: The Story of Prehistoric Australia and Its People.* HarperCollins Publishing, Sydney.

Flynn L. & Sabatier M, 1984. A muroid rodent of Asian affinity from the Miocene of Kenya. *Journal of Vertebrate Palaeontology.* 3: 160–165.

Foley, R.A. 1987. *Another Unique Species: Patterns in Human Evolutionary Ecology.* Longman Scientific and Technical, New York.

Foley, R.A. 1999. Evolutionary geography of Pliocene African hominids. In T.G. Bromage & F. Schrenk (eds.). *African Biogeography, Climate Change, & Human Evolution,* pp. 328–348. Oxford University Press, Oxford, UK.

Foley, R.A. 2002. The evolutionary consequences of increased carnivory in hominids. In C.B. Stanford & H.T. Bunn (eds.). *Meat-Eating and Human Evolution,* pp. 305–331. Oxford University Press, Oxford, UK.

Foley, R.A. & Lahr, M.M. 1997. Mode 3 technologies and evolution of modern humans. *Cambridge Archaeological Journal.* 7: 3–36.

Gabunia, L. & Vekua, A. 1995. A Plio-Pleistocene hominid from Dmanisi, East Georgia, Caucasus. *Nature.* 373: 509–512.

Gabunia, L., Vekua, A., Lordkipandze, D., Swisher, C.C., Ferring, R., Justus, A., Nioradze, M., Tvalchrelidze, M., Anton, S.C., Bosinski, G., Joris, O., de Lumley, M. A., Majsuradze, C. & Mouskhelishvili, A. 2000a. Earliest Pleistocene hominid cranial remains from Dmanisi, Republic of Georgia: Taxonomy, geological setting, and age. *Science.* 288: 1019–1025.

Gabunia, L., de Lumley, M.A. & Bérillion, G. 2000b. Morphologie et fonction du troisièmemétatarsien de Dmanissi, Géorgie orientale, *ERAUL.* 92: 29–41.

Gabunia, L., de Lumley, M.A., Vekua, A., Lordkipanidze, D. & de Lumely, H. 2002. Découverte d'un nouvel hominidé à Dmanissi (Tanscaucasie, Géorgie). *C. R. Palevol.* 1: 243–253.

Gamble, C. 1986. *The Palaeolitheic Settlement of Europe.* Cambridge University Press, Cambridge, UK.

Gamble, C. 1999. *The Palaeolithic Societies of Europe.* Cambridge University Press, Cambridge, UK.

Gargett, R.H. 1989. Grave shortcomings: The evidence for Neandertal burial. *Current Anthropology.* 30: 157–190.

Gargett, R.H. 1996. *Cave Bears and Modern Human Origins: The Spatial Taphonomy of Pod Hradem Cave, Czech Republic.* University Press of America.

Gargett, R.H. 1999. Middle Paleolithic burial is not a dead issue: The view from Qafzeh, Saint-Césaire, Kebara, Amud, and Dederiyeh. *Journal of Human Evolution.* 37: 27–90.

Gauthier, J., Kluge, A. & Rowe, T. 1988. Amniote phylogeny and the importance of fossils. *Cladistics.* 4: 105–209.

Geist, V. 1978. *Life Strategies, Human Evolution, Environmental Design.* Springer Verlag, New York.

Gentry, A. & Heizmann, E.P.J. 1996. Miocene ruminants of the Central and Eastern Paratethys. In R.L. Bernor, V. Fahlbusch & W. Mittmann (eds.). *The Evolution of Western Eurasian Neogene Mammal Faunas,* pp. 378–391. Columbia University Press. New York.

Goldberg, P., Weiner, S., Bar Yosef, O., Xu, Q. & Liu, J. 2001. Site formation processes at Zhoukoudian, China. *Journal of Human Evolution.* 41: 483–530.

Goodall, J. 1965. Chimpanzees of the Gombe stream Reserve. In I. DeVore (ed.). *Private Behaviour,* pp. 425–427. Chicago University Press.

Goodman, M. 1963. Man's Place in the Phylogeny of Primates as reflected in serum proteins S.L. Washburn (ed.). *Classification and Human Evolution,* pp. 204–234. Aldine, Chicago.

Goren-Inbar, N. & Belfer-Cohen, A. 1998. The technological abilities of the Levantine Mousterians: Cultural and mental capacities. In T. Akazawa, K. Aoki and O. Bar-Yosef (eds). *Neandertals and Modern Humans in Western Asia*, pp. 205–222. Plenum Press, New York.

Goodhart, C.B. 1964. A biological view of toplessness. *New Scientist*, 23: 558–560.

Gould, S.J. 2002. *The Structure of Evolutionary Theory*. Harvard University Press, Cambridge, MA.

Gould, S.J. & Eldredge, N. 1977. Punctuated equilibria: The tempo and mode of evolution reconsidered. *Palaeobiology*. 3: 115–151.

Grine, F.E. 1988. Evolutionary history of the "robust" Australopithecines: A summary and historical perspective. In F.E. Grine (ed.). *Evolutionary History of the "Robust" Australopithecines*, pp. 509–520. Aldine de Gruyter, New York.

Grine, F.E., Demes, B., Jungers, W.L. & Cole, T.M. 1993. Taxonomic affinity of the early *Homo* cranium from Swartkrans, South Africa. *American Journal of Physical Anthropology*. 92: 411–426.

Groves, C.P. 1989a. *A Theory of Primate and Human Evolution*. Oxford University Press, Oxford, UK.

Groves, C.P. 1989b. A regional approach to the problem of the origin of modern humans in Australia. In P. Mellars and C. Stringer (eds.). *The Human Revolution: Behavioral and Biological Perspectives on the Origins of Modern Humans*, pp. 274–285. Princeton University Press. Princeton, NJ.

Groves, C.P. 1991. New look at old remains. *ANU Reporter*, 23 Feb. 1990: 9.

Groves, C.P. 1994. The origin of modern humans. *Interdisciplinary Science Reviews*. 19: 23–34.

Groves, C.P. 1999. Nomenclature of African Plio-Pleistocene hominins. *Journal of Human Evolution*. 37: 869–872.

Groves, C.P. 2001. Time and taxonomy. *Ludus Vitalis*. 9: 91–96.

Groves, C.P. 2002. *Primate Taxonomy*. The Smithsonian Institution Press, Washington, DC.

Groves, C.P. (in Press). What is the Maba fossil? *Perspectives in Human Biology*.

Groves, C.P. & Mazak, V. 1975. An approach to the taxonomy for the Hominidae: Gracile Villafranchian hominids of Africa. *Casopis pro Mineralogii a Geologii*. 20: 225–246.

Groves, C.P. & Thorne, A.G. 1999. The terminal Pleistocene and early Holocene populations of northern Africa. *Homo*. 50: 249–262.

Grün, R. 1996. A re-analysis of electron spin resonance dating associated with the Petralona hominid. *Journal of Human Evolution*. 30: 227–241.

Grün, R. & Stringer, C.B. 1991. Electron spin resonance dating and the evolution of modern humans. *Archaeometry*. 33: 153–199.

Grün, R., Stringer, C.B. & Schwarcz, H.P. 1991. ESR dating of teeth from Garrod's Tabun Cave collection. *Journal of Human Evolution*. 20: 231–248.

Grün, R., Brink, J., Spooner, N., Taylor, A., Stringer, C., Franciscus, R. & Murray, A. 1996. Direct dating of the Florisbad hominid. *Nature*. 382: 500–501.

Gutiérrez, G., Sánchez, D. & Marín, A. 2002. A reanalysis of the ancient mitochondrial DNA sequences recovered from Neandertal bones. *Molecular Biology and Evolution*. 19: 1359–1366.

Guthrie, R.D. 1984. Mosaics, allelochemics and nutrients: An ecological theory of Late Pleistocene megafaunal extinctions. In P. Martin & R. Klein (eds.). *Quaternary Extinctions: A Prehistoric Revolution*, pp. 259–298. University of Arizona Press, Tucson.

Habgood, P. 1989. The origins of anatomically modern humans in Australasia. In P. Mellars and C.B. Stringer (eds.). *The Human Revolution: Behavioural and Biological Perspectives on the Origins of Modern Humans*, pp. 245–273. Princeton University Press, Princeton, NJ.

Haile-Selassie, Y. 2001. Late Miocene hominids from the Middle Awash. *Nature*. 412: 178–181.

Haldane, J.B.S. 1932. *The Causes of Evolution*. Longman, London.

Harrison, T. 1987. The phylogenetic relationships of the early catarrhine primates: A review of the current evidence. *Journal of Human Evolution*. 16: 41–80.

Harrison, T. 1988. A taxonomic revision of the small catarrhine primates from the early Miocene of East Africa. *Folia Primatologica*. 50: 59–108.

Harrison, T. 1992. A reassessment of the taxonomic and phylogenetic affinities of the fossil catarrhines from Fort Ternan, Kenya. *Primates*. 33: 501–522.

Harrison, T. 1993. Cladistic concepts and the species problem in hominoid evolution. In W.H. Kimbel and L.B. Martin (eds.). *Species, Species Concepts, and Primate Evolution*, pp. 345–371. Plenum Press, New York.

Harrison, T. 2002. Late Oligocene to middle Miocene catarrhines from Afro-Arabia. In W.C. Hartwig (ed.). *The Primate Fossil Record*, pp. 301–338. Cambridge University Press, Cambridge, UK.

Harrison, T. & Rook, L. 1997. Enigmatic anthropoid or misunderstood ape? The phylogenetic status of *Oreopithecus bambolii* reconsidered. In D.R. Begun, C.V. Ward & M.D. Rose (eds.). *Function, Phylogeny, and Fossils: Miocene Hominoid Evolution and Adaptations*, pp. 327–362, Plenum Press, New York.

Hartwig-Scherer, S. & Martin, R.D. 1991. Was "Lucy" more human than her "Child"? Observations on early hominid postcranial skeletons. *Journal of Human Evolution*. 21: 439–449.

Harvey, P.H. & Pagel, M.D. 1998. *The Comparative Method in Evolutionary Biology*. Oxford University Press, Oxford, UK.

Hawks, J., Oh, S., Humley, K., Dobson, S., Cabana, G., Dayalu, P. & Wolpoff, M.H. 2000. An Australasian test of the recent African origin theory using the WLH-50 calvarium. *Journal of Human Evolution*. 39: 1–22.

Heinrich, R.E., Rose, M.D., Leakey, R.E.F. & Walker, A.C. 1993. Hominid radius from the Middle Pliocene of Lake Turkana, Kenya. *American Journal of Physical Anthropology*. 92: 139–148.

Heinzelen, J. de., Clarke, J.D., White, T.D., Hart, W., Renne, P., WoldGabriel, G., Beyene, Y. & Vrba, E. 1999. Environment and behavior of 2.5 million-year-old Bouri hominids. *Science*. 284: 625–629.

Heizmann, E.P. & Begun, D.R. 2001. The oldest Eurasian hominoid. *Journal of Human Evolution*. 41: 463–482.

Hennig, W. 1966. *Phylogenetic Systematics*. University of Illinois Press, Urbana, IL.

Hill, A. 1994. Late Miocene and early Pliocene hominoids from Africa. In R.S. Corrucini & R.L. Ciochon (eds.). *Integrative Paths to the Past: Palaeoanthropological Advances in Honor of F. Clark Howell*, pp. 123–146. Prentice Hall, Englewood Cliffs, NJ.

Holloway, R.L. 1972. New australopithecine endocast, SK 1585, from Swartkrans, South Africa. *American Journal of Physical Anthropology*. 37: 173–186.

Holloway, R.L. 1981. The Indonesian *Homo erectus* brain endocasts revisited. *American Journal of Physical Anthropology*. 55: 503–521.

Holloway, R.L. 1988. "Robust" australopithecine brain endocasts: Some preliminary observations. In F.E. Grine (ed.). *Evolutionary History of the "Robust" Australopithecines*, pp. 97–106. New York: Aldine de Gruyter.

Howell, F.C. 1960. European and northwest African middle Pleistocene hominids. *Current Anthropology*. 1: 195–232.

Howell, F.C. 1978. Hominidae. In V.J. Maglio & H.B.S. Cooke (eds.). *Evolution of African Mammals*, pp. 154–248. Harvard University Press, Cambridge, MA.

Howell, F.C. 1998. Evolutionary implications of altered perspectives on hominine demes and populations in the later Pleistocene of western Eurasia. In T. Akazawa, K. Aoki & O. Bar-Yosef (eds.). *Neandertals and Modern Humans in Western Asia*, pp. 5–28. Plenum Press, New York.

Howell, F.C. & Coppens, Y. 1976. An overview of Hominidae from the Omo Succession, Ethiopia. In Y. Coppens, F.C. Howell, G.L. Isaac & R.E.F. Leakey (eds.). *Earliest Man and Environments in the Lake Rudolf Basin: Stratigraphy, Paleoecology, and Evolution*, pp. 522–532. University of Chicago Press, Chicago.

Howells, W.W. 1992. The dispersion of modern humans. In S. Jones, R. Martin & D. Pilbeam (eds.). *The Cambridge Encyclopaedia of Human Evolution*, pp. 389–401. Cambridge University Press, Cambridge, UK.

Hublin, J.J. 1985. Human fossils from the North African Middle Pleistocene and the origins of *Homo sapiens*. In E. Delson (ed.). *Ancestors: The Hard Evidence*, pp. 283–288. Alan R. Liss, New York.

Hublin, J.J. 1998. Climatic changes, paleogeography, and the evolution of the Neandertals. In T. Akazawa, K. Aoki & O. Bar-Yosef (eds.). *Neandertals and Modern Humans in Western Asia*, pp. 295–310. Plenum Press, New York.

Hublin, J.J. 2000. Modern–nonmodern hominid interactions: A Mediterranean perspective. In O. Bar Yosef & D. Pilbeam (eds.). *The Geography of Neanderthals and Modern Humans in Europe and the Greater Mediterranean*, pp. 157–182. Peabody Museum Bulletin No. 8. Harvard University, Cambridge, MA.

Hublin, J.J., Spoor, F., Braun, M., Zonneveld, F. & Condemi, S. 1996. A late Neandertal associated with Upper Palaeolithic artefacts. *Nature*. 231: 224–226.

Hylander, W.L. 1984. Stress and strain in the mandibular synphysis of primates: A test of competing hypotheses. *American Journal of Physical Anthropology*. 64: 1–46.

Hylander, W.L. 1988. Implications of *in vivo* experiments for interpreting the functional signifigance of "Robust" Australopithecine jaws. In F.E. Grine (ed.). *Evolutionary History of the Robust Australopithecines*, pp. 55–83. Aldine de Gruyter, New York.

Hylander, W.L. & Johnson, K.R. 1994. Jaw muscle function and wishboning of the mandible during mastication in macaques and baboons. *American Journal of Physical Anthropology*. 94: 523–547.

Hylander, W.L. & Ravosa, M.J. 1992. An analysis of the supraorbital region of primates: A morphometric and experimental approach. In P. Smith, & E. Tchernov (eds.). *Structure, Function and Evolution of Teeth*, pp. 223–255. Freund, London.

Hylander, W.L., Picq, P.G. & Johnson, K.L. 1991. Masticatory-stress hypothesis and the supraorbital region of primates. *American Journal of Physical Anthropology*. 86: 1–36.

Hylander, W.L., Ravosa, M., Ross, C.F., Wall, C.E. & Johnson, K.R. 2000. Symphyseal fusion and jaw-adductor muscle force: An EMG study. *American Journal of Physical Anthropology*. 112: 469–492.

Isaac, G.L. 1976. East Africa as a source of fossil evidence for human evolution. In G. Isaac & E.R. McCown (eds.). *Human Origins: Louis Leakey and the East African Evidence*, pp. 121–138. W.A. Benjamin, Menlo Park, CA.

Isaac, G.L. 1977. *Olorgesailie*. University of Chicago Press, Chicago.

Isaac, G.L. 1978. The food sharing behavior of protohuman hominids. *Scientific American*. 238: 90–108.

Isaac, G.L. 1986. Foundation stones: Early artefacts as indicators of activities and abilities. In G.N. Bailey & P. Callow (eds.). *Stone Age Prehistory*, pp. 221–241. Cambridge University Press, Cambridge, UK.

Isaac, G.L. 1989. Cutting and carrying: Archaeology and the emergence of the genus *Homo*. In J.R. Durant (ed.). *Human Origins*, pp. 106–122. Oxford University Press, Oxford, UK.

Isaac, G.L. & Behrensmeyer, A.K. 1997. Geological context and paleoenvironments. In G.L. Isaac & B. Isaac (eds.). *Koobi Fora Research Project*, Vol. 5: *Plio-Pleistocene Archaeology*, pp. 12–70. Oxford University Press, Oxford, UK.

Isaac, G.L. & Isaac, B. 1997. *Koobi Fora Research Project*, Vol. 5: *Plio-Pleistocene Archaeology*. Oxford University Press, Oxford, UK.

Isaac, G.L. & McCown, E.R. 1976. *Human Origins: Louis Leakey and the East African Evidence*. W.A. Benjamin, Inc. Menlo Park, California.

Ishida, H. & Pickford, M. 1997. A new late Miocene hominoid from Kenya: *Samburupithecus kiptalami* gen. et sp. nov. *C. R. Academy of Science, Paris*. 325: 823–829.

Ishida, H., Pickford, M., Nakaya, H. & Nakano, Y. 1984. Fossil anthropoids from Nachola and Samburu Hills, Samburu District, Kenya. *African Study Monograph Supplement Issue*. 2: 73–85.

Ishida, H., Kunimatsu, Y., Nakatukasa, M. & Nakano, Y. 1999. New hominoid genus from the middle Miocene of Nachola, Kenya. *Anthropological Science.* 197: 189–191.

Jablonski, N.G. 2002. Fossil Old World monkeys: The late Neogene radiation. In W.C. Hartwig (ed.). *The Primate Fossil Record*, pp. 255–299. Cambridge University Press, Cambridge, UK.

Jernvall, J. 1995. Mammalian molar cusp patterns: Developmental mechanisms of diversity. *Acta Zoologica Fennica.* 98: 1–61.

Jia, L. 1985. China's earliest Palaeolithic assemblages. In R. Wu & J.W. Olsen (eds.). *Palaeoanthropology and Palaeolithic Archaeology in the People's Republic of China*, pp. 135–146. Academic Press, New York.

Johanson, D.C. & Edey, M.A. 1981. *Lucy: The Beginnings of Humankind.* Penguin Books, London.

Johanson, D.C. & Taieb, M. 1976. Plio-Plesitocene hominid discoveries in Hadar, Ethiopia. *Nature.* 260: 293–297.

Johanson, D.C., White, T.D. & Coppens, Y. 1978. A new species of the genus *Australopithecus* (Primates: Hominidae) from the Pliocene of eastern Africa. *Kirtlandia.* 28: 1–14.

Johanson, D.C., Masao, F.T., Eck, G.G., White, T.D., Walter, R.C., Kimbel, W.H., Asfaw, B., Manega, P., Ndessokia, P. & Suwa, G. 1987. New partial skeleton of *Homo habilis* from Olduvai Gorge, Tanzania. *Nature.* 327: 205–209.

Jolly, C.J. 1970. The seed-eaters: A new model of hominid differentiation based on a baboon analogy. *Man.* 5: 5–26.

Jones, M. 1999. *The Molecular Hunt: Archaeology and the Search for Ancient DNA.* Allen Lane, Penguin Press, UK.

Jones, R. 1977. Man as an element of a continental fauna: The case of the sundering of the Bassian Bridge. In J.A. Golson & R. Jones (eds.). *Sunda & Sahul: Prehistoric Studies in Southeast Asia, Melanasia and Australia*, pp. 317–386. Academic Press, New York.

Jones, R. 1989. East of Wallace's line: Issues and problems in the colonization of the Australian continent. In P. Mellars & C. Stringer (eds.). *The Human Revolution: Behavioral and Biological Perspectives on the Origins of Modern Humans*, pp. 743–782. Princeton University Press. Princeton, NJ.

Jones, R.M. 1999. Marine invertebrate (chiefly foraminiferal) evidence for the palaeogeography of the Oligocene-Miocene of western Eurasia, and consequences for terrestrial vertebrate migration. In J. Agusti, L. Rook & P.J. Andrews (eds.). *Hominoid Evolution and Climatic*

Change in Europe, Vol. 1: *The Evolution of Neogene Terrestrial Ecosystems in Europe*, pp. 273–308. Cambridge University Press, Cambridge, UK.

Jordan, P. 1999. *Neanderthal: Neanderthal Man and the Story of Human Origins*. Sutton Publishing.

Kahn, P. & Gibbons, A. 1997. DNA from an extinct human. *Science.* 277: 176–178.

Kamminga, J. & Wright, R.V.S. 1988. The Upper Cave at Zhoukoudian and the origins of the mongoloids. *Journal of Human Evolution.* 17: 739–767.

Kappelman, J., Richmond, B.G., Seiffert, E.R., Maga, A.M. & Ryan, T.M. 2003. Hominoidea (Primates). In M. Fortelius, J. Kappleman, S. Sen and R. Bernor (eds.). *Geology and Paleontology of the Miocene Sinap Formation, Turkey*, pp. 90–124. Columbia University Press, New York.

Kay, R.F. & Grine, F.E. 1988. Tooth morphology, wear and diet in *Australopithecus* and *Paranthropus* from southern Africa. In F.E. Grine (ed.). *Evolutionary History of the "Robust" Australopithecines*, pp. 427–448. Aldine de Gruyter, New York.

Kay, R.F. & Ungar P.S. 1997. Dental evidence for diet in some Miocene catarrhines with comments on the effects of phylogeny on the interpretation of adaptation. In D.R. Begun, C.V. Ward & M.D. Rose (eds.). *Function, Phylogeny, and Fossils: Miocene Hominoid Evolution and Adaptations*, pp. 131–151. Plenum Press, New York.

Kay, R.F., Cartmill, M. & Balow, M. 1998. The hypoglossal canal and the origin of human vocal behavior. *Proceedings of the National Academy of Science, USA.* 95: 5417–5419.

Kelley, J. 2002. The hominid radiation in Asia. In W.C. Hartwig (ed.). *The Primate Fossil Record*, pp. 369–384. Cambridge University Press, Cambridge, UK.

Kemp, T.S. 1999. *Fossils and Evolution*. Oxford University Press, Oxford, UK.

Kennedy, A.R.K., Sonakia, A., Chiment, J. & Verma, K.K. 1991. Is the Narmada hominid an Indian *Homo erectus*? *American Journal of Physical Anthropology.* 86: 475–496.

Kennett, J.P. 1995. A review of polar climatic evolution during the Neogene, based on the marine sediment record. In E.S. Vrba, G.H. Denton, T.C. Partridge & L.H. Buckle (eds.). *Paleoclimate and Evolution with Emphasis on Human Origins*, pp. 49–64. Yale University Press, New Haven, CT.

Keyser, A.W. 2000. The Drimolen skull: the most complete australopithecine cranium and mandible to date. *South African Journal of
Science.* 96: 189–197.

Kimbel, W.H. 1984. Variation in the pattern of cranial venous sinuses and
hominid phylogeny. *American Journal of Physical Anthropology.* 63:
243–263.

Kimbel, W.H. 1988. Identification of a partial cranium of *Australopithecus afarensis* from Koobi Fora Formation, Kenya. *Journal of
Human Evolution.* 17: 647–656.

Kimbel, W.H. & Martin, L.B. (eds.). 1993. *Species, Species Concepts, and
Primate Evolution.* Plenum Press, New York.

Kimbel, W.H. & Rak, Y. 1985. Functional morphology of the asterionic
region in extant hominioids and extinct hominids. *American Journal of
Physical Anthroplogy.* 66: 337–388.

Kimbel, W.H., White, T.D. & Johanson, D.C. 1984. Cranial morphology
of *Australopithecus afarensis*: A comparative study based on a composite reconstruction of the adult skull. *American Journal of Physical
Anthropology.* 64: 337–388.

Kimbel, W.H., White, T.D. & Johanson, D.C. 1988. Implications of KNM-
WT 17000 for the Evolution of "robust" *Australopithecus.* In F.E. Grine
(ed.). *Evolutionary History of the "Robust" Australopithecines*, pp. 259–
268. Aldine de Gruyter, New York.

Kimbel, W.H., Johanson, D.C. & Rak, Y. 1994. The first skull and other new
discoveries of *Australopithecus afarensis* at Hadar, Ethiopia. *Nature.* 368:
449–251.

Kimbel, W.H., Johanson, D.C. & Rak, Y. 1997. Systematic assessment of
a maxilla of *Homo* from Hadar, Ethiopia. *American Journal of Physical
Anthropology.* 103: 235–262.

Kingdon, J. 1997. *Self-Made Man and His Undoing.* Simon & Schuster,
London.

Kingdon, J. 2003. *Lowly Origin: Where, When, and Why Our Ancestors
First Stood Up.* Princeton University Press, Princeton, NJ.

Kingston, J.D. 1999. Environmental determinants in early hominid evolu-
tion: Issues and evidence from the Tugen Hills, Kenya. In P.J. Andrews &
P. Banhan (eds.). *Late Cenozoic Environments and Hominoid
Evolution: A Tribute to Bill Bishop,* pp. 69–84. Geological Society,
London.

Klein, R.G. 1999. *The Human Career: Human Biological and Cultural
Origins,* 2nd ed. University of Chicago Press, Chicago.

Köhler, M. & Moyà-Solà, S. 1997. Ape-like or hominid-like? The positional behavior of *Oreopithecus bambolii* reconsidered. *Proceedings of the National Academy of Sciences USA.* 94: 11747–11750.

Köhler, M., Moyà-Solà, S. & Alba, D.M. 2001. Eurasian hominoid evolution in the light of recent *Dryopithecus* findings. In L. de Bonis, G.D. Koufos & P.J. Andrews (eds.). *Hominoid Evolution and Climate Change in Europe*, Vol. 2: *Phylogeny of the Neogene Hominoid Primates of Eurasia*, pp. 192–212. Cambridge University Press, Cambridge, UK.

Koppe, T. & Ohkawa, Y. 1999. Pneumatization of the facial skeleton in catarrhine primates. In T. Koppe, H. Nagai, & K.W. Alt (eds.). *The Paranasal Sinuses of Higher Primates: Development, Function and Evolution*, pp. 77–119. Chicago: Quintessence.

Kordos, L. & Begun, D.R. 1997. A new reconstruction of RUD 77, a partial cranium of *Dryopithecus brancoi* from Rudabánya, Hungary. *American Journal of Physical Anthropology.* 103: 277–294.

Kornet, D.J. 1993. *Reconstructing Species: Demarcations in Geological Networks*. Instituut voor Theoretsche Biologie, Rijksherbarium, Leiden, Netherlands.

Koufos, G. 1993. A mandible of *Ouranopithecus macedoniensis* from the late Miocene of Macedonia (Greece). *American Journal of Physical Anthropology.* 91: 225–235.

Koufos, G. 1995. The first female maxilla of the hominoid *Ouranopithecus macedoniensis* from the late Miocene of Macedonia (Greece). *Journal of Human Evolution.* 29: 385–399.

Kramer, A. 1986 Hominid-pongid distinctiveness in the Miocene-Pliocene fossil record: The Lothagam mandible. *American Journal of Physical Anthropology.* 70: 457–473.

Krings, M., Stone, A., Schmitz, R.M., Krainitzki, H., Stoneking, M. & Pääbo, S. 1997. Neanderthal DNA sequences and the origin of modern humans. *Cell.* 90: 19–30.

Krings, M., Capelli, C., Tschentscher, F., Geisert, H., Meyer, S., von Haeseler, A., Grossschmidt, K., Possnert, G., Paunovic, M., & Paabo, S. 2000. A view of Neandertal genetic diversity. *Nature Genetics.* 26: 144–146.

Kuman, K. & Clarke, R.J. 2000. Stratigraphy, artefact industries and hominid associations for Sterkfontein, Member 5. *Journal of Human Evolution.* 38: 827–847.

Larick, R. & Ciochon, R.L. 1996. The African emergence and early Asian dispersals of the genus *Homo*. *American Scientist.* 84: 538–551.

Leakey, L.S.B. 1962. A new lower Pliocene fossil primate from Kenya. *Annals of the Magazine of Natural History,* Series 13(4): 689–696.

Leakey, L.S.B., Tobias, P.V. & Napier, J.R. 1964. A new species of the genus *Homo* from Olduvai Gorge. *Nature.* 202: 7–9.

Leakey, M.D. 1971. *Olduvai Gorge. Excavations in Beds I and II, 1960–1963,* Vol. 3. Cambridge University Press, Cambridge, UK.

Leakey, M.D. 1994. *Olduvai Gorge. Excavations in Beds III, IV, and the Masek Beds, 1968–1971,* Vol. 5 Cambridge University Press, Cambridge, UK.

Leakey, M.D. & Hay, R.L. 1979. Pliocene footprints in the Laetoli Beds at Laetoli, northern Tanzania. *Nature.* 278: 317–323.

Leakey, M.G. & Walker, A.C. 1997. *Afropithecus* function and phylogeny. In D.R. Begun, C.V. Ward & M.D. Rose (eds.). *Function, Phylogeny, and Fossils: Miocene Hominoid Evolution and Adaptations,* pp. 225–239. Plenum Press, New York.

Leakey, M.G. & Walker, A.C. 2003. The Lothagam hominids. In M.G. Leakey & J.M. Harris (eds.). *Lothagam: The Dawn of Humanity in Eastern Africa,* pp. 249–257. Columbia University Press, New York.

Leakey, M.G., Feilbel, C.S., McDougall, I. & Walker, A.C. 1995. New four-million-year-old hominid species from Kanapoi and Allia Bay, Kenya. *Nature.* 376: 565–571.

Leakey, M.G., Spoor, F., Brown, F.H., Gathogo, P.N., Kiarie, C., Leakey, L.N. & McDougall, I. 2001. New hominin genus from eastern Africa shows diverse middle Pliocene lineages. *Nature.* 410: 433–440.

Leakey, R.E.F. 1973a. Further evidence of Lower Pleistocene hominids from East Rudolf, North Kenya, 1972. *Nature.* 242: 170–173.

Leakey, R.E.F. 1973b. Australopithecines and hominins: A summary on the evidence from the early Pleistocene of eastern Africa. In S. Zuckerman (ed.). *Concepts of Human Evolution,* pp. 53–69. *Symposium Zoological Society.* Number 33. London.

Leakey, R.E.F. 1974. Further evidence of Lower Pleistocene hominids from East Rudolf, North Kenya, 1973. *Nature.* 248: 653–656.

Leakey, R.E.F, Leakey, M.G, Richtsmeier, J.T, Simons, E.L. & Walker, A.C. 1991. Similarities in *Aegyptopithecus* and *Afropithecus* facial morphology. *Folia Primatologica.* 56: 65–85.

Lieberman, D.E. 1995. Testing hypotheses about recent human evolution from skulls. *Current Anthropology* (including comments). 36: 159–195.

Lieberman, D.E. 1996. How and why recent humans grow thin skulls: experimental data on systemic cortical robusticity. *American Journal of Physical Anthropology.* 101: 217–236.

Lieberman, D.E. 1998. Sphenoid shortening and the evolution of modern human cranial shape. *Nature.* 393: 158–162.

Lieberman, D.E. 2000. Ontogeny, homology, and phylogeny in the hominid craniofacial skeleton: The problem of the browridge. In P.O. Higgins, M. Cohn (eds.). *Development, Growth and Evolution: Implications for the Study of the Hominid Skeleton*, pp. 85–122. Academic Press, New York.

Lieberman, D.E. 2001. Another face in our family tree. *Nature.* 410: 419–420.

Lieberman, D.E. & Crompton, A.W. 1998. Responses of bone to stress. In E. Wiebel, D.R. Taylor & L. Bolis (eds.). *Principles of Biological Design: The Optimization and Symmorphosis Debate*, pp. 78–86. Cambridge University Press. Cambridge, UK.

Lieberman, D.E., Pilbeam, D.R. & Wood, B.A. 1988. A probabilistic approach to the problem of sexual dimorphism in *Homo habilis*: A comparison of KNM-ER 1470 and KNM-ER 1813. *Journal of Human Evolution.* 17: 503–511.

Lieberman, D.E., Ross, C.F. Ravosa, M.J. 2000. The primate cranial base: Ontogeny, function, and integration. *Yearbook of Physical Anthropology.* 43: 117–170.

Lieberman, D.E., Wood, B.A. & Pilbeam, D.R. 1996. Homoplasy and early *Homo*: An analysis of the evolutionary relationships of *H. habilis sensu stricto* and *H. rudolfensis. Journal of Human Evolution.* 30: 97–120.

Lieberman, D.E. Pearson, O.M. & Mowbray, K.M. 2000. Basicranial influence on overall cranial shape. *Journal of Human Evolution.* 38: 291–315.

Lieberman, D.E., McBratney, B.M. & Krovitz, G. 2002. The evolution and development of cranial form in *Homo sapiens. Proceedings of the National Academy of Sciences, USA.* 99: 1134–1139.

Lieberman, P. 1989. The origins of some aspects of human language and cognition. In P. Mellars and C. Stringer (eds.). *The Human Revolution: Behavioral and Biological Perspectives on the Origins of Modern Humans*, pp. 391–414. Princeton University Press, Princeton, NJ.

Lieberman, P. 1991. *Uniquely Human: The Evolution of Speech, Thought, and Selfless Behavior.* Harvard University Press, Cambridge, MA.

Lovejoy, C.O. 1974. The gait of australopithecines. *Yearbook of Physical Anthropology.* 17: 147–161.

Lovejoy, C.O. 1981. The origin of man. *Science.* 211: 341–350.

Lumly, de, H. & Lumley, de, M.A. 1971. Découverte de restes humaines anténéandertaliens datés au debut de Riss a la Caune d'Arago (Tautavel, Pyrénées-Orientales). *Comptes Rendus de é Academie des Sciences Paris.* 272: 1729–1742.

Lumley, de, M.A. & Sonakia, A. 1985. Première découverte d'un *Homo erectus* sur le continent Indien a Hathnora, dans la Moyenne Vallée de la Narmada. *L'Anthropologie* (Paris). 89: 13–61.

Macintosch, N.W.G. & Larnach, S.L. 1972. The persistence of *H. erectus* traits in Australian aboriginal crania. *Archaeology and Physical Anthropology in Oceania.* 7: 1–7.

Macintosch, N.W.G. & Larnach, S.L. 1976. Aboriginal affinities looked at in world context. In R.L. Kirk & A.G. Thorne (eds.). *The Origin of the Australians*, pp. 113–126. Australian Institute of Aborigiunal Studies, Canberra Humanities Press, Canberra.

MacLarnon, A.M. 1993. The vertebral canal. In A.C. Walker & R.E.F. Leakey (eds.). *The Nariokotome* Homo erectus *Skeleton*, pp. 359–390. Harvard University Press, Cambridge, MA.

MacLarnon, A.M. & Hewitt, G.P. 1999. The evolution of human speech: The role of enhanced breathing control. *American Journal of Physical Anthropology.* 109: 341–363.

Maier, W. & Nkini, A. 1984. Olduvai hominid 9: New results of investigation. *Courier Forschung-Institut Senckenberg.* 69: 123–130.

Mallegni, F., Carnieri, E., Bisconti, M., Tartarelli, G., Ricci, S., Biddittu, I. & Segre, A. 2003. *Homo cepranensis sp. nov.* and the evolution of African-European Middle Pleistocene hominids. *Compter Rendus Palévol.* 2: 153–159.

Marks, A.E. & Volkman, P.W. 1983. Changing core reduction strategies: A technological shift from the Middle to the Upper Paleolithic in the Southern Levant. In E. Trinkaus (ed.). *The Mousterian Legacy: Human Biocultural Change in the Upper Pleistocene*, pp. 13–34. BAR International Series 164. Oxford, UK.

Marquez, S., Mowbray, K., Swayer., G.J, Sawyer., Jacob, T. & Silvers, A. 2001. New fossil hominid calvaria from Indonesia — Sambungmacan 3. *Anatomical Record.* 262: 344–368.

Martin, L.B. & Andrews, P.J. 1984. The phyletic position of *Graecopithecus freybergi* Koenigswald. *Courier Forschungsinstitut Senckenberg.* 69: 25–40.

Martin, R.B., Burr, D.B. & Sharkey, N.A. 1999. *Skeletal Tissue Mechanics.* Springer Verlag, New York.

Martin, R.D. 1990. *Primate Origins and Evolution: A phylogenetic Reconstruction*. Princeton University Press, Princeton, NJ.

McCollum, M.A. 1997. Palatal thickening and facial form in *Paranthropus*: Examination of alternative developmental models. *American Journal of Physical Anthropology*. 103: 375–392.

McCollum, M.A. & Ward, S.C. 1997. Subnasoalveolar anatomy and hominoid phylogeny: Evidence from comparative ontogeny. *American Journal of Physical Anthropology*. 102: 377–405.

McCollum, M.A., Grine, F.E., Ward, S.C. & Kimbel, W.H. 1993. Subnasal morphological variation in extant hominoids and fossil hominids. *Journal of Human Evolution*. 24: 87–111.

McCrossin, M.L. & Benefit, B.R. 1997. On the relationships and adaptations of *Kenyapithecus*, a large-bodied hominoid from the Middle Miocene of Eastern Africa. In D.R. Begun, C.V. Ward & M.D. Rose (eds.). *Function, Phylogeny, and Fossils: Miocene Hominoid Evolution and Adaptations*, pp. 241–268. Plenum Press, New York.

McDougal, I. & Feibal, C.S. 1999. Numerical age control for the Miocene-Pliocene succession at Lothagam, a hominoid-bearing sequence in the northern Kenya Rift. *Journal of the Geological Society, London*. 156: 731–745.

McGowan, C. 1999. *A Practical Guide to Vertebrate Mechanics*. Cambridge University Press. Cambridge, UK.

McHenry, H.M. 1986. The first bipeds: A comparison of the *A. afarensis* and *A. africanus* postcranium and implications for the evolution of bipedalism. *Journal of human Evolution*. 15: 177–191.

McHenry, H.M. 1988. New estimates of body weight in early hominids and their significance to encephalization and megadontia in "robust" australopithecines. In F.E. Grine (ed.). *Evolutionary History of the "Robust" Australopithecines*, pp. 133–148. Aldine de Gruyter, New York.

McKee, J.K. 1999. The autocatalytic nature of hominid evolution in African Plio-Pleistocene environments. In T.G. Bromage & F. Schrenk (eds.). *African Biogeography, Climate Change and Human Evolution*, pp. 57–67. Oxford University Press, Oxford, UK.

Mellars, P. 1989. Technological changes at the Middle–Upper Palaeolithic transition: Economic, social and cognitive perspectives. In P. Mellars & C. Stringer (eds.). *The Human Revolution: Behavioral and Biological Perspectives on the Origins of Modern humans*, pp. 338–365. Princeton University Press. Princeton, NJ.

Mellars, P. 1996. *The Neanderthal Legacy: An Archaeological Perspective from Western Europe.* Princeton University Press. Princeton, NJ.

Mellars, P. 2000. The archaeological records of the Neanderthal–modern human transition in France. In O. Bar Yosef & D. Pilbeam (eds.). *The Geography of Neanderthals and Modern Humans in Europe and the Greater Mediterranean*, pp. 35–48. Peabody Museum Bulletin No. 8, Harvard University, Cambridge, MA.

Miller, G. 2000. *The Mating Mind: How Sexual Choice Shaped the Evolution of Human Nature.* William Heinemann, London.

Morgan, E. 1990. *The Scars of Evolution.* Souvenier Press, London.

Morgan, E. 1997. *The Aquatic Ape Hypothesis.* Souvenier Press, London.

Morant, G. 1938. The form of the Swanscombe skull. *Journal of the Royal Anthropological Institute.* 68: 67–97.

Moss, M.L. & Young, R.W. 1960. A functional approach to craniology. *American Journal of Physical Anthropology.* 18: 281–292.

Moyà-Solà, S. & Köhler, M. 1993. Recent discoveries of *Dryopithecus* shed new light on evolution of great apes. *Nature.* 365: 543–545.

Moyà-Solà, S. & Köhler, M. 1995. New partial cranium of *Dryopithecus lartet* 1863 (Hominoidea, Primates) from the upper Miocene of Can Llobateres, Barcelona, Spain. *Journal of Human Evolution.* 29: 101–139.

Moyà-Solà, S. & Köhler, M. 1996. The first *Dryopithecus* skeleton: Origins of great ape locomotion. *Nature.* 379: 156–159.

Mulvaney, J. & Kamminga, J. 1999. *Prehistory of Australia.* Allan and Unwin, Sydney.

Nakatskasa, M., Yamanaka, A., Kumimatsu, Y., Shimizu, D. & Ishia, H. 1998. A newly discovered *Kenyapithecus* skeleton and its implications for the evolution of postcranial behavior in Miocene East African hominoids. *Journal of Human Evolution.* 34: 657–664.

Newby, J. 1997. *The Pact for Survival.* Sydney: ABC Books.

Niewoehner, W.A. 2001. Behavioral inferences from the Skul/Qafzeh early modern human hand remains. *Proceedings of the National Academy of Sciences, USA.* 98: 2979–2984.

Noble, W. & Davidson, I. 1996. *Human Evolution, Language and Mind: A Psychological and Archaeological Inquiry.* Cambridge University Press, Cambridge, UK.

Noreen, E.W. 1989. *Computer-Intensive Methods for Testing Hypotheses: An introduction.* John Wiley & Sons, New York.

Novacek, M.J. 1992. Fossils, topologies, missing data, and the higher level phylogeny of eutherian mammals. *Systematic Biology*. 41: 58–73.

Oakley, K.P. & Hoskins, C.R. 1950. New evidence on the antiquity of Piltdown. *Nature*. 165: 379–382.

Olsen, T.R. 1981. Basicranial morphology of the extant hominoids and Pliocene hominids: The new material from the Hadar Formation, Ethiopia, and its significance in early human evolution and taxonomy. In C.B. Stringer (ed.). *Aspects of Human Evolution*, pp. 99–128. Taylor and Francis, London.

Olsen, T.R. 1985. Cranial morphology and systematics of the Hadar Formation hominids and *"Australopithecus" afarensis*. In E. Deslon (ed.). *Ancestors: The Hard Evidence*, pp. 94–101. Alan R. Liss, New York.

Oppenheimer, S. 2003. *Out of Eden: The Peopling of the New World*. Constable, London.

Ovchinnikov, I.V., Götherström, A., Romanova, G.P., Kharltonov, V.M., Lidén, K. & Goodwin, W. 2000. Molecular analysis of Neanderthal DNA from the northern Caucasus. *Nature*. 404: 490–493.

Owen-Smith, N. 1999. Ecological links between African savanna environments, climate change, and early hominid evolution. In T.G. Bromage & F. Schrenk (eds.). *African Biogeography, Climate Change & Human Evolution*, pp. 138–149. Oxford University Press, Oxford, UK.

Pagel, M. & Bodmer, W. 2003. A naked ape would have fewer parasites. *Proceedings of the Royal Society of London*, B, *Biological Sciences*. 270: S117–S119.

Pardoe, C. 1991. Competing paradigms and ancient human remains: The state of the discipline. *Archaeology in Oceania*, 26: 79–85.

Parés, J.M. & Pérez-Gonzáles, A. 1995. Paleomagnetic age for hominid fossils at Atapuerca archaeological site, Spain. *Science*. 269: 830–832.

Partridge, T.C., Wood B.A. & deMenocal, P.B. 1995. The influence of global climatic change and regional uplift on large-mammalian evolution in East and Southern Africa. In E.S. Vrba, G.H. Denton, T.C. Partridge & L.H. Burckle (eds.). *Paleoclimate and Evolution, with Emphasis on Human Origins*, pp. 331–355. Yale University Press, New Haven, CT.

Pickering, T.R. 2001. Taphonomy of the Swatkrans hominid postcrania and its bearing on issues of meat-eating and fire management. In C.B. Stanford & H.T. Bun (eds.). *Meat-Eating and Human Evolution*, pp. 33–72. Oxford University Press, Oxford, UK.

Pickford, M., Senut, B., Gommery, D. & Treil, J. 2002. Bipedalism in *Orrorin tugenensis* revealed by its femora. *C.R.Palevolution.* 1: 191–203.

Pilbeam, D.R. 1996. Genetic and morphological records of the Hominoidea and hominid origins: Asynthesis. *Molecular Phylogenetics and Evolution.* 5: 155–168.

Pilbeam, D.R. 1997. Research on Miocene hominoids and hominid origins: The last three decades. In D.R. Begun, C.V. Ward & M.D. Rose (eds.). *Function, Phylogeny, and Fossils: Miocene Hominoid Evolution and Adaptations*, pp. 13–28. Plenum Press, New York.

Pilbeam, D.R. 2002. Perspectives on the Miocene Hominoidea. In W.C. Hartwig (ed.). *The Primate Fossil Record*, pp. 303–310. Cambridge University Press, Cambridge, UK.

Pilbeam, D.R. & Bar Yosef, O. 2000. Afterward. In O. Bar Yosef & D.R. Pilbeam (eds.). *The Geography of Neanderthals and Modern Humans in Europe and the Greater Mediterranean*, pp. 186–189. Peabody Museum Bulletin No. 8. Harvard University, Cambridge, MA.

Pitts, M. & Roberts, M. 1998. *Fairweather Eden: Life in Britain Half a Million Years Ago as Revealed by the Excavations at Boxgrove.* Arrow Books, London.

Plavcan, J.M. 1993. Catarrhine dental variability and species recognition in the fossil record. In W.H. Kimbel & L.B. Martin (eds.). *Species, Species Concepts, and Primate Evolution*, pp. 239–263. Plenum Press, New York.

Potts, R. 1986. Temporal span of bone accumulations at Olduvai Gorge and implications for early hominid foraging behavior. *Paleobiology.* 12: 25–31.

Potts, R. 1988. *Early Hominid Activities at Olduvai.* Aldine de Gruyter, New York.

Potts, R. 1996. *Humanity's Descent: The Consequences of Ecological Instability.* William Morrow and Company, New York.

Potts, R. & Shipman, P. 1981. Cutmarks made by stone tools on bones from Olduvai Gorge, Tanzania. *Nature.* 291: 577–580.

Prothero, D.R. 1994. *The Eocene–Oligocene Transition: Paradise Lost.* Columbia University Press, New York.

Quade, J., Cerling, T.E. & Bowman, J.R. 1989. Development of Asian monsoon revealed by marked ecological shifts during the latest Miocene in northern Pakistan. *Nature.* 342: 163–166.

Rae, T.C. 1999. The maxillary sinus in primate palaeontology and systematics. In T. Koppe, H. Nagai & K.W. Alt (eds.). *The Paranasal Sinuses of*

Higher Primates: Development, Function, and Evolution, pp. 177–190. Quintessence Publishing, Chicago.

Raff, R.A. 1996. *The Shape of Life: Genes, Development, and the Evolution of Animal Form.* University of Chicago Press, Chicago.

Rak, Y. 1978. The functional significance of the squamosal suture in *Australopithecus boisei. American Journal of Physical Anthropology.* 49: 71–78.

Rak, Y. 1983. *The Australopithecine Face.* Academic Press, New York.

Rak, Y. & Kimbel, W.H. 1991. On the squamosal suture of KNM-WT 17000. *American Journal of Physical Anthropology.* 85: 1–6.

Rak, Y. & Kimbel, W.H. 1993. Reply to Drs. Walker, Brown and Ward. *American Journal of Physical Anthropology.* 90: 506–507.

Ravosa, M.J. 1988. Browridge development in Cercopithecidae: A test of two models. *American Journal of Physical Anthropology.* 76: 535–555.

Reed, K.E. 1997. Early hominid evolution and ecological change through the African Plio-Pleistocene. *Journal of Human Evolution.* 32: 289–322.

Relethford, J.H. 2001. *Genetics and the Search for Modern Human Origins.* Wiley-Liss, New York.

Richmond, B.G. & Strait, D.S. 2000. Evidence that humans evolved from a knuckle-walking ancestor. *Nature.* 404: 382–385.

Rightmire, P.G. 1984. Comparison of *Homo erectus* from Africa and southeast Asia. In P.A. Andrews & L. Franzen (eds.). *The Early Evolution of Man with Special Emphasis on Southeast Asia and Africa. Courier Forschungsinstitut Senckenberg.* 69: 83–98.

Rightmire, P.G. 1985. The tempo of change in the evolution of mid-Pleistocene *Homo.* In E. Delson (ed.). *Ancestors: The Hard Evidence,* pp. 255–264. Alan R. Liss, New York.

Rightmire, P.G. 1986. Body size and encephalization in *Homo erectus. Anthropos (Brno).* 23: 139–150.

Rightmire, P.G. 1990. *The Evolution of* Homo erectus: *Comparative Anatomical Studies of an Extinct Human Species.* Cambridge University Press. Cambridge, UK.

Rightmire, P.G. 1993. Variation among early *Homo* crania from Olduvai Gorge and the Koobi Fora region. *American Journal of Physical Anthropology.* 90: 1–33.

Rightmire, P.G. 1996. The human cranium from Bodo, Ethiopia: Evidence for speciation in the Middle Pleistocene? *Journal of Human Evolution.* 31: 21–39.

Robinson, J.T. 1954. The genera and species of the Australopithecinae. *American Journal of Physical Anthropology.* 12: 181–200.

Robinson, J.T. 1958. Cranial cresting patterns and their significance in the Hominoidea. *American Journal of Physical Antropology.* 16: 397–428.

Robinson, J.T. 1962. The origin and adaptive radiation of the australopithecines. In G. Kurth (ed.). *Evolution und Hominisation*, pp. 120–140. Gustav Fischer, Stuttgart.

Robinson, J.T. 1972. *Early Hominid Posture and Locomotion.* University of Chicago Press, Chicago.

Rodman, P.S. & McHenry, H.M. 1980. Bioenergetics and the origin of hominid bipedalism. *American Journal of Physical Anthropology.* 52: 103–106.

Rögl, F. 1999. Mediterranean and Paratethys palaeogeography during the Oligocene and Miocene. In J. Agusti, L. Rook & P.J. Andrews (eds.). *Hominoid Evolution and Climatic Change in Europe: The Evolution of Neogene Terrestrial Ecosystems in Europe*, Vol. 1, pp. 8–22. Cambridge University Press, Cambridge, UK.

Rook, L., Bondioli, L., Köhler, M., Moyà-Solà, S. & Macchiarelli, R. 1999. *Oreopithecus* was a biped after all: Evidence from the iliac cancellous architecture. *Proceedings of the National Academy of Science.* 96: 8795–8799.

Rosen, B.R. 1999. Palaeoclimate implications of the energy hypothesis from Neogene corals of the Mediterranean region. In J. Agusti, L. Rook & P.J. Andrews (eds.). *Hominoid Evolution and Climatic Change in Europe.* Vol. 1: *The Evolution of Neogene Terrestrial Ecosystems in Europe*, pp. 309–327. Cambridge University Press, Cambridge, UK.

Ruff, C.B. & Walker, A.C. 1993. Body size and body shape. In A.C. Walker & R.E.F. Leakey (eds.). *The Nariokotome* Homo erectus *Skeleton*, pp. 234–265. Harvard University Press, Cambridge, MA.

Ruff, C., Trinkaus, E. & Holliday, T. 1997. Body mass and encephalization in Pleistocene *Homo. Nature.* 387: 173–176.

Ruvolo, M. 1994. Molecular evolutionary processes and conflicting gene trees: The hominoid case. *American Journal of Physical Anthropology.* 94: 89–114.

Ruvolo, M. 1997. Molecular phylogeny of the hominoids: Inference from multiple independent DNA sequence data sets. *Molecular Biology and Evolution.* 14: 248–265.

Santa Luca, A.P. 1978. A re-examination of presumed Neanderthal-like fossils. *Journal of Human Evolution.* 7: 619–636.

Santa Luca, A.P. 1980. *The Ngandong Fossil Hominids: A Comparative Study of a Far Eastern* Homo erectus *group.* Yale University Publications in Anthropology No. 78. New Haven, CT.

Sawada, Y., Pickford, M., Senut, B., Itaya, T., Hyodo, M., Miura, T. Kashine, C., Chuho., T. & Fujii, H. 2002. The age of *Orrorin tugenensis*, an early hominid from the Tugen Hills, Kenya. *C. R. Palevolution.* 1: 293–303.

Schick, K.D. & Toth, N. 1993. *Making Silent Stones Speak: Human Evolution and the Dawn of Technology.* Weidenfeld & Nicholson, London.

Schoch, R.M. 1986. *Phylogeny Reconstruction in Palaeontology.* Van Nostrand Reinhold, New York.

Schoeninger, M.J., Bunn, H.T., Murray, S., Pickering, T. & Moore, J. 2002. Meat-eating by the fourth African ape. In C.B. Stanford & H.T. Bunn (eds.). *Meat-Eating and Human Evolution*, pp. 179–195. Oxford University Press, Oxford, UK.

Schwartz, J.H. & Tattersall, I. 2002. *The Human Fossil Record*, Vol. 1: *Terminology and Craniodental Morphology of Genus* Homo *(Europe).* Wiley-Liss, New York.

Schwartz, J.H. & Tattersall, I. 2003. *The Human Fossil Record*, Vol. 2: *Craniodental Morphology of Genus* Homo *(Africa and Asia).* Wiley-Liss, New York.

Semaw, S., Renne, P., Harris, J.W.K., Feibel, C.S., Bernor, R.L., Fesseha, N. & Mowbray, K. 1997. 2.5-million-year-old stone tools from Gona, Ethiopia. *Nature.* 385: 333–336.

Senut, B. & Tardieu, C. 1985. Functional aspects of Plio-Pleistocene hominid limb bones: Implications for taxonomy and phylogeny. In E. Delson (ed.). *Ancestors: The Hard Evidence*, pp. 193–201. Alan R. Liss, New York.

Senut, B., Pickford, M., Gommery, D., Mein, P., Cheboi, K. & Coppens, Y. 2001. First hominid from the Miocene (Lukeino Formation, Kenya). *Comptes Rendus de l'Académie des Sciences de Paris.* 332: 137–144.

Shea, B.T. 1985. On aspects of skull form in african apes and orangutans, with implications for hominoid evolution. *American Journal of Physical Anthropology.* 68: 329–342.

Shea, B.T. 1993. Bone growth and primate evolution. In B.K. Hall (ed.). *Bone*, Vol. 7: *Bone — B*, pp. 133–158. CRC Press, London.

Shea, B.T., Leigh, S.R. & Groves, C.P. 1993. Multivariate craniometric variation in chimpanzees. Implications for species identification in palaeoanthropology. In W.H. Kimbel & L.B. Martin (eds.). *Species, Species Concepts, and Primate Evolution*, pp. 265–296. Plenum Press, New York.

Shen, G-J., Wang, W., Wang, Q., Zhao, J.-X., Collerson, K., Zhao, C.-L., & Tobias, P.B. 2002. U-series dating of Liujiang hominid site in Guangxi, Southern China. *Journal of Human Evolution.* 43: 817–829.

Sherwood, R.J. 1999. Pneumatic processes in the temporal bone of chimpanzee (*Pan troglodytes*) and gorilla (*Gorilla gorilla*). *Journal of Morphology.* 241: 127–137.

Shreeve, J. 1995. *The Neandertal enigma: Solving the mystery of modern human origins.* Viking Press, London.

Singer, P. & Cavalier, P. (eds.). 1994. *The Great Ape Project.* London: Fourth Estate.

Singleton, M. 2000. The phylogenetic affinities of *Otavipithecus namibiensis. Journal of Human Evolution.* 38: 537–574.

Skelton, R.R. & McHenry, H.M. 1992. Evolutionary relationships among early hominids. *Journal of Human Evolution.* 23: 309–349.

Skelton, R.R. & McHenry, H.M. 1998. Trait list bias and a reappraisal of early hominid phylogeny. *Journal of human Evolution.* 34: 109–113.

Skelton, R.R., McHenry, H.M. & Drawhorn, G.M. 1986. Phylogenetic analysis of early hominids. *Current Anthropology.* 94: 307–325.

Smith, A.B. 1994. *Systematics and the Fossil Record: Documenting Evolutionary Patterns.* Blackwell Scientific Publications, London.

Smith, F.H. 1983. A behavioral interpretation of changes in craniofacial morphology across the archaic/modern *Homo sapiens* transition. In E. Trinkaus (ed.). *The Mousterian Legacy: Human Biocultural Change in the Upper Pleistocene,* pp. 141–164. BAR International Series 164. Oxford, UK.

Smith, F.H. 2002. Migrations, radiations and continuity: Patterns in the evolution of Middle and Late Pleistocene humans. In W.G. Hartwig (ed.). *The Primate Fossil Record,* pp. 437–456. Cambridge University Press, Cambridge, UK.

Smith, F.H. & Raymond, G.L. 1980. Evolution of the supraorbital region in Upper Pleistocene fossil hominids from South-Central Europe. *American Journal of Physical Anthropology.* 53: 589–610.

Smith, F.H., Simek, J.F. & Harrill, M.S. 1989. Geographic variation in supraorbital torus reduction during the Later Pleistocene (c. 80,000–15,000 Bp). In P. Mellars & C. Stringer (eds.). *The Human Revolution: Behavioral and Biological Perspectives on the Origins of Modern Humans,* pp. 172–193. Princeton University Press. Princeton, NJ.

Smith, H.B. 1993. The physiological age of KNM-WT 15000. In A.C. Walker & R.E.F. Leakey (eds.). *The Nariokotome* Homo erectus *Skeleton,* pp. 195–220. Harvard University Press, Cambridge, MA.

Smith-Woodward, A. 1914. On the lower jaw of an anthropoid ape (*Dryopithecus*) from the upper Miocene of Lerida (Spain). *Quarterly Journal of the Geological Society.* 70: 316–320.

Solounias, N., Plavcan, M.J., Quade, J. & Witmer, L. 1999. The palaeoecology of the Pikermian Biome and the savanna myth. In J. Agusti, L. Rook & P.J. Andrews (eds.). *Hominoid Evolution and Climatic Change in Europe: The Evolution of Neogene Terrestrial Ecosystems in Europe,* Vol. 1, pp. 436–453. Cambridge University Press, Cambridge, UK.

Spencer, F. 1990. *Piltdown: A Scientific Foragery.* Oxford University Press, Oxford UK.

Spencer, M.A. 1998. Force production in the primate masticatory system: Electrographic tests of biomechanical hypotheses. *Journal of Human Evolution.* 34: 25–54.

Spencer, M.A. 1999. Constraints on masticatory system evolution in anthropoid primates. *American Journal of Physical Anthropology.* 108: 483–506.

Sponheimer, M. & Lee-Thorpe, J.A. 1999. Isotopic evidence for diet of an early hominid, *Australopithecus africanus. Science.* 283: 368–370.

Spoor, F. & Zonneveld. F. 1999. Computed tomography–based three-dimensional imaging of hominid fossils: Features of the Broken Hill 1, Wadjak 1 and SK 47 crania. In T. Koppe, H. Nagai, & K. W. Alt (eds.). *The Paranasal Sinuses of Higher Primates: Development, Function, and Evolution,* pp. 207–226. Quintessence Publishing, Chicago.

Stanley, S.M. 1978. Chonospecies' longevities, the origin of genera, and the punctuated model of evolution. *Paleobiology.* 4: 26–40.

Stanley, S.M. 1979. *Macroevolution: Pattern and Process.* Johns Hopkins University Press, Baltimore.

Stanley, S.M. 1996. *Children of the Ice Age.* W.H. Freeman and Company, New York.

Stern, J.T. & Susman, R.L. 1983. The locomotor anatomy of *Australopithecus afarensis. American Journal of Physical Anthropology.* 60: 279–317.

Stewart, C. & Disotell, T.R. 1998. Primate evolution — in and out of Africa. *Current Biology.* 8: 582–588.

Stiner, M.C. 1991. The faunal remains at Grotta Guattari: A taphonomic perspective. *Current Anthropology.* 32: 103–117.

Stiner, M.C. 1994. *Honor Among Thieves: A Zooarchaeological Study of Neandertal Ecology.* Princeton University Press, Princeton, NJ.

Stone, T. & Cupper, M.L. 2003, Last glacial maximum ages for robust humans at Kow Swamp, Southern Australia. *Journal of Human Evolution,* 45: 1–13.

Storm, P. 1995. The evolutionary significance of the Wajak skulls. *Scripta Geology*. 110: 1–247.

Strait, D.S. 2001. Integration, phylogeny, and the hominid cranial base. *American Journal of Physical Anthropology*. 114: 273–297.

Strait, D.S. & Grine, F.E. 1998. Trait list bias? A reply to Skeleton and McHenry. *Journal of Human Evolution*. 34: 115–118.

Strait, D.S. & Grine, F.E. 2001. The systematics of *Australopithecus garhi*. *Ludus Vitalis*. 9: 109–135.

Strait, D.S. & Ross, C. 1999. Kinematic data on primate head and neck posture: Implications for the evolution of basicranial flexion and an evaluation of registration planes used in palaeoanthropology. *American Journal of Physical Anthropology*. 108: 205–222.

Strait, D.S. & Wood, B.A. 1999. Early hominid biogeography. *Proceedings of the National Academy of Science USA*. 96: 9196–9200.

Strait, D.S., Grine, F.E. & Moniz, M.A. 1997. A reappraisal of early hominid phylogeny. *Journal of Human Evolution*. 32: 17–82.

Stringer, C.B. 1983. Some further notes on the morphology and dating of the Petralona hominid. *Journal of Human Evolution*. 12: 731–742.

Stringer, C.B. 1984. The definition of *Homo erectus* and the existence of the species in Africa and Europe. In P.A. Andrews & L. Franzen (eds.). *The Early Evolution of Man, with Special Emphasis on Southeast Asia and Africa. Courier Forschungsinstitut Senckenberg* 69: 131–143.

Stringer, C.B. 1985. Middle Pleistocene hominid variability and the origin of the Late Pleistocene humans. In E. Delson (ed.). *Ancestors: The Hard Evidence*, pp. 289–295. Alan R. Liss, New York.

Stringer, C.B. 1986. The credibility of *Homo habilis*. In B. Wood, L. Martin & P.A. Andrews (eds.). *Major Topics in Primate and Human Evolution*, pp. 266–294. Cambridge University Press, Cambridge, UK.

Stringer, C.B. 1989. The origins of early modern humans: A comparison of the European and non-European Evidence. In P. Mellars & C. Stringer (eds.). *The Human Revolution: Behavioral and Biological Perspectives on the Origins of Modern Humans*, pp. 232–244. Princeton University Press. Princeton, NJ.

Stringer, C.B. 1994. Out of Africa: A personal history. In M.H. Nitecki & D.V. Nitecki (eds.). *Origins of Anatomically Modern Humans*, pp. 149–174. Penum Press, New York.

Stringer, C.B. 1998. Chronological and biogeographic perspectives on later human evolution. In T. Akazawa, K. Aoki, & O. Bar-Yosef (eds.). *Neandertals and Modern Humans in Western Asia*, pp. 29–37. Plenum Press, New York.

Stringer, C.B. 1999. Evolution of early humans. In S. Jones, R. Martin & D. Pilbream (eds.). *The Cambridge Encyclopedia of Human Evolution*, pp. 241–251. Cambridge University Press, Cambridge, UK.

Stringer, C.B. 2000a. Modern human origins: Out of Africa. In E. Delson, I. Tattersall, J.A. Van Couvering & A.S. Brooks (eds.). *Encyclopedia of Human Evolution and Prehistory*, 2nd ed., pp. 429–432. Garland, New York.

Stringer, C.B. 2000b. Neanderthals. In E. Delson, I. Tattersall, J.A. Van Couvering & A.S. Brooks (eds.). *Encyclopedia of Human Evolution and Prehistory*, 2nd ed., pp. 469–474. Garland, New York.

Stringer, C.B. 2003. Out of Ethiopia. *Nature*. 423: 692–695.

Stringer, C.B. & Andrews, P.A. 1988. Genetic and fossil evidence for the origin of modern humans. *Science*. 239: 1263–1268.

Stringer, C.B. & Gamble, C. 1993. *In Search of the Neanderthals: Solving the Puzzle of Human Origins*. Thames & Hudson, London.

Stringer, C.B. & McKie, R. 1996. *African Exodus: The Origins of Modern Humanity*. Pimlico, London.

Stringer, C.B. & Hublin, J-J. 1999. New age estimates for the Swanscombe hominid, and their significance for human evolution. *Journal of Human Evolution*. 37: 873–877.

Stringer, C.B. & Trinkaus, E. 1981. The Shanidar Neanderthal crania. In C.B. Stringer (ed.). *Aspects of Human Evolution*, pp. 129–165. Taylor & Francis, London.

Stringer, C.B., Howell, F.C. & Melentis, J. 1979. The significance of the fossil hominid skull from Petralona, Greece. *Journal of Archaeological Science*. 6: 235–253.

Stringer, C.B., Hublin, J.J. & Vandermeersch, B. 1984. The origins of anatomically modern humans in western Europe. In F.H. Smith & F. Spencer (eds.). *The Origins of Modern Humans*, pp. 51–135. Alan R. Liss, New York.

Stringer, C.B., Dean, M.C. & Martin, R.D. 1990. A comparative study of cranial and dental development within a recent British sample and among Neanderthals. In C.J. DeRousseau (ed.). *Primate Life History and Evolution*. Wiley-Liss. New York.

Stringer, C.B., Humphrey, L.T. & Crompton, T. 1997. Cladistic analysis of dental traits in recent humans using a fossil outgroup. *Journal of Human Evolution*. 32: 389–402.

Susman, R.L. 1979. Comparative and functional morphology of hominoid fingers. *American Journal of Physical Anthropology*. 50: 215–235.

Susman, R.L. 1988. New postcranial remains from Swartkrans and their bearing on the functional morphology and behavior of *Paranthropus robustus*. In F.E. Grine (ed.). *Evolutionary History of the "Robust" Australopithecines*, pp. 149–174. Aldine de Gruyter, New York.

Susman, R.L., Stern, J.T. & Jungers, W.L. 1984. Arboreality and bipedality in Hadar (Ethiopia) hominids. *Folia Primatologica*. 43: 113–156.

Suwa, G., Asfaw, B., Beyene, Y., White, T.D., Katoh, S., Nagaoka, S., Nakaya, H., Uzawa, K., Rene, P. & WoldeGabriel, G. 1997. The first skull of *Australopithecus boisei*. *Nature*. 389: 489–492.

Swisher, C.G., Curtis, G.H., Jacob, T., Getty, A.G., Suprijo, A. & Widiasmoro. 1994. Age of the earliest known hominids in Java, Indonesia. *Science*. 263: 1118–1121.

Swisher, G.G., Rink, W.J., Antón, S.C., Schwarcz, H.P., Curtis, G.H., Suprijo, A. & Widiasmoro, 1996. Latest *H. erectus* of Java: Potential contemporaneity with *H. sapiens* in Southeast Asia. *Science*. 274: 1870–1874.

Swisher, C.G., Curtis, G.H. & Lewin, R. 2000. *Java Man: How Two Geologists' Dramatic Discoveries Changed Our Understanding of the Evolutionary Path to Modern Humans*. Scribner, New York.

Swofford, D.L. 1998. *PAUP: Phylogenetic Analysis Using Parsimony (Version 4)*. Sinauer Associates, MA.

Sykes, B. 2001. *The Seven Daughters of Eve: The Science That Reveals Our Genetic Ancestry*. W.W. Norton & Company, New York.

Tattersall, I. 1986. Species recognition in human paleontology. *Journal of Human Evolution*. 15: 165–175.

Tattersall, I. 1995. *The Fossil Trail: How We Know What We Think We Know About Human Evolution*. Oxford University Press, Oxford, UK.

Tattersall, I. 1999. *The Last Neanderthal: The Rise, Success, and Mysterious Extinction of Our Closest Human Relatives*. Westview Press, New York.

Tattersall, I. & Eldredge, N. 1977. Fact, theory and fantasy in human paleontology. *American Scientist*. 65: 204–211.

Tattersall, I. & Schwartz, J. 2000. *Extinct Humans*. Westview Press. New York.

Taylor C.R. & Rowntree, V.J. 1970. Running on two or four legs: Which consumes more energy? *Science*. 179: 186–187.

Tchernov, E. 1998. The faunal sequence of the Southwest Asian Middle Paleolithic in relation to hominid dispersal events. In T. Akazawa, K. Aoki & O. Bar-Yosef (eds.). *Neandertals and Modern Humans in Western Asia*, pp. 77–90. Plenum Press, New York.

Teaford, M.F. 2000. Primate dental functional morphology revisited. In M. F. Teaford, M.M. Smith & M.W.J. Ferguson (eds.). *Development, Function and Evolution of Teeth*, pp. 290–304. Cambridge University Press, Cambridge, UK.

Thomas, H. 1985. The early and middle Miocene land connections of the Afro-Arabian plate and Asia: A major event for hominoid dispersal? In E. Delson (ed.). *Ancestors: The Hard Evidence*, pp. 42–50. Alan R. Liss, New York.

Thorne, A.G. 1976. Morphological contrasts in Pleistocene Australians. In R.L. Kirk & A.G. Thorne (eds.). *The Origin of the Australians*, pp. 95–112. Australian Institute of Aborigiunal Studies, Canberra Humanities Press, Canberra.

Thorne, A.G. 1977. Separation or reconciliation? Biological clues to the development of Australian Society. In J. Allen, J. Golson & R. Jones (eds.). *Sunda and Sahul: Prehistoric Studies in Southeast Asia, Melanesia and Australia*, pp. 187–204. Academic Press, New York.

Thorne, A.G. 1984. Australia's human origins — how many sources? *American Journal of Physical Anthropology*. 63: 227.

Thorne, A.G. 2000. Modern human origins: Multiregional evolution. In E. Delson, I. Tattersall, J.A. Van Couvering & A.S. Brooks (eds.). *Encyclopedia of Human Evolution and Prehistory*, 2nd ed., pp. 425–429. Garland, New York.

Thorne, A.G. & Wilson, S.R. 1977. Pleistocene and recent Australians. *Journal of Human Evolution*. 6: 393–402.

Thorne, A.G. & Wolpoff, M. 1981. Regional continuity in Australasian Pleistocene hominid evolution. *American Journal of Physical Anthropology*. 55: 337–349.

Thorne, A.G. & Wolpoff, M. 1992. The multiregional evolution of humans. *Scientific American*. 266: 28–33.

Thorne, A.G., Grün, R., Mortimer, G., Spooner, N.A., Simpson, J.J., McCulloch, M., Taylor, L. & Curnoe, D. 1999. Australia's oldest human remains: Age of the Lake Mungo 3 skeleton. *Journal of Human Evolution*. 36: 591–612.

Tiller, A. 1998. Ontogenetic variation in Late Pleistocene *Homo sapiens* from the Near East. In T. Akazawa, K. Aoki & O. Bar-Yosef (eds.). *Neandertals and Modern Humans in Western Asia*, pp. 381–389. Plenum Press, New York.

Tobias, P.V. 1967. *Olduvai Gorge*, Vol. 2: *The Cranium and Maxillary Dentition of* Australopithecus (Zinjanthropus) boisei. Cambridge University Press, Cambridge, UK.

Tobias, P.V. 1980. *"Australopithecus afarensis"* and *A. africanus*: Critique of an alternative hypothesis. *Palaeontologica Africana.* 23: 1–17.

Tobias, P.V. 1991. *Olduvai Gorge*, Vol. 4: *The Skulls, Endocasts, and Teeth of* Homo habilis. Cambridge University Press, Cambridge, UK.

Tong, H. & Jaeger, J.J. 1993. Muroid rodents from the middle Miocene Fort Ternan locality (Kenya) and their contribution to the phylogeny of muroids. *Palaeontographica.* 229: 51–73.

Torre, de la, I., Mora, R., Dominguez-rodrigo, M., Luque, de, L. & Alcala, L. 2003. The Oldowan industry of Peninj and its bearing on the reconstruction of the technological skills of Lower Pleistocene hominids. *Journal of Human Evolution.* 44: 203–224.

Trinkaus, E. 1983. *The Shanidar Neanderthals.* Academic Press, New York.

Trinkaus, E. 1986. The Neanderthals and modern human origins. *Annual Review of Anthropology.* 15: 193–218.

Trinkaus, E. & Shipman, P. 1993. *The Neandertals: Changing the Image of Mankind.* Alfred A. Knopf, New York.

Trinkaus, E., Moldovan, O., Milota, S., Bilgar, A., Sarcina, L., Athreya, S., Bailey, S.E., Rodrigo, R., Mircea, G., Higham, T., Ramsey, C.B. and van der Plicht, J. 2003. An early modern human from the Pestera cu Oase, Romania. *Proc. Natl. Acad. Sci. USA.* 100: 11231–11236.

Turner, A. 1984. Hominids and fellow travellers: Human migration into high latiudes as part of a large mammal community. In R. Foley (ed.). *Hominid Evolution and Community Ecology*, pp. 193–217. Academic Press, London.

Turner, A. & Antón, M. 1997. *The Big Cats and Their Fossil Relatives.* Columbia University Press, New York.

Turq, A. 1989. Exploitation des matières premières lithiques et occupation du sol: l'exemple du Moustérien entre Dordogne et Lot. In H. Laville (ed.). *Variations des Paléomilieux et Peuplement Préhistorique*, pp. 179–204. Centre National de la Recherche Scientifique, Paris.

Turq, A. 1992. Raw material and technological studies of the Quina Mousterian in Périgord. In H.L. Dibble & P.A. Mellar (eds.). *The Middle Paleolithic: Adaptation, Behavior, and Variability*, pp. 75–85. University of Pennsylvania Press, Philadelphia.

Tzedakis, P.C. & Bennett, K.D. 1995. Interglacial vegetation succession: A view from southern Europe. *Quaternary Science Reviews.* 14: 967–982.

Ungar, P.S. 2002. Reconstructing the diets of fossil primates. In M.J. Plavcan, R.F. Kay, W.L. Jungers & C.P. Schaik (eds.). *Reconstructing Behavior in the Primate Fossil Record*, pp. 261–296. Kluwer Academic/Plenum Publishers, New York.

Ungar, P.S. & Kay, R.F. 1995. The dietary adaptations of European Miocene catarrhines. *Proceedings of the National Academy of Sciences USA.* 92: 5479–5481.

Valladas, H., Mercier, N., Joron, J.L. & Reyss, J.L. 1988. GIF laboratory dates for Middle Palaeolithic Levant. In T. Akazawa, K. Aoki & O. Bar-Yosef (eds.). *Neandertals and Modern Humans in Western Asia*, pp. 69–76. Plenum Press, New York.

Van Valen, L. 1973. A new evolutionary law. *Evolutionary Theory.* 1: 1–30.

Vekua, A., Lordkipanidze, D., Rightmire, G.P., Agusti, J., Ferring, R., Maisuradze, G., Mouskhelishili, A., Nioradze, M., de Leon, M.P., Tappen, M., Tvalchrelidze, M. & Zollikofer, C. 2002. A new skull of early *Homo* from Dmanisi, Georgia. *Nature.* 297: 85–89.

Vignaud, P., Duringer, P., Mackaye, H.T., Likius, A., Blondel, C., Boisserie, J.R., de Bonis, L., Eisenmann, V., Etienne, M.E., Geraads, D., Guy, F., Lehmann, T., Lihoreau, F., Lopez-Martinez, N., Mourer-Chauvire, Otero, O., Rage, J.C., Schuster, M., Viriot, L., Zazzo, A. & Brunet, M. 2002. Geology and palaeontology of the Upper Miocene Tores-Menalla hominid locality, Chad. *Nature.* 418: 152–155.

Vrba, E.S. 1980. Evolution, species and fossils: How does life evolve? *South African Journal of Science.* 76: 61–84.

Vrba, E.S. 1985. Ecological and adaptive changes associated with early hominid evolution. In E. Delson (ed.). *Ancestors: The Hard Evidence*, pp. 63–71. Alan R. Liss., New York.

Vrba, E.S. 1999. Habitat theory in relation to the evolution in African Neogene biota and hominids. In T.G. Bromage & F. Schrenk (eds.). *African Biogeography, Climate Change and Human Evolution*, pp. 19–34. Oxford University Press, Oxford, UK.

Walker, A.C. 1976. Remains attributed to *Australopithecus* in the East Rudolf succession. In Y. Coppens, F.C. Howell, G.L. Isaac & R.E.F. Leakey (eds.). *Earliest Man and Environments in the Lake Rudolf Basin: Stratigraphy, Paleoecology, and Evolution*, pp. 484–489. University of Chicago Press, Chicago.

Walker, A.C. 1993. Perspectives on the Nariokotome discovery. In A.C. Walker & R.E.F. Leakey (eds.). *The Nariokotome* Homo erectus *skeleton*, pp. 411–430. Harvard University Press, Cambridge, MA.

Walker, A.C. 1994. Early *Homo* from 1.8–1.5 million year deposits at Lake Turkana, Kenya. *Courier Forschungs-Institut Senckenberg.* 171: 167–173.

Walker, A.C. 1997. *Proconsul*: Function and phylogeny. In D.R. Begun, C.V. Ward & M.D. Rose (eds.). *Function, Phylogeny, and Fossils: Miocene Hominoid Evolution and Adaptations*, pp. 209–224. Plenum Press, New York.

Walker, A.C. & Leakey, R.E.F. 1988. The evolution of *Australopithecus boisei*. In F.E. Grine (ed.) *Evolutionary History of the "Robust" Australopithecines*, pp. 247–258. Aldine de Gruyter, New York.

Walker, A.C. & Leakey, R.E.F. 1993a. The skull. In A.C. Walker & R.E.F. Leakey (eds.). *The Nariokotome* Homo erectus *Skeleton*, pp. 63–94. Harvard University Press, Cambridge, MA.

Walker, A.C. & Leakey, R.E.F. 1993b. The postcranial bones. In A.C. Walker & R.E.F. Leakey (eds.). *The Nariokotome* Homo erectus *Skeleton*, pp. 95–160. Harvard University Press, Cambridge, MA.

Walker, A.C. & Ruff, C.B. 1993. The reconstruction of the pelvis. In A.C. Walker & R.E.F. Leakey (eds.). *The Nariokotome* Homo erectus *Skeleton*, pp. 221–233. Harvard University Press, Cambridge, MA.

Walker, A.C. & Shipman, P. 1996. *The Wisdom of Bones: In Search of Human Origins*. Weidenfeld & Nicholson, London.

Walker, A.C., Zimmerman, M. & Leakey, R.E.F. 1982. A possible case of hypervitaminosis A in *Homo erectus*. *Nature*. 266: 248–250.

Walker, A.C., Leakey, R.E.F., Harris, J.M. & Brown, E.H. 1986. 2.5 Myr *Australopithecus boisei* from west of Lake Turkana, Kenya. *Nature*. 322: 517–522.

Walker, A.C., Brown, B. & Ward, S.C. 1993. Squamosal suture of cranium KNM-WT 17000. *American Journal of Physical Anthropology*. 90: 501–505.

Ward, C.V., Leakey, M.G. & Walker, A.C. 2001. Morphology of *Australopithecus anamensis* from Kanapoi and Allia Bay, Kenya. *Journal of Human Evolution*. 41: 255–368.

Ward, R. & Stringer, C. 1997. A molecular handle on the Neanderthals. *Nature*. 388: 225–226.

Ward, S. 1997. The taxonomy and phylogenetic relationships of *Sivapithecus* revisited. In D.R. Begun, C.V. Ward & M.D. Rose (eds.). *Function, Phylogeny, and Fossils: Miocene Hominoid Evolution and Adaptations*, pp. 269–290. Plenum Press, New York.

Ward, S. & Duren, D.A. 2002. Middle and late Miocene African hominoids. In W.C. Hartwig (ed.). *The Primate Fossil Record*, pp. 385–397. Cambridge University Press, Cambridge, UK.

Webb, S.G. 1989. *The Willandra Lakes Hominids*. Department of Prehistory, Research School of Pacific Studies. Australian National University, Canberra.

Webb, S.G. 1995. *Palaeopathology of Aboriginal Australians: Health and Disease Across a Hunter-Gatherer Continent*. Cambridge University Press, Cambridge, UK.

Weidenreich, F. 1937. The dentition of *Sinanthropus pekinensis*: A comparative odontography of the hominids. *Palaeont. Sin., N.S.D, No. 1*.

Weidenreich, F. 1939. On the earliest representatives of modern mankind recovered on the soil of East Asia. *Peking Natural History Bulletin*. 13: 161–174.

Weidenreich, F. 1941. The brain and its role in the phylogenetic transformation of the human skull. *Transactions of the American Philological Society*. 31: 321–442.

Weidenreich, F. 1943. The skull of *Sinanthropus pekinensis*: A comparative study on a primitive hominid skull. *Palaeontology Sinica*. [NS 10] 127: 1–134.

Weidenreich, F. 1946. *Apes, Giants and Man*. University of Chicago Press, Chicago.

Weidenreich, F. 1949. Interpretations of the fossil material. In W.W. Howells (ed.). *Early Man in the Far East: Studies in Physical Anthropology*. American Association of Physical Anthropologists, Detroit.

Wells, S. 2002. *The Journey of Man: A Genetic Odyssey*. Princeton University Press, USA.

Whallon, R. 1989. Elements of cultural change in the later Paleolithic. In P. Meller & C. Stringer (eds.). *The Human Revolution: Behavioral and Biological Perspectives on the Origins of Modern Humans*, pp. 433–454. Princeton University Press, Princeton, NJ.

Wheeler, P.E. 1984. The evolution of bipedality and loss of functional body hair in hominids. *Journal of Human Evolution*. 13: 91–98.

Wheeler, P.E. 1991. The influence of bipedalism in the energy and water budgets of early hominids. *Journal of Human Evolution*. 23: 379–388.

Wheeler, P.E. 1993. The influence of stature and body form on hominid energy and water budgets: A comparison of *Australopithecus* and early *Homo* physiques. *Journal of Human Evolution*. 24: 13–28.

White, R. 1983. Changing land-use patterns across the Middle/Upper Paleolithic transition: the complex case of the Perigord. In E. Trinkaus (ed.). *The Mousterian Legacy: Human Biocultural Change in the Upper Pleistocene*, pp. 113–122. BAR International Series 164. Oxford, UK.

White, P.J. & O'Connell, J.F. 1982. *A Prehistory of Australia, New Guinea and Sahul.* Academic Press, Sydney.

White, T.D. 1980. Evolutionary implications of Pliocene hominid footprints. *Science.* 208: 175–176.

White, T.D. 1986. *Australopithecus afarensis* and the Lothagam mandible. *Fossil Man: New Facts, New Ideas (Anthropos, Brno),* pp. 79–90.

White, T.D. 1992. *Prehistoric Cannibalism at Mancos 5MTUMR-2346.* Princeton University Press, Princeton, NJ.

White, T.D. 2002. Earliest hominids. In W.G. Hartwig (ed.). *The Primate Fossil Record,* pp. 407–417. Cambridge University Press, Cambridge, UK.

White, T.D. 2003. Early hominids — diversity or distortion? *Science.* 299: 1994–1997.

White, T.D., Johanson, D.C. & Kimbel, W.H. 1983. *Australopithecus africanus:* Its phyletic position reconsidered. In R.L. Ciochon & R.S. Corruccini (eds.). *New Interpretations of Ape and Human Ancestry,* pp. 721–780. Plenum Press, New York.

White, T.D., Suwa, G. & Asfaw, B. 1994. *Australopithecus ramidus,* a new species of early hominid from Aramis, Ethiopia. *Nature.* 371: 306–312.

White, T.D., Suwa, G. & Asfaw, B. 1995. Corrigendum: *Australopithecus ramidus,* a new species of early hominid from Aramis, Ethiopia. *Nature.* 375: 88.

White, T.D., Asfaw, B., DeGusta, D., Gilbert, H., Richards, G.D., Suwa, G. & Howell, F.C. 2003. Pleistocene *Homo sapiens* from the Middle Awash, Ethiopia. *Nature.* 423: 742–747.

Wiley, E.O. 1981. *Phylogenetics: The Theory and Practice of Phylogenetic Systematics.* Wiley-Liss, New York.

Williams, M., Dunkerley, D., de Decker, P., Kershaw, P. & Chappell, J. 1998. *Quaternary Environments.* Arnold, London.

Wilson, A.C., Stoneking, M., Cann, R.L., Prager, E.M., Ferris, S.D., Wrischnik, L.A. & Higuchi, R.G. 1987. Mitochondrial clans and the age of our common mother. In F. Vogel & K. Sperling (eds.). *Human Genetics: Proceedings of the 7th International Congress,* pp. 158–164. Springer Verlag, Berlin.

Wilson, M.V.H. 1992. Importance for phylogeny of single and multiple stem-group fossil species with examples from freshwater fishes. *Systematic Biology.* 41: 462–470.

Wilson, R.C.L., Drury, S.A. & Chapman, J.L. 2000. *The Great Ice Age: Climate Change and Life.* Routledge, London.

WoldeGabriel, G., White, T.D., Suwa, G., Renne, P., De Heinzelin, J., Hart, W.K. & Heiken, G. 1994. Ecological and temporal placement of early Pliocene hominids at Aramis, 1995. Ethiopia. *Nature.* 371: 330–333.

Wolpoff, M.H. 1971. *Metric trends in hominid dental evolution.* Case Western Reserve University Studies in Anthropology, 2. Cleveland, Ohio.

Wolpoff, M.H. 1989. Multiregional evolution: The fossil alternative to Eden. In P. Mellars & C.B. Stringer (eds.). *The Human Revolution: Behavioral and Biological Perspectives on the Origins of Modern Humans,* pp. 62–109. Princeton University Press. Princeton, NJ.

Wolpoff, M.H. 1996. *Human Evolution.* Mc Graw-Hill, New York.

Wolpoff, M.H. 1999. *Palaeoanthropology.* McGraw-Hill, Boston.

Wolpoff, M.H. & Caspari, R. 1997. *Race and Human Evolution: A Fatal Attraction.* Simon & Schuster, New York.

Wolpoff, M.H., Wu, X.Z. & Thorne, A.G. 1984. Modern *Homo sapiens* origins: A general theory of hominid evolution involving the fossil evidence from East Asia. In F.H. Smith & F. Spencer (eds.). *The Origins of Modern Humans,* pp. 411–484. Alan R. Liss, New York.

Wolpoff, M.H., Hawks. J., Frayer, D.W. & Hunley, K. 2001. Modern human ancestry at the peripheries: A test of the replacement theory. *Science.* 291: 293–297.

Wood, B.A. 1976. Remains attributable to *Homo* in the East Rudolf Succession. In Y. Coppens, F.C. Howell, G.L. Isaac & R.E.F. Leakey (eds.). *Earliest Man and Environments in the Lake Rudolf Basin: Stratigraphy, Paleoecology, and Evolution,* pp. 490–506. University of Chicago Press, Chicago.

Wood, B.A. 1984. The origins of *Homo erectus. Courier Forschungsinstitut Senckenberg.* 69: 99–112.

Wood, B.A. 1988. Are "Robust" australopithecines a monophyletic group? In F.E. Grine (ed.). *Evolutionary History of the "Robust" Australopithecines,* pp. 269–284. Aldine de Gruyter, New York.

Wood, B.A. 1991. *Hominid Cranial Remains. Koobi Fora Research Project,* Vol. 4. Oxford University Press, Oxford, UK.

Wood, B.A. 1992. Origin and evolution of the genus *Homo. Nature.* 355: 783–790.

Wood, B.A. 2002. Hominid revelations from Chad. *Nature.* 418: 133–135.

Wood, B.A. & Chamberlain, A.T. 1986. *Australopithecus*: Grade or clade? In B.A. Wood, L. Martin & P.J. Andrews (eds.). *Major Topics in*

Primate and Human Evolution, pp. 220–248. Cambridge University Press, Cambridge, UK.

Wood, B.A. & Collard, M. 1999. The Human Genus. *Science.* 284: 65–71.

Wood, B.A. & Richmond, B.G. 2000. Human evolution: Taxonomy and paleobiology. *Journal of Anatomy.* 196: 19–60.

Wood, C. 1994. The correspondence between diet and masticatory morphology in a range of extant primates. *Zoological Morphology Anthropologica.* 80: 19–50.

Woodburne, M.O., Bernor, R.L. & Swisher, C.C. 1996. An appraisal of the stratigraphic and phylogenetic bases for the "Hipparion" datum in the Old World. In R.L. Bernor, V. Fahlbusch & W. Mittmann (eds.). *The Evolution of Western Eurasian Neogene Mammal Faunas*, pp. 124–136. Columbia University Press, New York.

Wrangham, R.W. 2001. Out of the *Pan*, into the fire: How our ancestors' evolution depended on what they ate. In F.B.M. de Waal (ed.). *Tree of Origin: What Primate Behavior Can Tell Us About Human Social Evolution*, pp. 121–143. Harvard University Press, Cambridge, MA.

Wright, R.V.S. 1976. Evolutionary process and semantics: Australian prehistoric tooth size as a local adjustment. In R.L. Kirk & A.G. Thorne (eds.). *The Origin of the Australians*, pp. 265–274. Australian Institute of Aboriginal Studies, Canberra.

Wu, X. 1961. On the racial types of the Upper Cave Man of Choukoutien. *Scientia Sinica.* 10: 998–1005.

Wu, X. 1980. Palaeoanthropology in the new China. In L.K. Koenigsson (ed.). *Current Arguments on Early Man*, pp. 182–206. Pergamon Press, London.

Wu X. 2000. Longgupo hominoid mandible belongs to ape. *Acta Anthropologica Sinica.* 19: 1–10.

Wu, X. & Poirier, F.E. 1995. *Human Evolution in China: A Metric Description of the Fossils and a Review of the Sites.* Oxford University Press. Oxford, UK.

Wu, X. & Zhang, Z. 1983. *Homo sapiens* remains from Late Palaeolithic and Neolithic China. In R. Wu & J.W. Olsen (eds.). *Palaeoanthropology and Palaeolithic Archaeology in the People's Republic of China*, pp. 107–134. Academic Press, New York.

Wunderlich, R.E., Walker, A.C. & Jungers, W.L. 1999. Rethinking the positional repertoire of *Oreopithecus. American Journal of Physical Anthropology.* 28: 282.

Yi, S. & Clark, G.A. 1983. Observations on the Lower Paleolithic of northeast Asia. *Current Anthropology.* 24: 181–202.

Yokoyama, Y. & Nguyen, H.V. 1981. Datation directe de homme de Tautavel par la spectrométrie gamma, non-destructive, du crâne humain fossile Arago XXI. *C. R. Academy of. Science Paris, II.* 292: 927–930.

INDEX